Nanoscale Materials for Warfare Agent Detection: Nanoscience for Security

NATO Science for Peace and Security Series

This Series presents the results of scientific meetings supported under the NATO Programme: Science for Peace and Security (SPS).

The NATO SPS Programme supports meetings in the following Key Priority areas: (1) Defence Against Terrorism; (2) Countering other Threats to Security and (3) NATO, Partner and Mediterranean Dialogue Country Priorities. The types of meeting supported are generally "Advanced Study Institutes" and "Advanced Research Workshops". The NATO SPS Series collects together the results of these meetings. The meetings are co-organized by scientists from NATO countries and scientists from NATO's "Partner" or "Mediterranean Dialogue" countries. The observations and recommendations made at the meetings, as well as the contents of the volumes in the Series, reflect those of participants and contributors only; they should not necessarily be regarded as reflecting NATO views or policy.

Advanced Study Institutes (ASI) are high-level tutorial courses to convey the latest developments in a subject to an advanced-level audience.

Advanced Research Workshops (ARW) are expert meetings where an intense but informal exchange of views at the frontiers of a subject aims at identifying directions for future action.

Following a transformation of the programme in 2006, the Series has been re-named and re-organised. Recent volumes on topics not related to security, which result from meetings supported under the programme earlier, may be found in the NATO Science Series.

The Series is published by IOS Press, Amsterdam, and Springer, Dordrecht, in conjunction with the NATO Emerging Security Challenges Division.

Sub-Series

A.	Chemistry and Biology	Springer
B.	Physics and Biophysics	Springer
C.	Environmental Security	Springer
D.	Information and Communication Security	IOS Press
E.	Human and Societal Dynamics	IOS Press

http://www.nato.int/science
http://www.springer.com
http://www.iospress.nl

Series A: Chemistry and Biology

Nanoscale Materials for Warfare Agent Detection: Nanoscience for Security

edited by

Carla Bittencourt
Université de Mons - UMONS
Mons, Belgium

Chris Ewels
Inst. des Materiaux Jean Rouxel (IMN)
CNRS / University of Nantes
Nantes, France

and

Eduard Llobet
Universitat Rovira i Virgili
Tarragona, Spain

 Springer

Published in Cooperation with NATO Emerging Security Challenges Division

Proceedings of the NATO Advanced Research Workshop on Nanoscale
Materials for Warfare Agent Detection: Nanoscience for Security
Levi, Finland
13–16 February 2017

Library of Congress Control Number: 2019934443

ISBN 978-94-024-1622-0   (PB)
ISBN 978-94-024-1619-0   (HB)
ISBN 978-94-024-1620-6   (e-book)
https://doi.org/10.1007/978-94-024-1620-6

Published by Springer,
P.O. Box 17, 3300 AA Dordrecht, The Netherlands.

www.springer.com

Printed on acid-free paper

Preface

The current global political climate places ever-increasing pressure on civil and military society to rapidly detect and characterise toxic and dangerous agents in our environment. Nanomaterials offer a new range of tools capable of providing low-cost passive sensing solutions, with possibilities for novel deployment and detection routes.

This book comes out of a NATO Advanced Research Workshop organised in February 2017. The workshop brought together a diverse group of international civilian researchers focused on nanoscience and nanotechnology problems relevant to chemical defence needs. The reliable detection of chemical warfare agents to protect civilians in public places remains one of the most significant topics open for research. Despite recent advancements in detection of toxic gases, the existent available techniques do not necessarily meet the requirements for on-site chemical warfare agent detection.

The Workshop provided a forum for leading scientists in Nanotechnology and Devices, and this mix is reflected in the chapters in this current book. The first chapter introduces the topic and provides an industrialist's view of the feasibility of graphene-based defence applications. There are a wide range of nanomaterials with potential application in detection of chemical warfare agents, with different electronic, mechanical, and chemical behaviour. The next chapters explore some of these key nanomaterial families: carbon nanomaterials, oxides, polymers, tuneable molecular structures, layered transition metal dichalcogenides, and hollow nanoparticles.

Since carbon nanomaterials are the subject of major international research effort at present, we next have a series of chapters exploring carbon nanomaterials in more detail, with chapters discussing synthesis routes, methods for doping and chemical modification to tune the material properties, and key characterisation approaches as applied to gas-reactive nanocarbon surfaces.

Finally, there is much potential beyond direct sensing for nanomaterials technology. We finish this collection with a chapter exploring one such area, photoactivation of nanomaterials for eliminating biohazards. Such technology also has great promise in civilian applications such as cancer treatment.

We hope you find this collection a useful reference source.

Mons, Belgium   Carla Bittencourt
Nantes, France   Chris Ewels
Tarragona, Spain   Eduard Llobet

Contents

Chapter 1
The Feasibility of Graphene-Based Defense Applications: An Industry Perspective

Antonio Miramontes

Abstract Not all scientific research is innovation. New knowledge only becomes an innovation when it is transformed into products, processes and services that add value to the customers. If a new idea cannot be produced economically, or not be produced at all, it is just that: a new idea.

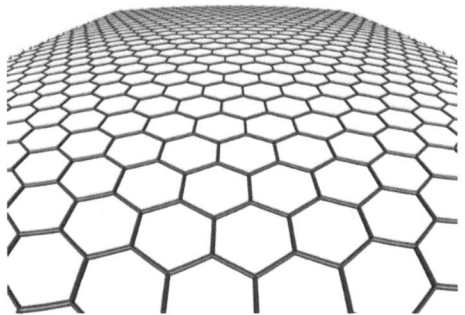

Schematic of a graphene sheet, showing the hexagonally arranged monolayer of carbon atoms (c) www.ewels.info

Of course, scientific research does not need to reach the market to be highly valuable, but for an idea to be manufactured and fulfill a practical need it is important to take an industry perspective.

That is especially true for graphene, a new nanoscale material that is supposed to change many products and industries. However, while graphene research is gaining momentum by the day, graphene applications are lagging behind. In spite of the huge amount of resources invested worldwide and the thousands of patents filed each year, most applications are failing to deliver when it comes to scale up production and adjust to manufacturing conditions.

A. Miramontes (✉)
Graphenemex, Ciudad de México, Mexico
e-mail: jam@graphenemex.com

© Springer Nature B.V. 2019
C. Bittencourt et al. (eds.), *Nanoscale Materials for Warfare Agent Detection: Nanoscience for Security*, NATO Science for Peace and Security Series A: Chemistry and Biology, https://doi.org/10.1007/978-94-024-1620-6_1

In particular, graphene-based defense applications show great potential. There is a lot of research going on and each result brings many new developments. The use of graphene for warfare agent detection could set a new standard for sensor devices. That is not only an opportunity, but also a major challenge for filling the gap between lab research and industry processes.

This chapter reviews some of the most important defense applications that could be radically improved by graphene, it also intends to map out some of the main constraints that are preventing graphene applications from being produced and commercialized.

Keywords Graphene · Industry · Defense

1.1 Graphene Potential in Defense Applications

The unique properties of graphene materials offer potential advantages for enabling advances in multiple applications, such as electronics, medical, disease detection, polymer materials, photovoltaic cells, water desalination, electrode materials in batteries, gas sensors, flexible display screens and nanofiltration membranes.

By the same token, there is a great span of potential graphene applications that are relevant to defense needs:

Bulletproofing for Combat Vehicles and Body Armor These applications require the development of composites that are lighter and stronger. Graphene is 10 times better than steel at dissipating kinetic energy, the amount of dissipated energy reaches up to 0.92 megajoules of energy per kilogram of graphene, compared to the 0.08 megajoules per kilogram that steel exhibits [1].

Stealth Systems Graphene is useful for camouflage, signature reduction and detection avoidance. For example, vanadium dioxide, coated with graphene and carbon nanotubes, allows adaptive camouflage in thermal imaging [2].

Special Polymer Composites and Coatings Reinforced plastics with increased compression and tension resistance, conductivity, fire retardation, corrosion protection and ultraviolet irradiation resistance.

Energy Pulse Generation Graphene and silicon have been used for femtosecond laser propagation [3] or modulation [4].

Wearable Technology Flexible and stretchable wearable electronic devices can be improved through graphene-based gas and chemical sensors [5] or printed graphene for wireless communications, such as antennas [6].

Other graphene applications that could be used for defense purposes are anti-fog crystals, anti-electric shock coatings, fuel additives, EMI interference shielding, wave absorbers, interface systems, electronic displays, antibacterial solutions, electrostatic charge dissipation and diffusion barriers.

1.2 Graphene Potential in Warfare Agent Detection

Current agent detection technologies were adapted from traditional chemistry techniques and face several limitations: they are not sensitive enough; involve complex processes; lack portability; vary under different environmental conditions; are either too selective (adjusted to only pre-determined targets), or not selective at all (affected by interferent compounds); require high power requirements; are not consistent and lack accuracy.

Fortunately, graphene is expected to have impact on high performance sensors for threat detection of explosives, chemical agents and biological agents.

Due to increasing terrorist threats, the role of detection is no longer related only to attacks on soldiers on the battlefield, but also to attacks on civilians on any location. It is imperative to develop sensor devices that are able to efficiently detect, identify and monitor warfare agents for both military and civilian defense.

In that regard, providing a solution to opportune and effective warfare agents detection should be a priority for graphene-based sensors. Several of the inherently unique properties of graphene show potential to enhance sensors and fulfill their performance requirements:

- High electron mobility.
- Thermal conductivity.
- Mechanical robustness.
- High electron-transfer rates.
- High surface area.
- Lightweight.
- Optical transparency.
- Flexibility.
- Environmental stability.

In addition, graphene materials are highly adaptable and can be implemented to construct sensors based on various sensing methods such as optical (fluorescence) and electrochemical. For instance, reduced graphene oxide has been applied in sensors with different levels of chemical reduction. These sensors were capable of detecting 10 s exposures to simulants of the three main classes of chemical-warfare agents and an explosive at parts-per-billion concentrations [7].

There are several sensing solutions that have been researched: enzyme-based sensors, chemically reactive sensors, supramolecular sensors, mass-sensitive sensors, chemicapacitors, chemiresistors and field-effect transistors, just to name a few. New sensors are based on different graphene materials (CVD graphene, graphene oxide or reduced graphene oxide), increasing sensitivity, processing capability and miniaturization.

According to recent research on graphene-based sensors, they are suitable to improve detection of most warfare agents that have been used as weapons of mass destruction. Graphene-based sensors can be adapted to the characteristics of each warfare agent type, either chemical or biological, and each detection should be treated in a different way.

Chemical warfare agents are powerful noxious chemicals that are usually categorized as nerve agents (sarin, soman, tabun), blister agents (sulfur mustard, nitrogen mustard), arsenicals (lewisite, phosgene oxime), blood agents (hydrogen cyanide, cyanogen chloride, arsine), choking-pulmonary agents (chorine, phosgene, diphosgene, chloropicrin), incapacitating agents and tear agents. In general, these agents are relatively stable while deployed to the target, easily dispersed and highly toxic.

Several studies have found that graphene-based sensors effectively detect different agent simulants, such as dimethyl-methylphosphonate (DMMP) [8–10] or dimethyl chlorophosphonate (DCP) [11] for nerve agents, as well as ipropylene glycol monomethyl ether (DPGME) [8] or thiodiglycol (TDG) [12] for mustard gas. Detection must be also selective and graphene improves discrimination capabilities of sensors among different simulants and background interferents [13].

On the other hand, biological warfare agents comprise bacteria and viruses. These are living organisms with short incubation periods, a high ability to infect or intoxicate with small doses, easiness to be dispersed and limited immunity among the population. As oppose to chemical weapons, biological agents are not stable when released into the environment as they are affected by factors such as sun and rain exposure, they have the ability to reproduce and even to mutate. These agents are becoming more lethal and sophisticated through genetic engineering. Various researches have shown that graphene is useful for detecting biological agents; for example, graphene functionalized with single strand deoxyribonucleic acid aptamers [14] or modified with a macromolecule complex [15] is able to detect the bacillus anthracis.

In general, warfare agents show high lethality, immediate symptoms, resistance to treatments with common medications and strong ability to attack under extreme environmental conditions. These agents are also easily stored, transported and propagated, and can be reproduced in high volumes with a small laboratory infrastructure.

There is a need for stand-alone, field-deployable sensor devices for the detection of chemical/biological warfare agents as sensitive and selective measures are critical to prevention, disease diagnosis and therapy. However, current sensors face some problems such as slow response time, low detecting volume and complex processes. It is imperative to develop cheaper, faster and more accurate sensors.

Graphene-based sensors enable several characteristics that are crucial to improve sensor technology. The following would be a list of the 'Ten Commandments' of warfare agent detection:

1. **High detection sensitivity.** Detect with confidence even the lowest agent concentrations.
2. **Selective detection.** Distinguish target from similar compounds. Sensors must differentiate specific agents and not be affected by coexisting substances or other chemicals present in the environment.

3. **Rapid response.** Require short time for the detector to respond to targeted agents. It is critical to identify warfare agents as quickly and as accurately as possible, providing advanced warning, once positive identification is obtained, in order to reduce immediate damage, especially when there may be no effective therapy available.
4. **Performance stability.** Maintain sensor's designated functions regardless of the environmental conditions.
5. **Portability.** Compact devices and miniaturize systems, assuring low power requirements, to replace standard laboratory diagnostics in order to detect warfare agents in situ.
6. **Linearity.** Facilitate detection through a linear relationship between cause and effect (inputs and outputs), so that measures are homogeneous and additive.
7. **Repeatability.** Achieve consistency so that results remain the same for the same input with every measure.
8. **Easy to use.** Design a simple interface and user-friendly process.
9. **Low calibration requirements.** Reduce verification requirements of capacity to perform using simulants.
10. **Accuracy.** Minimize false alarm rates, either false positives (if a detector responds when an agent is not present) or false negatives (if the detector fails to respond to an agent that is present). Sometimes, to confirm results it is useful to use another detector, based on a different technology.

Needless to say, if everything goes ahead as planned, there would be soon an 11th commandment to be added: use graphene.

1.3 An Industry Perspective

All market research reports project a considerable expansion of the graphene market in the next 5–10 years, is just a matter of time for the graphene market to explode, as there are large-scale investments, both in the public and private sectors, committed to developing graphene applications. However, there is still a preoccupying lack of commercial applications based on graphene materials. As of today, the "killer application" that will increase graphene demand is still missing and, in particular, no graphene-based warfare agent detectors are a commercial reality.

The ability to commercialize graphene applications, to move from lab concepts to market, is crucial in light of the fast changes in nanotechnology. In doing so, it is critical to bridge the gap between research, real world application and commercialization. It is clear that there is a lot of work that has to be carried out in order to develop practical solutions. New applications must consider both technology that is advanced enough to satisfy market requirements and the existence of clear demand/market for the innovation.

Past experience has shown that even the most promising investigations can fail because of production/commercialization challenges. In order to convert ideas into viable solutions, products must be designed to be readily manufactured at low cost and high quality. While maintaining the desired functionality, the core technology must be integrated into an optimal product architecture and production process.

The graphene market faces several challenges:

- High price/production cost.
- Limited production volume.
- Lack of consistency.
- Lack of industry standards.
- Difficulty to get incorporated into industrial processes and products.
- Market constraints.
- Gap between research and industry.

Most graphene applications either do not reach the market, fall short of expectations or are never profitable at all. It is possible that, even when the solution is useful and efficient, there is not a market whatsoever (sometimes, solutions are found to non-existent problems).

In many cases, graphene is set to create brand new markets with no competition. However, graphene applications must compete frequently against one or various substitutes that are well established in the market. For an industry to switch from traditional materials to graphene, there must be a combination of several factors, such as low costs, a clear improvement in properties and easiness to adjust to current processes. In addition, current producers of competing materials will never give up without a fight and will certainly reduce prices or improve features in order to protect their market.

It is common for research work to show disdain for the technical, production and market requirements of the particular applications or, if commercialization is foreseen in some way, it is assumed that the required implementation adjustments will be carried out by the industry. Therefore, most research prototypes are not suitable to achieve the lowest possible cost or are not robust enough to be replicated in production environments and perform well under different operating conditions. Product design must consider product optimization, cost reductions and manufacturing constraints, minimizing subsequent changes and requalification.

High Price/Production Cost The high price of graphene is still one of the major barriers to its widespread adoption for commercial applications. Graphene costs have been following a sustained declining trend, but the material is still lost out on price. It is common for industries to compete in a low value, high demand elasticity, market where any price increase draws customers away. In such a situation, graphene will prevail as long as it is more effective and cheaper than other materials.

For graphene applications to gain momentum, it is imperative for producers to reduce prices even further. Improving manufacturing methods and making available high quality-low cost graphene is crucial for moving from a materials market to an applications market.

Prices vary according to the quality and type of graphene material (CVD graphene, exfoliated graphene, graphene oxide, reduced graphene oxide or any of its functionalized forms). Every application requires a different type of graphene material and, not necessarily, the best quality one. Frequently, some defects are required for some particular applications. In order to accelerate the commercialization of graphene applications it is critical, from design and the initial graphene material selection, to consider the best price/performance ratio.

Limited Production Volume Research results are usually obtained under very controlled environments and using small sample sizes. The subsequent challenge is to obtain similar results in large quantities in production facilities, with quality and reliability. Most current methods for producing graphene, either bottom-up and top-down, fail to maintain quality in large quantities for scaling-up is non-linear and most chemical proportions change. Sometimes, a completely new process must be developed.

At larger scales, obtaining raw materials and identifying appropriate and cost-effective manufacturing represents a significant challenge. Research design for commercialization should established whether the chemical technology can be scaled up and the required output can be produced for an acceptable cost.

Lack of Consistency Graphene demand could be discouraged by lack of consistency between different production batches, in such a situation it would be difficult to build relationships between industry and graphene suppliers.

To be consistent, results must be replicable, but there is a clear difference between scientific replicability and industrial replicability. The former is a principle that calls for precision in measurement and accuracy of procedure, the latter is a consequence of solid engineering that covers all stages of production, from raw materials to final products, so that results are similar given similar inputs and conditions.

Industrial consistency ensures products to meet quality expectations and gives customers production predictability.

Lack of Industry Standards Development of graphene standards is an essential aspect of building the infrastructure needed to support commercialization.

Standards promote consistency and the lack of standards has been casting a long shadow over the introduction of graphene materials. For buyers it is difficult to depend on the quality of the acquired graphene. What is sold as graphene by some producer can be different from what other producer offers as the same material or, just the opposite, similar types of graphene can be commercialized as different products. Even worse, the "graphene" sold may not even be graphene, but some kind of expanded graphite.

Difficulty to Get Incorporated into Industrial Processes and Products Inflexible product design approaches and manufacturing processes often limit the introduction of new materials. Frequently, industries use processes that are very difficult to adapt to new materials, especially when current systems are not compatible with the specific properties and characteristics required.

Industries face costs when switching supplier's products to a new material (for example, new equipment, employee training and technical support). It is difficult to convince potential customers to bear the costs required to make a change.

In order to overcome this barrier, graphene suppliers should adjust their product offering so that the specific needs of industrial processes are fulfilled. It is possible to produce or functionalize different types of graphene depending on each application. Companies could require graphene powder or graphene dispersed in some liquid, like water or ethanol, and with different concentrations. Some processes in the plastic industry require prefabricated graphene polymer composites with different features.

Market Constraints The commercialization of new technologies faces additional difficulties, as they are new to the end users, they lack standardization and third-party certifications. Graphene is an emerging technology and the market is immature and fragmented, there is a disarray regarding price, quality, consistency and what kind of graphene material is most suitable for the particular application.

Most industries do not know what is the best graphene material for them. Should they use CVD graphene, exfoliated graphene, graphene oxide or reduced graphene oxide? Do they require material produced with or without defects? The material is required to be functionalized or doped in some way? How much graphene should be added and in what way? There is no a straight answer, there is not best or worse solution, material selection should depend on the best price/performance ratio for each particular application. In the face of this lack of knowledge it is difficult for graphene producers to accelerate the testing of products by the potential buyers.

Another constraint is that it is common for industries to have long term supply contracts, with pre-determined prices, delivery conditions, quality requirements and minimum annual quantities of the competing materials.

Finally, most graphene producers are not integrated in a larger supply chain or are located far from the industry clusters related to the application. Graphene needs to be integrated into the supply chains of manufacturers to really become a true game-changer.

Gap Between Research and Industry Research institutions need to increase their relationship with industry, developments should be targeted towards real world applications. A great deal of research fails to get through the real-world sieve and to be transformed into real industrial processes that are efficient and profitable enough to be introduced into industry. As a result, most research results end up in paper publications, patent filings and reports to the organizations that provided the funds.

Frequently, research is too specialized on a specific issue and fails to go beyond that. An industry ready application has to consider the whole system, not a single component. There are many implications on a real-world development that are missed when an individual component view prevails rather than a systemic one.

In general, researches lack the appropriate skills to consider real industrial processes, economic feasibility and market demand. To improve development efforts, it is necessary to put together lab teams with a wide range of skills and it is crucial to assure the interaction among experts from different fields. For industries, it is important to combine in-house and external resources, with an open innovation approach and to orient research objectives towards market needs.

Science and business objectives usually differ and researchers tend to investigate issues out of intellectual curiosity without any regard to real world applications, market needs and commercialization requirements. In addition, incentives are different among research institutions and industries. For instance, researchers tend to time projects according to the duration of grants while industries seek to get commercial results as quickly as possible.

It is common that project funding does not consider operational tests of products in real industrial conditions and activities leading to commercialization. In such a case, industry should provide the funds required for final tests, adjustments, norm compliance and commercialization.

Clearly, there is a need for graphene producers and applications developers, such as Graphenemex, to ensure proper communication and understanding between research and industry.

1.4   Conclusion

In general, it is difficult to transit from the lab to the real world as a lot of research miss important lessons from real operating conditions. In spite of the huge investment in graphene applications development and the myriad of studies, there are serious deficiencies in rapidly moving technology from research to products.

Many times, researchers answer a question that solves the wrong problem or economic concerns are frequently underestimated or not considered at all. Generally, research does not take into account process efficiency, industrial scale and costs. It is important to include detailed analysis of actual costs and to determine technical feasibility. In addition, it is necessary to determine how easy it is to adapt the new technologies to current production processes.

Graphene-based defense applications should lead to substantial improvement over competing technologies and is critical to determine whether enhanced properties or cost savings are sufficient to justify the adoption of the new technology. Graphene applications research should also consider product design approaches to reflect the unique properties of the new material, as well as to any environmental and use changes that the application might experience as a result of added capability.

More than ever, it is necessary to look for the challenges where they are. In so doing, graphene applications development must follow several steps:

- Identify potential value of research, differentiating pure science aspects from business.

- Identify research suitable to be converted into commercial solutions.
- Develop graphene applications according to innovation cycles, technology maturity and economic feasibility.
- Produce graphene material tailored to the needs of each application.
- Determine changes required to adjust findings to practical requirements.
- Determine requirements for: mass production, supply chains, distribution, logistics, pricing, production processes, quality control, marketing, technology certification, regulations compliance, human resources, finance, etc.
- Determine whether those requirements are able to be fulfilled.
- Ensure reliable production scale-up.
- Determine rights and ownership in order to protect the intellectual property (patents, licenses and copyrights).
- Secure the necessary regulatory approvals, if required.
- Identify time to commercialization.
- Perform technical proof-of-concept studies to demonstrate that the application can be effective in a real-world setting.
- Develop the technology into a commercially attractive proposition.
- Support continuous improvement.

Graphene has the potential to push performance boundaries, transform multiple industries and render previous technologies obsolete. The defense industry can play an important role in pushing graphene applications ahead.

Acknowledgement I would like to thank for support and cooperation to my colleagues from Graphenemex.

References

1. Lee J-H, Loya PE, Lou J, Thomas EL (2014) Dynamic mechanical behavior of multilayer graphene via supersonic projectile penetration. Science 346(6213):1092–1096. https://doi.org/10.1126/science.1258544
2. Lin X, He M, Liu J et al (2015) Fast adaptive thermal camouflage based on flexible VO2/graphene/CNT thin films. Nano Lett 15(12):8365–8370. https://doi.org/10.1021/acs.nanolett.5b04090
3. Liu K, Zhang JF, Xu W et al (2015) Ultra-fast pulse propagation in nonlinear graphene/silicon ridge waveguide. Sci Rep. https://doi.org/10.1038/srep16734
4. Yu H, Chen X, Zhang H et al (2010) Large energy pulse generation modulated by graphene epitaxially grown on silicon carbide. CS Nano 4(12):7582–7586. https://doi.org/10.1021/nn102280m
5. Singh E, Meyyappan M, Nalwa HS (2017) Flexible graphene-based wearable gas and chemical sensors. ACS Appl Mater Interfaces 9(40):34544–34586. https://doi.org/10.1021/acsami.7b07063
6. He H, Akbari M, Sydänheimo L et al (2017) 3D-Printed graphene antennas and interconnections for textile rfid tags: fabrication and reliability towards humidity Hindawi. Int J Antennas Propag 2017, Article ID 1386017:5
7. Robinson JT, Perkins FK, Snow ES et al (2008) Naval research lab, Washington DC reduced graphene oxide molecular sensors defense technical information center. Accession number: ADA540573. Nano Lett 8(10):3137–3140

8. Sayago I, Matatagui D, Fernández MJ et al (2016) Graphene oxide as sensitive layer in love-wave surface acoustic wave sensors for the detection of chemical warfare agent simulants. Talanta 148(1):393–400

9. Ha S, Lee M, Seo HO et al (2017) Structural effect of thioureas on the detection of chemical warfare agent simulants. ACS Sens 2(8):1146–1151. https://doi.org/10.1021/ acssensors.7b00256. Publication Date (Web): August 4, 2017

10. Hwang HM et al (2016) Mesoporous non-stacked graphene-receptor sensor for detecting nerve agents. Sci Rep 6:33299. https://doi.org/10.1038/srep33299

11. Olguin C, Laguarda-Miro N, Vidal Lluis PI et al (2014) An electronic nose for the detection of Sarin, Soman and Tabun mimics and interfering agents. Sensors Actuators B Chem 202:31–37. https://doi.org/10.1016/j.snb.2014.05.060

12. Singh VV, Nigam AK, Yadav SS et al (2013) Graphene oxide as carboelectrocatalyst for in situ electrochemical oxidation and sensing of chemical warfare agent simulant. Sensors Actuators B Chem 188:1218–1224

13. Wiederoder MS, Nallon EC, Weiss M et al Graphene nanoplatelet-polymer chemiresistive sensor arrays for the detection and discrimination of chemical warfare agent simulants. ACS Sens, Just Accepted Manuscript 2:1669–1678. https://doi.org/10.1021/acssensors.7b00550

14. Quinton, Matthew J. Air Force Institute of Technology, Wright-Patterson Air Force Base, Ohio. Optimization of graphene biosensors to detect biological warfare agents. Accession Number: ADA598854. Thesis Mar 27, 2014

15. Ryu J, Lee E, Lee K, Jang J (2015) A graphene quantum dots based fluorescent sensor for anthrax biomarker detection and its size dependence. J Mater Chem B 24(3):4865–4870

Chapter 2
Carbon Nanomaterials Integrated in Rugged and Inexpensive Sensing Platforms for the In-Field Detection of Chemical Warfare Agents

Juan Casanova-Chafer and Eduard Llobet

Abstract Fast, real-time and reliable detection of Chemical Warfare Agents (CWAs) is one of the most important challenges for our societies. The common techniques used to detect CWAs are expensive and require trained personnel, such as gas or liquid chromatography coupled to mass spectrometry or ion mobility spectrometry. For that reason, a lot of research has been conducted in recent years to achieve low-cost and portable technology for monitoring the presence of CWAs in the local environment. According to reported results, carbon nanomaterials have been found to be promising sensitive materials due to their excellent electronic properties and high possibilities for being functionalized, decorated or modified, this achieving tailored surface chemistry. In this chapter we study the use of different carbon nanomaterials (carbon nanotubes, graphene, carbon nanofibers and carbon black) to detect compounds with intrinsic potential to be used as a chemical weapon against military or civilian's targets. The available technology usually employs chemiresistive, electrochemical, gravimetric and optical sensors to measure the concentration of CWAs at ppm or ppb levels in the environmental or in biological samples. The suitable detection of chemical agents employing inexpensive sensors constitutes an important strategy to enhance population security.

Keywords Chemical warfare agents · Carbon nanomaterials · Sensors · Graphene · Carbon nanotubes

J. Casanova-Chafer · E. Llobet (✉)
MINOS-EMaS, Universitat Rovira i Virgili, Tarragona, Spain
e-mail: eduard.llobet@urv.cat

© Springer Nature B.V. 2019
C. Bittencourt et al. (eds.), *Nanoscale Materials for Warfare Agent Detection: Nanoscience for Security*, NATO Science for Peace and Security Series A: Chemistry and Biology, https://doi.org/10.1007/978-94-024-1620-6_2

13

2.1 Introduction

The reliable detection of chemical warfare agents (CWAs) such as sulfur mustard, lewisite or nerve agents has been driving intensive research efforts in the last two decades. The two main applications are, on the one hand, the sensitive analysis of CWAs and their decomposition products in environmental samples [1] and, on the other hand, the biomonitoring of CWAs [2]. The most evolved technologies currently available are best suited for the retrospective detection exposure to CWAs. Retrospective detection aims at diagnosing exposure to CWAs and dosimetry of casualties in order to decide medical treatment [3]. In the event of a terrorist attack, a correct diagnose of casualties becomes as important as confirming non-exposure to worried individuals. In addition, these methods can be of help for verifying an alleged nonadherence to the Chemical Weapons Convention [4] but also in health surveillance of workers in facilities for the destruction of CWAs or for biomonitoring potential terrorists who are involved in the production and/or handling of CWAs. While urinary metabolites are readily accessible biomarkers, the rapid elimination of which limits their use for retrospective detection [5], adducts with macromolecules such as proteins offer long-lived biological markers of the exposure to CWAs [6]. Gas or liquid chromatography combined with tandem mass spectrometry are the methods of choice for the unequivocal identification of these adducts or metabolites at trace levels. However, these techniques are costly, require well trained personnel and remain difficult to apply in field laboratories.

In the analysis of CWAs in air, current procedures comprise the sampling by sorption tubes, the thermal desorption or extraction with solvent and analysis using GC-MS [7]. In some cases, reagents are added to sorption tubes for the derivatization of trapped CWAs into products that can be more easily identified or quantified [8, 9]. The time needed for performing such an analysis ranges between 5 min to half an hour. Since environmental analysis is best performed in situ, portable instruments have been developed. These generally consist of a gas sampling device, an analyte concentrator, a thermal desorption stage, a chromatographic column, a detector (mass or ion mobility spectrometers or capillary electrophoregraphs) and a computer for controlling the whole system and data analysis [10–13].The attained limits of detection with portable equipment are useful and they respond in minutes, however, the main drawbacks of these instruments are the generation of false alarms, and the formation of analyte residues, which are difficult and take time to remove [14]. Nowadays research efforts are conducted towards the simplification of sampling methods, the avoidance of derivatization and the miniaturization and monolithic integration of system components. These efforts should help achieving good enough detection limits with improved selectivity (e.g. avoiding false alarms) and lower analysis times.

In parallel to the aforementioned instrumental techniques, there have been considerable research efforts from the research community aimed at developing other innovative detection methodologies. These alternative methodologies should provide high sensitivity and selectivity to CWAs, short analysis times and ability to perform correctly in real environments [15]. In addition, these techniques should be portable, lightweight, less expensive and simpler to operate than analytical techniques, yet reliable, thus providing added dimensions of analytical information to the combatant or to security agents in a timely manner. This is of high importance in the current climate of terrorism awareness. A great deal of research has been applied to the synthesis of acidic [16] or molecularly-imprinted polymers [17–19] for selective agent adsorption. These are often used as coatings in resonant transducers such as bulk or surface acoustic wave devices. The tailoring of novel sensor surfaces (e.g. metal oxides, carbon nanomaterials, porous materials) [20–24] has been addressed as well and, the adaptation of enzymatic assays coupled to colorimetry, fluorescence, or electrochemistry [25–28]. A valuable, recent review on electrochemical sensors for detecting CWAs can be found in [29]. Another strategy has consisted of studying the chemical derivatization of nerve agents to generate easily monitored colorimetric or fluorescent species [30–32].

This chapter critically reviews the latest developments in sensors that employ carbon nanomaterials (carbon black, carbon nanofibers and nanotubes and graphene) for detecting CWAs both in the gas phase and in liquid samples. Different transducing schemes such as chemoresistors, electrochemical sensors and biosensors, resonant and optical sensors are discussed, because these could be integrated in portable, miniaturized analyzers or even in cost effective smart tags with on-board gas sensing and communication capabilities that would be deployed as an unattended sensing network throughout the area to be monitored and able to provide a rapid, real-time, on-site alarm. Achievements are comprehensively and critically reviewed, shortcomings and bottlenecks identified and new research directions unveiled.

2.2 Carbon-Based Warfare Agent Sensors

2.2.1 Chemiresistive Sensors

Chemiresistive gas sensors interact with an analyte to produce a change in their resistance. When the gas-sensitive material of a carbon-based chemo-resistor reacts with electron donor or acceptor species, the result is a charge transfer between them, which eventually can be measured as a variation in the electrical resistance of the sensor. Such sensors present some advantages like simplicity in the driving and measuring electronics, low cost of the transducer (any standard rigid or flexible substrate with a pair of electrodes suffices) and high sensitivity [33, 34]. However, the main limitation of chemiresistive sensors is their poor selectivity [35]. Most of the research conducted for developing chemiresistive sensors aimed at the detection of warfare agents has involved the use of carbon nanomaterials. These include carbon black, carbon nanofibers and nanotubes and, more recently, graphene and graphene-based materials.

Most of carbon nanomaterial-based chemiresistive sensors have been devised for detecting, dimethyl methyl phosphonate (DMMP), a simulant of Sarin gas. Fewer studies report on the detection of other CWAs such as dimethylacetamide (DMA) or dipropylene glycol methyl ether (DPGME), simulants of distilled and nitrogen mustard gas, respectively. A reason explaining this is the polarity of DMMP, because in general CNTs are more sensitive to polar compounds. In consequence, other CWAs with lower polarity (e.g. DPGMA) result in a significantly lower charge transfer upon adsorption to the carbon nanomaterial film [36], and consequently, in a lower resistance change.

One of the first carbon-based materials considered for being integrated in chemiresistive sensors to detect CWAs was carbon black [37]. Birglin and co-workers [38] performed an array of carbon black-polymer composite to detect DNT at ppb range in seconds. The DMMP and the disopropyl methyl phosphonate (DIMP) were studied by Hopkins [39] using arrays of carbon black/organic polymer hybrids, demonstrating that carbon black sensors have some advantages like low power consumption and high possibilities to create arrays with hundreds of compositionally different detectors. The long-term stability of these sensors is affected both by polymer degradation and by migration of carbon black particles within the polymeric matrix after repeated detection cycles (e.g. these involve swelling of the polymer).

Carbon nanotube gas sensors have been extensively researched and used for the detection of airborne Chemical Warfare Agents (CWAs). Both single-wall carbon nanotubes (SWCNTs) and multi-wall carbon nanotubes (MWCNTs) have been employed. These materials present good chemical stability, low noise levels, large specific surface area and excellent mechanical and electronic properties [40]. However, pristine carbon nanotubes show low response and selectivity towards target gases. For that reason, many studies have focused on the functionalization, by physical or chemical methods, of the surface of CNTs in order to enhance their sensing performance. Substitutional doping, generation of controlled defects, grafting of metal/ metal oxide nanoparticles or functional groups to their outermost wall are different techniques that have been explored for increasing sensitivity and tuning selectivity.

Unlike other types of materials, like for example metal oxides, carbon nanotube chemiresistors can work at room temperature [36]. However, after their resistance has been altered due to the interaction of CNTs with a target gas, it is often difficult for these sensors to fully regain their baseline resistance during a cleaning phase in pure air. Sometimes, it is a matter of hours for CNTs to desorb target molecules at room temperature, and thus for sensors to come back to their initial resistance. Many studies have reported that heating the sensor or applying ultraviolet light during a cleaning phase can help desorbing analytes and speed up recovery. Alternatively, there is the possibility of functionalizing carbon nanotubes to improve their response and recovery processes, as it will be discussed later.

A critical parameter is the density of carbon nanotubes in the sensor film. Wang and co-workers [41] proved that too low a density of nanotubes can induce rather high noise levels due to CNTs being sparsely distributed and loosely interconnected.

However, too high density of nanotubes within the film decreases its baseline resistance and, as consequence, the resistance response is reduced too. In summary, the study reported an optimal SWCNTs density of about 30–40 tubes μm^{-3}. In addition, SWNTs were deposited onto an oxidized silicon surface functionalized with 3-aminopropyltrimethysilane (APS) [41]. This functionalization provides amino-terminations on Si/SiO2 wafer, which create an electrostatic interaction with carbon nanotubes, resulting in good stabilization of the film on the substrate and improved electrical contact between SWCNTs and gold electrodes.

Even though CNT chemiresistors can offer high sensitivity to CWAs, these must be made insensitive to potential interfering species. To solve the characteristic poor selectivity of carbon nanotube chemiresistors, Novak and co-workers [20] used a chemoselective polymer filter composed by a strong hydrogen-bonding compound (HC). This filter was designed to improve the specificity against a potential interfering species, such as ammonia, because this is a strong electron donor like dimethyl methyl phosphonate (DMMP). Cattanach and co-workers [42] reported the use of polyisobutylene (PIB) as chemoselective polymer with great improvements in the selectivity of SWCNTs in front of hydrocarbons. These polymers constitute an option to filter out potential interfering species such as ambient humidity, aliphatic hydrocarbons (hexanes) and aromatic hydrocarbons (xylenes), simulants of gasoline and diesel fuel, respectively.

A well-known approach for increasing selectivity is to employ an array of chemiresistors with partially overlapping selectivity to better differentiate among analytes and discriminate interfering species. For instance, Chuang and co-workers [43] designed an array of multi-wall carbon nanotubes employing 15 different polymers to detect different CWAs at room temperature. A dissolution of CNTs and a given polymer was prepared and ultrasonicated for 30 minutes. Then, the dissolution was spray-deposited onto one of the array transducer element and heated at 40 °C in vacuum for 12 hours. This process was repeated for the different polymers used. This study measured four different species in the ppm range (from 43 ppm to 356 ppm): DMMP, dichloromethane, acetonitrile and 2-chloroethyl ethyl sulfide, obtaining different fingerprints. However, the concentrations tested are too high for the application sought.

Enhanced CNTs sensitivity has been reported employing organic polymers to functionalize carbon nanotubes, such as hexafluoroisopropanol (HFIP) attached to polythiophene [33], N-4-Hexafluoroisopropanolphenyl-1-pyrenebutyramide (HFIPP) [44] and p-hexafluoroisopropanol phenyl (HFIPP) [45]. HFIP binds to the SWCNTs surfaces through non-covalent $\pi-\pi$ stacking interactions. Moreover, when DMMP gas is applied, a strong hydrogen-bonding is formed between HFIP groups and phosphate esters (common structural element in nerve agents), providing a high increase in sensitivity and selectivity of carbon-based sensors, lowering down to 50 ppb the limit of detection [44]. These results are summarized in Fig. 2.1. More recently, Fennell and co-workers [46] reported wrapped SWCNTs by poly (3,4-ethylenedioxythiophene) (PEDOT). The hexafluoroisopropanol group of the polymer improves the sensitivity to DMMP until a few ppm. However, the presence of ambient humidity decreases the chemiresistive response, probably

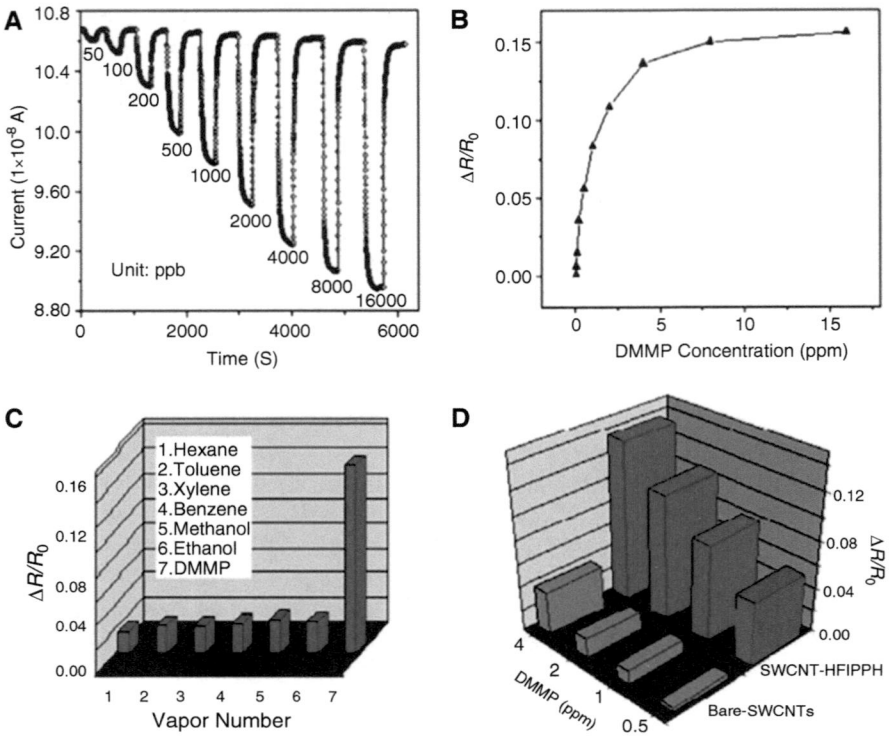

Fig. 2.1 (**a**) Response curves of the sensor upon exposure to different concentrations of DMMP. (**b**) Resistance changes of the sensor towards different concentrations of DMMP. (**c**) Resistance changes of the sensor upon exposure to different species and DMMP diluted to 1% of saturated vapor conditions. (**d**) Resistance changes of the SWCNT-HFIPPH and bare SWCNT devices were compared at different concentrations of DMMP. Reproduced from [45], ©Elsevier with permission

because ambient moisture competes as a hydrogen bond donor with DMMP gas. Wei and co-workers [47] functionalized single-wall carbon nanotubes by tetrafluorohydroquinone (TFQ) to decrease the response time (shorter than 2 minutes) and to lower down to 20 ppt the detection limit of DMMP. This solid organic acid promotes hydrogen-bonding with the surface of SWCNTs. TFQ improves the sensitivity and selectivity due to several reasons. First, TFQ and DMMP interact by strong hydrogen bonding. Second, DMMP is a strong electron donor, so was adsorbed to SWCNTs surface. Third, high hole doping of carbon nanotubes by TFQ is a result of this functionalization, which increases the binding affinity for DMMP.

It has been reported that single stranded-DNA decorated carbon nanotubes are suitable for detecting, at room temperature, dimethyl methyl phosphonate (DMMP) and di-nitrotoluene (DNT), with limits of detection at ppb and ppm level, respectively [48]. DNA binds onto CNTs by non-covalent $\pi-\pi$ interactions, improving the sensitivity towards CWAs. Upon adsorption of CWAs, the resistance of the sensor increases due to the negative charge of DNA being transferred towards CNTs, thus compensating carbon nanotube holes and decreasing the number of free charge carriers.

Other alternatives to carbon nanotubes have been explored to perform chemiresistive carbon-based sensors. One option is carbon nanofibers (CNFs), the production of this material is less costly than that of CNTs and shows some of its promising properties. One example is the work presented by Seop Lee and co-workers [49]. They decorated the surface of carbon nanofibers with metal oxides (ZnO and SnO_2) reporting a detection limit of about 0.1 ppb for DMMP. However, this material has not been used as intensively as carbon nanotubes as transducer of chemiresistive sensors, for that reason some of its properties remain unexplored such as selectivity towards potential interfering species or the effect of relative humidity in electrical response.

Recently, graphene has become one of the most intensively researched materials for designing resistive gas sensors. It has been demonstrated that graphene oxide (with multiple oxygen functional groups) is a promising candidate because this material has a diverse range of surface sites, whose density can be controlled easily. In addition, graphene presents in theory larger charge carrier mobility and lower noise levels than CNTs, which should help improving the signal-to-noise ratio and detection limits [50].

Hu and co-workers [51] reported graphene oxide reduced by p-phenylenediamine (PPD) as sensing material to detect DMMP. The lowest concentration successfully measured was 5 ppm, and the effect of potential interfering species was analyzed as well. Gases such as hexane, xylenes and chloroform cannot be considered interfering gases because sensor response is too low. However, dichloromethane and methanol trigger reasonable sensor responses (1/3 of the response towards DMMP), in consequence the selectivity of this reduced graphene oxide should be improved. In addition, relative humidity (R.H.) has an important effect in sensor response and, because of that, drying the gas flow input to the sensor chamber is a requirement. More recently, Alizadeh and Soltani [52] designed a gas sensor array to detect DMMP. The array was constructed using different reduced graphene oxides (RGO) based on three reducing agents, hydrazine hydrate, ascorbic acid and sodium borohydride. The effect on sensitivity related to the reducing agent used was reported. GO reduced with ascorbic acid and sodium borohydride was found to be more sensitive to DMMP than the one reduced with hydrazine hydrate. The reason is the negative effect of nitrogen atoms introduced by hydrazine hydrate as reducing agent, due to the fact that nitrogen atoms act as n-dopant, interfering with the p-type RGO. However, hydrazine hydrate and sodium borohydride present better efficiency to discriminate DMMP from other vapors such as methanol, acetone, and diethyl ether. In addition, it was found that ambient moisture had significant effect in sensor responses.

In summary, carbon nanomaterials have been extensively used in chemiresistive sensors due to their wide options for being functionalized in order to improve their electrical response, stability, sensitivity and selectivity. However, when applied to the detection of CWAs, the detection of DMMP has been by far the most studied case, and fewer results have been reported for other CWAs such as DPGME or DMA. For that reason, it is still necessary to explore new molecules and strategies to functionalize carbon materials in order to improve their sensitivity and selectivity towards non-polar CWAs. Moreover, the cross-sensitivity effects of many potential interfering species have been reported. These is the case of gases or vapors such as hydrocarbons or alcohols. In some cases, selectivity is improved via hydrogen bonding specific to organophosphorus compounds, however, carbon nanomaterials remain very sensitive to oxidizing gases such as nitrogen dioxide or ozone and this is generally overlooked. In consequence, considering CWA detection in real applications, carbon nanomaterial chemiresistive sensors are still far from reaching all the specifications needed.

2.2.2 *Electrochemical Sensors*

Electrochemical techniques are gaining popularity because of their advantages of portability, simple instrumentation, fast analysis, environmental friendliness, low cost, limited necessity of skilled personnel, ease of miniaturization, and high sensitivity and selectivity [29, 53]. Due to the non-electrochemical activity of CWAs, it is difficult to perform an electrochemical determination [54]. For that reason, it is necessary to modify the electrodes and, carbon nanomaterials such as carbon black, graphene oxide or carbon nanotubes offer high potential for achieving interesting results. Electrochemical sensors have been developed for detecting CWAs in liquids. The most common technique is the combination of cyclic voltammetry, in order to research the anodic and cathodic peaks, with amperometric measurements to quantify the response towards target species.

Periyakaruppan and co-workers [55] reported a biosensor to detect ricin employing nanoelectrode arrays (NEA) composed by vertically aligned carbon nanofibers (30–50 nm in diameter). Ricin was bonded to both, ricin-A antibody and Aricin RNA aptamer, demonstrating that aptamer probes gave higher electron transfer resistance (R_{ct}) than antibody probes. Probably this is due to ricin not being well captured by the antibody. Cyclic voltammetry analysis was performed to vertically aligned CNFs electrodes in a solution of 5 mM $K_3Fe(CN)_6$ in 0.1 M phosphate buffer at pH 7.4, and impedance spectroscopy was applied with a solution of 5 mM $K_3Fe(CN)_6$ in 0.1 M KCl. The concentration of ricin applied was 5 μL of 0.55 μ/mL to both, aptamer and antibody probe. In addition, the decreasing of R_{ct} observed suggests either antibody degradation or the formation of non-covalently bound layers on the electrode surface. Unlike the unstable antibody probe, the aptamer probe remains stable and can be used in several detection cycles, giving a reusable property to the biosensor.

Cinti and co-workers [56] reported an integrated paper-based, screen-printed device. They used carbon black/Prussian blue nanocomposite as working electrode. The carbon black acted as a substrate to grow ferric hexacyanoferrate (known as Prussian Blue) nanoparticles that enhanced the surface-to-volume ratio and improved electrochemical performance. The measurements are based on enzymatic activity of butyrylcholinesterase (BChe) towards butyrylthiocholine, with and without exposure to paraoxon, a nerve simulant agent. BChe hydrolyzes butyrylthiocholine yielding butyric acid and thiocholine, but paraoxon inhibits the enzymatic activity of BChe. Cinti et al. showed a detection of paraoxon down to 3 μg/L thanks to thiocholine electrochemical detection by Prussian Blue nanoparticles. The performed cyclic voltammetry's and chronoamperometric detection were tested in aqueous solution with 0.05 M phosphate buffer containing 0.1 M KCl at pH 7.4. In addition, the electrode was tested in water river and wastewater matrix to demonstrate the suitability for in field applications. However, the main limitation of this method is the repeatability, since an 11% standard deviation was observed in the determination of 25 μg/L of paraoxon.

Another alternative to detect paraoxon was proposed by Lee and co-workers [57]. They prepared a glass carbon electrode with an anodic layer composed by mesoporous carbon (MC) and carbon black (CB). This biosensor is based on the immobilization of organophosphorus hydrolase (OPH) onto screen-printed carbon electrodes that detects p-nitrophenol (PNP), a product of organophosphorus hydrolase reaction in presence of nerve agents, such as paraoxon. Cyclic voltammetry was used to obtain the best operating potential and, after that, chronoamperometry was applied to determine the PNP concentration. In both electro-analytical measurements, a potassium phosphate buffer (0.05 M) diluted with deionized water was used. Several concentrations of paraoxon were measured, being the lower 2 μM and establishing a detection of limit of 0.12 μM (estimated from employing the signal/noise ratio approach in which detection limit is estimated when S/N = 3). From this study, it was derived that a hybrid electrode with MC and CB improved the sensitivity and stability of measurements thanks to the resulting microstructure with high specific pore volume and surface area.

The capability of graphene to detect thiodiglycol (TDG, bis (2-hydroxyethyl) sulfide), the main hydrolysis product of sulfur mustard (bis (2-chloroethyl) sulfide) has been reported. Singh and co-workers [58] used graphene oxide (GO) to modify the surface of gold electrodes. GO acts as carboelectrocatalyst for TDG oxidation at room temperature, revealing the formation of 2-(2-hydroxy-ethanesulfinyl)-aldehyde as main oxidation product. A Mili-Q water with 0.05 M phosphate buffer solution (pH 7) as electrolyte was used in cyclic voltammetry for characterizing the electrodes. Differential pulse voltammetry (DPV) was employed to establish the relationship between the anodic peak current and concentration of TDG. The detection limit was estimated to be 0.2 μM (S/N = 3). Mahyari and co-workers [59] created an electrode with graphene nanosheets, silver nanoparticles and ionic liquid octylpyridinium hexafluorophosphate (GNs-AgNPs-ILE). The electrochemical solution prepared was composed by Britton-Robinson buffer 0.1 M (pH 4) diluted with distilled water. The electrocatalytic behaviour was investigated by cyclic voltammetry and finally, an amperometric determination of TDG down to 6 μM was performed. It was proved that GNs-AgNPs-ILE showed good repeatability and reproducibility. In addition, interfering compounds caused low cross-sensitivity effects, exception made of catechol, which can be removed from aqueous solution by different methods in a pre-treatment step.

Carbon nanotubes present optimal properties for performing electrochemical analysis, such as large surface area and high electrocatalytic activity. Joshi and co-workers [28] reported a biosensor based on the enzyme-catalyzed hydrolysis of nerve agents. The screen-printed electrodes consisted of multiwall carbon nanotubes that immobilized organophosphorus hydrolase (OPH) thanks to their large surface area and hydrophobicity, enhancing the hydrolysis of P-S bonds towards demeton-S, used as a nerve agent mimic. The authors performed an amperometric detection (50 mM phosphate buffer solution with 0.1 M KCl at pH 7.4) of 2-(Diethylamino)ethanethiol (DEAET) and 2-(dimethylamino)ethanethiol (DMAET), thiol-containing hydrolysis products of R-VX and VX (thiophosphonate CWAs), respectively. The limit of detection (S/N = 3) could be estimated in 2 μM DMAET and 0.8 μM DEAET due to low noise levels. This electrode presented good reproducibility and several commonly employed pesticides were measured to evaluate the interference effect, concluding that compounds such as malathion caused minimum interference.

Khan and co-workers [60] explored an electrochemical sensor using multiwalled carbon nanotubes on indium tin oxide (ITO) surfaces and a recognition layer composed by ferrocene-amino acid conjugates non-covalently attached to MWCNTs for the detection of a CWA mimic. 5 mL of 2 M NaClO$_4$ was used as electrolyte solution to set up the alternating current voltammetry and capacitance measurements. In this study, methylposphonic acid (MPA), Diethyl Cyanophosphonate (DECP), Ethylmethylphosphonate (EMP) and Pinacolyl Methylphosphonate (PMP) were analyzed due to their reasonable mimic properties of tabun (nerve agent) and degradation products. The detection limits were found to be at subnanomolar level for a rather long detection time ranging between 60 and 90 minutes. Finally, in order to test selectivity, the sensor was exposed to diethyl ethylsulfide (DEES), a mustard gas mimic, which could not be detected and confirmed the high specificity towards tabun of the sensor developed.

Suresh and co-workers [61] developed an electrochemical method to detect ricin in water samples for in field applications. Two different electrodes were implemented, one of them using graphite powder (CPE) and the other employing multi-wall carbon nanotubes (MWCNTs). Both nanomaterials were mixed with paraffin oil. The morphology of these nanomaterials is shown in Fig. 2.2. A linear response to ricin antibodies in the range 0.625–25 ng/mL for MWCNTs and 2.5–25 ng/mL for CPE was recorded. The limit of detection was found to be 562 pg/mL for MWCNTs and 1.7 ng/mL in the case of CPE. Taking into consideration these results, MWCNTs had higher sensitivity to ricin than CPE. In other words, MWCNTs adsorbs antibodies better than CPE. However, due to non-specific binding of antibodies, this higher sensitivity can also be attributed to the electrochemical properties of carbon nanotubes.

Electrochemical sensors have demonstrated high potential for developing new methods to detect a wide range of CWAs in the liquid phase. Despite the known problems to effectively enhance selectivity and specificity to target analytes using

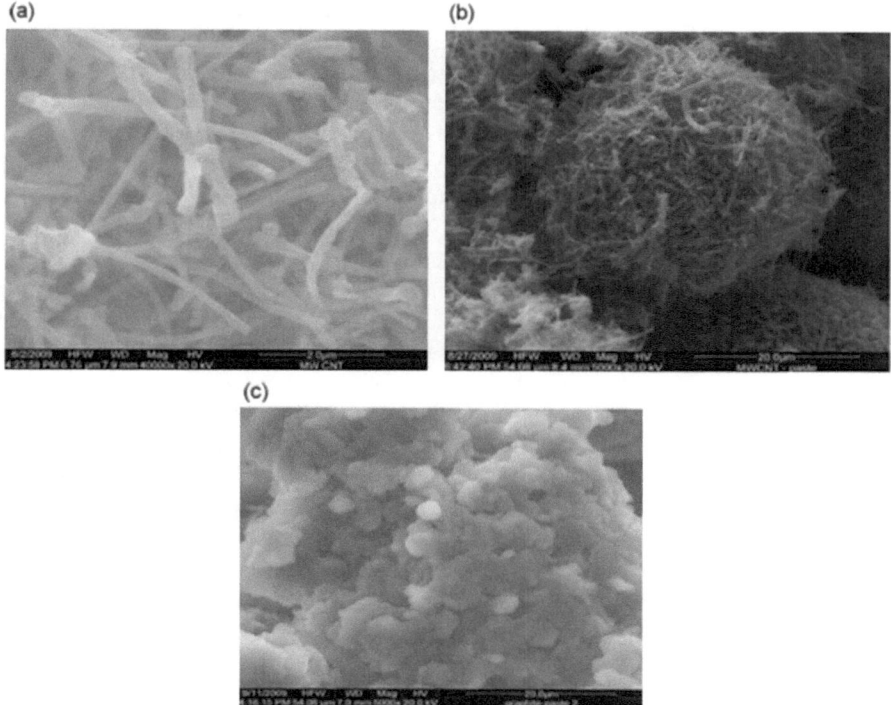

Fig. 2.2 (**a**) SEM image of MWCNT, (**b**) SEM of MWCNT/oil mixture, and (**c**) SEM image of graphite paste/oil mixture. (Reproduced from [61], ©Elsevier with permission)

electrochemical methods, in the recent years some remarkable advancements have been achieved by developing carbon nanomaterial based electrochemical biosensors.

2.2.3   Gravimetric Sensors

Gravimetric sensors consist either of bulk acoustic wave (BAW) or surface acoustic wave (SAW) resonators. Commercially available BAW devices consist of a piezoelectric quartz crystal between two gold electrodes or quartz crystal microbalance (QCM). In QCM devices, the upper electrode is modified with a coating that is able to adsorb target analytes. The detection mechanism comprises increased mass effects upon analyte adsorption, which translate in a shift in the resonance frequency of the device. SAW chemical sensors are attractive because of their small size, ruggedness, high sensitivity and large operational gas concentration dynamic range [62]. In surface acoustic devices, the propagation channel lying between the excitation and reading electrodes is coated with a sensitive film. Thin chemoselective polymer films are typically used as coatings on the sensors to reversibly sorb chemical species at the sensor's surface. The adsorption of gas/ vapor molecules on the polymer coating results in mass and/or stiffness changes, which lead to a variation in the resonance frequency of the acoustic wave propagating between the electrodes of the sensor. These changes (i.e., sensor response) are read with a frequency counter. The nature of the interaction between the coating and the gaseous molecules determines the selectivity and influences the sensitivity of the sensor and its reversibility [63].

McGill and co-workers [64] reported in the use of a SAW sensor for detecting DMMP. They employed fluoropolyol, which is a hydrogen-bond acidic polymer because it contains both hydroxyl groups and fluoro substitution [65] as DMMP-sensitive coating. They demonstrated that the SAW sensor was useful at evaluating the lifetime of a carbon-based sorbent. For doing so, they performed several filter breakthrough experiments, showing that the SAW sensor could respond in real time as DMMP advanced through the filter. Given the aim of this paper, no information on the limit of detection nor selectivity towards DMMP is available.

Wang and co-workers developed a SAW sensor for detecting DMMP and ethanol vapors [66]. They employed air-brushing for coating SAW substrates with a hybrid film consisting of acid-treated multiwalled carbon nanotubes dispersed in a polymer matrix. The polymer was poly (n, n dimethylamino propylsilsequioxane): SXNR polymer for short. The acid treatment of carbon nanotubes produces carbonyl and carboxyl functional groups, which help achieving a more homogeneous dispersion in the polymer matrix. When compared to coatings consisting exclusively of SXNR polymer, carbon-nanotube-SNXR polymer hybrids show increased reversibility and significantly ameliorated response times (of about 10 s). Concentrations of 200 ppm DMMP can be detected with moderate baseline drift. The sensor is operated at room temperature both during detection and recovery phases. A decrease in resonance frequency is observed during the detection events, which is indicative of a mass increase upon absorption of DMMP. Even though the signal to noise ratio for 200 ppm DMMP is good enough to assume that lower DMMP concentrations could be detected, the limit of detection would be probably higher than desired for the application. In addition, these sensors are rather sensitive to ethanol and cross-sensitivity to ambient moisture has not been checked, so selectivity issues can be expected.

Sayago and co-workers reported recently the use of graphene oxide (GO) coated SAW devices for detecting simulants of sarin and nitrogen mustard gases [67]. GO was selective as gas-sensitive material because it combines the two-dimensional structural features of graphene with the presence of oxygen-containing functional groups (mainly, epoxide, hydroxyl and carbonyl groups), which can potentially interact with a great variety of analytes, in particular via hydrogen bonding. GO was obtained employing Hummer's method, ultrasonically dispersed in ethanol and air brushed onto the SAW transducer. These sensors, which are operated at room temperature, show some of the highest sensitivities ever reported to DMMP and DPGME with theoretical limits of detection of 9.7 and 33.9 ppb, respectively.

However, their response and recovery times are rather slow (tens of minutes) and given the nature of the sensitive film used, important cross-sensitivity issues can be anticipated (e.g. to ammonia, nitrogen oxides, ozone, alcohols and ambient moisture).

Hwang and co-workers have reported the use of a functionalized graphene oxide coating in QCM sensors for the selective detection of DMMP [68]. In their paper, they consider using a non-stacked graphene oxide (ns-GO), because it offers a surface area and pore volume that are significantly higher than those characterizing other carbon nanomaterials [69]. In addition, non-stacked graphene oxide was functionalized with hexafluorohydoroxypropanyl benzene (HFHPB). HFHPB is considered to be a good compound for selectively detecting DMMP. It possesses an acidic proton of the hydroxy group, due to a strong electron-withdrawing trifluoromethyl group and is highly sensitive and selective to the organophosphate group of DMMP via strong hydrogen-bonding [70]. Indeed, a QCM sensor coated with an ns-GO-HFHPB film was found to be highly responsive to DMMP and almost no responsive to acetone, hexane, methanol, ethanol or moisture. The limit of detection for DMMP is close to 4 ppm, but this limit could be further decreased, for example by using a SAW instead of a QCM transducer. The main concern about these results is that target and interfering species were measured in an inert atmosphere (nitrogen). It is possible that the oxygen present in air, which is the standard background atmosphere in warfare agent detection scenarios, could affect the intensity of response and also the long-term stability of the adsorptive coating. In a closely related approach, the same group reported the functionalization of graphene quantum dots with HFHPB for detecting DMMP via hydrogen bonding [71]. While the primary sensing mechanism was monitoring changes in the photoluminescence spectrum of HFHPB-functionalized quantum dots upon DMMP adsorption, these were also employed for coating QCM transducers to prove the selective interaction with this organophosphorus compound.

2.2.4   Optical Sensors

Fluorescent biosensors have been reported for the optical detection of CWAs. Fluorescence spectroscopy is a detection method based on samples being excited at a given irradiation wavelength (generally in the visible-ultraviolet range) and, as a result, certain molecules present in the samples emit radiation at a different wavelength (generally in the visible range). By monitoring the intensity of fluorescence it is possible to determine the presence and quantify a species in the sample. This technique is fast, highly sensitive, low cost and simple [72].

Xiao and co-workers [73] have reported the use of graphene oxide (GO) as fluorescence anisotropy (FA) amplifier. Exonuclease III (Exo III) was added in order to co-amplify the fluorescence polarization detected and to recycle the process of blocker-assisted digestion of probe DNAs. The latter is an aptamer that acted as recognition element for the targeted CWA, which was ricin B-chain. Aptamer and blocker were hybridized and attached on magnetic beads (MBs), in consequence, when ricin was added, the blocker was released and hybridized with the dye-modified probe DNA on the surface of GO. The formed blocker–probe DNA duplex triggered the Exo III-assisted cyclic signal amplification by re-starting the hybridization and digestion of probe DNA and releasing the fluorophore with several nucleotides, which resulted in a low FA value [73]. In the presence of ricin B-chain, the fluorescence anisotropy decreased significantly, giving a high sensitivity towards this CWA. The excitation and emission wavelengths used, were 560 nm (green) and 581 nm (yellow), respectively. Figure 2.3 summarizes the detection pathway. In addition, the suitability of this quantitative method towards ricin was proved in orange juice. The linear range measured spanned from 1.0 μg/mL to 13.3 μg/mL and the theoretical limit of detection (LOD) was estimated to be 400 ng/mL.

Li and co-workers [74] reported a biosensor to detect ricin, specifically ricin B-chain (RTB) based on isothermal strand-displacement polymerase reaction (ISDPR). Graphene oxide (GO) was used as a substrate and an aptamer as recognition element. Ricin-binding aptamer (RBA) was first hybridized with a short blocker and subsequently immobilized on the surface of streptavidin-coated magnetic beads (MBs). In absence of CWA, the background signal by GO remains an extremely low fluorescence due to the fact that hairpin probes can be adsorbed on the GO surface through π-π stacking interactions. However, the addition of ricin releases the blocker, which hybridizes with hairpin probes and triggers the ISDPR, producing a high fluorescence intensity signal. Similarly, to the previously

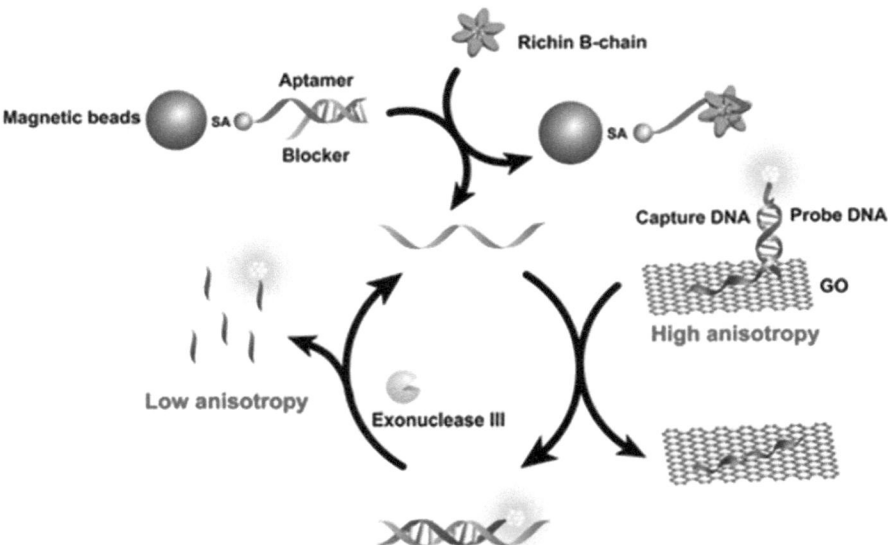

Fig. 2.3 Scheme of ricin B-chain detection using Exo III assisted, graphene oxide amplified fluorescence anisotropy. Reproduced from [73], ©Elsevier with permission

reported approach, the excitation and emission wavelengths were 560 and 580 nm, respectively. During the ISDPR reaction, a complementary DNA is synthetized and the blocker is displaced. The carboxytratramethylrhodamine (TAMRA)-labeled duplex DNA formed cannot adsorb onto the surface of GO, increasing the fluorescence signal. Besides, in order to confirm the suitability of this method, it was applied to detect RTB in different real matrices such as human urine, orange juice and castor beans. The linear range of response spanned from 0.75 μg/mL to 100 μg/mL and the limit of detection was established at 0.6 μg/mL.

Recent studies indicate that this approach could be employed for selectively detecting any target, provided that a specific aptamer was used [75]. However, this technique is still at an early research stage and seems far from meeting the requirements of ruggedness, reversibility and durability for in field CWA detection.

2.3 Conclusions and Outlook

The technology for detecting CWAs in environmental and biological samples is evolving rapidly. Currently, the best technologies available for the univocal detection of CWAs consist of gas or liquid chromatography combined with tandem mass spectrometry or ion mobility spectrometry. These instrumental techniques employ bulky and expensive equipment. Furthermore, the operation and data interpretation requires well-trained, expert personnel. More recently portable equipment employing these instrumental techniques has been developed for in field applications. However, these instruments remain cumbersome, costly and involved since they comprise sampling device, analyte concentrator, thermal desorption stage, chromatographic column and detector (mass or ion mobility spectrometers or capillary electrophoregraphs) and a computer. Research efforts are aimed at simplifying sampling methods, and miniaturizing system components. Advances towards a monolithic integration of the whole system by applying standard microsystem technologies (MEMS) could help to significantly shorten response times and lower fabrication costs. Yet, this approach needs to show it can deal with the formation of and find ways to easily remove analyte residues, avoiding the generation of false alarms.

In contrast, other innovative detection methodologies are attracting remarkable research interest because they should enable developing lightweight, inexpensive and simple to operate equipment for the real-time, in field, high sensitivity detection of CWAs. Nanotechnology is fueling the development of new functional nanomaterials with tailored interfacial properties for the sensitive and selective detection of CWAs. In particular, intensive research activity has been devoted to the synthesis, functionalization and integration onto standard transducing platforms of carbon nanomaterials such as carbon black, carbon nanotubes and graphene. These carbon nanomaterials are often dispersed into or functionalized with polymers. Polymers constitute an option to filter out potential interfering species such as ambient humidity or aliphatic and aromatic hydrocarbons. In addition, they provide hydrogen bonding specific to organophosphorus compounds, thus boosting sensitivity and selectivity towards nerve agents. These hybrid nanomaterials have been integrated into resistive or gravimetric transducing platforms for developing gas sensors. Although results are, in general, very encouraging, there are still important aspects to be addressed in future research. There is a gap to be filled between the results achievable in a laboratory and those to be expected in field and under real conditions.

- Cross-sensitivity effects caused by the presence in the ambient of oxidizing species are unknown and largely overlooked.
- Long term stability of gas sensors employing carbon nanotubes and graphene is still an issue.
- Further research is needed for achieving the mass-production of functionalized carbon nanomaterials with highly reproducible surface properties, which will lead eventually, to highly reproducible sensors.
- More research in the functionalization of carbon nanomaterials would be necessary to achieve a selective detection of the whole range of CWAs.
- In most cases the experimentally verified detection limits are still too high. The design and use of pre-concentration units should be considered.

Besides detecting CWAs in the gas phase, detecting these compounds in the liquid phase (e.g. in contaminated waters or in body fluids such as urine or plasma) is also of paramount importance. Functionalized carbon nanomaterials have been employed for developing electrochemical sensors and biosensors or fluorescent biosensors for detecting CWAs in aqueous matrices. The selectivity and limits of detection achieved for biosensors are considerably better than those of electrochemical sensors, since the former comprise carbon nanomaterials functionalized with specific receptors for target analytes. However, there are a few aspects that still need to be addressed before biosensors could be used in field detectors.

- Fabrication methods are far from being standardized. Device to device reproducibility is still an issue.
- Reversibility of the biosensor is complicated or sometimes impossible.
- The durability of a biosensor system is also an issue and ideal storage conditions before use are difficult to be met while under real, in field operation.

Despite these issues, functionalized carbon nanomaterials have potential for becoming, in the next few years, the sensitive coatings of inexpensive, miniaturized, and wearable CWA detectors. These characteristics would facilitate a widespread deployment and use within an area to be monitored. If combined with geo-localization and communication capabilities, even if such detectors would not reach the accuracy and selectivity of instrumental equipment, they would remain highly valuable for providing a rapid, real-time, on-site first alarm for improving situational awareness and identifying potential areas of hazard. Once an alarm is raised, the reliable identification of the CWA risk for alarm confirmation and hazard identification could be performed, possibly employing instrumental equipment.

Acknowledgements J.C-C. gratefully acknowledges a Martí i Franquès PhD-Fellowship from Universitat Rovira i Virgili. E.L. is supported by the Catalan Institution for Research and Advanced Studies (ICREA) via the ICREA Academia Award.

References

1. Kientz CE (1998) Chromatography and mass spectrometry of chemical warfare agents, toxins and related compounds: state of the art and future prospects. J Chromatogr A 814:1–23
2. Noort D, Benschop HP, Black RM (2002) Biomonitoring of exposure to chemical warfare agents: a review. Toxicol Appl Pharmacol 184:116–126
3. Murray V, Goodfellow F (2002) Mass casualty chemical incidents – towards guidance for public health management. Public Health 116:2–14
4. Hay A (2000) Old dogs or new tricks: chemical warfare at the millennium. Med Confl Surviv 16:37–41
5. Black RM, Read RW (1995) Improved methodology for the detection and quantitation of urinary metabolites of sulphur mustard using gas chromatography-tandem mass spectrometry. J Chromatogr B Biomed Appl 665:97–105
6. Black RM, Harrison JM, Read RW (1999) The interaction of sarin and soman with plasma proteins: the identification of a novel phosphonylation site. Arch Toxicol 73:123–126
7. Schneider JF, Boparai AS, Reed LL (2001) Screening for sarin in air and water by solid-phase microextraction-gas chromatography-mass spectrometry. J Chromatogr Sci 39:420–424
8. Muir B, Quick S, Slater BJ, Cooper DB, Moran MC, Timperley CM, Carrick WA, Burnell CK (2005) Analysis of chemical warfare agents: II Use of thiols and statistical experimental design for the trace level determination of vesicant compounds in air samples. J Chromatogr A 1068:315–326
9. Muir B, Slater BJ, Cooper DB, Timperley CM (2004) Analysis of chemical warfare agents: I Use of aliphatic thiols in the trace level determination of Lewisite compounds in complex matrices. J Chromatogr A 1028:313–320
10. Makas AL, Troshkov ML (2004) Field gas chromatography–mass spectrometry for fast analysis. J Chromatogr B 800:55–61
11. Buryakov IA (2004) Express analysis of explosives, chemical warfare agents and drugs with multicapillary column gas chromatography and ion mobility increment spectrometry. J Chromatogr B Anal Technol Biomed Life Sci 800:75–82
12. Nagashima H, Kondo T, Nagoya T, Ikeda T, Kurimata N, Unoke S, Seto Y (2015) Identification of chemical warfare agents from vapor samples using a field-portable capillary gas chromatography/membrane-interfaced electron ionization quadrupole mass spectrometry instrument with Tri-Bed concentrator. J Chromatogr A 1406:279–290
13. Sekiguchi H, Matsushita K, Yamashiro S, Sano Y, Seto Y, Okuda T, Sato A (2006) On-site determination of nerve and mustard gases using a field-portable gas chromatograph-mass spectrometer. Forensic Toxicol 24:17–22
14. Witkiewicz Z, Sliwka E, Neffe S (2016) Chromatographic analysis of chemical compounds related to the chemical weapons convention. TrAC – Trends Anal Chem 85:21–33
15. Giordano BC, Collins GE (2007) Synthetic methods applied to the detection of chemical warfare nerve agents. Curr Org Chem 11:255–265
16. Hartmann-Thompson C, Hu J, Kaganove SN, Keinath SE, Keeley DL, Dvornic PR (2004) Hydrogen-bond acidic Hyperbranched polymers for surface acoustic wave (SAW) sensors. Chem Mater 16:5357–5364
17. Jenkins AL, Bae SY (2005) Molecularly imprinted polymers for chemical agent detection in multiple water matrices. Anal Chim Acta 542:32–37
18. Harvey SD (2005) Molecularly imprinted polymers for selective analysis of chemical warfare surrogate and nuclear signature compounds in complex matrices. J Sep Sci 28:1221–1230

19. Zhou Y, Yu B, Eric Shiu A, Levon K (2004) Potentiometric sensing of chemical warfare agents: surface imprinted polymer integrated with an indium tin oxide electrode. Anal Chem 76:2689–2693
20. Novak JP, Snow ES, Houser EJ, Park D, Stepnowski JL, McGill RA (2003) Nerve agent detection using networks of single-walled carbon nanotubes. Appl Phys Lett 83:4026–4028
21. Snow E, Perkins F, Houser E, Badescu S, Reinecke T (2005) Chemical detection with a single-walled carbon nanotube capacitor. Science 307:1942–1945
22. Menzel ER, Menzel LW, Schwierking JR (2005) Rapid fluorophosphate nerve agent detection with lanthanides. Talanta 67:383–387
23. Yang Y, Ji H-F, Thundat T (2003) Nerve agents detection using a Cu^{2+}/l-cysteine bilayer-coated microcantilever. J Am Chem Soc 125:1124–1125
24. Zheng Q, Fu YC, Xu JQ (2010) Advances in the chemical sensors for the detection of DMMP a simulant for nerve agent sarin. Procedia Eng 7:179–184
25. Sabelle S, Renard P-Y, Pecorella K, de Suzzoni-Dézard S, Créminon C, Grassi J, Mioskowski C (2002) Design and synthesis of chemiluminescent probes for the detection of cholinesterase activity. J Am Chem Soc 124:4874–4880
26. Liu G, Riechers SL, Mellen MC, Lin Y (2005) Sensitive electrochemical detection of enzymatically generated thiocholine at carbon nanotube modified glassy carbon electrode. Electrochem Commun 7:1163–1169
27. Joshi K, Tang J, Haddon R, Wang J, Chen W, Mulchandani A (2005) A disposable biosensor for organophosphorus nerve agents based on carbon nanotubes modified thick film strip electrode. Electroanalysis 17:54–58
28. Joshi KA, Prouza M, Kum M, Wang J, Tang J, Haddon R, Chen W, Mulchandani A (2006) V-type nerve agent detection using a carbon nanotube-based amperometric enzyme electrode. Anal Chem 78:331–336
29. Singh VV (2016) Recent advances in electrochemical sensors for detecting weapons of mass destruction. A review. Electroanalysis 28:920–935
30. Wallace KJ, Morey J, Lynch VM, Anslyn EV (2005) Colorimetric detection of chemical warfare simulants. New J Chem 29:1469
31. Zhang SW, Swager TM (2003) Fluorescent detection of chemical warfare agents: functional group specific Ratiometric Chemosensors. J Am Chem Soc 125:3420–3421
32. Wang J, Zima J, Lawrence NS, Chatrathi MP, Mulchandani A, Collins GE (2004) Microchip capillary electrophoresis with electrochemical detection of thiol-containing degradation products of V-type nerve agents. Anal Chem 76:4721–4726
33. Wang F, Gu H, Swager TM (2008) Carbon nanotube/Polythiophene Chemiresistive sensors for chemical warfare agents. J Am Chem Soc 113:5392–5393
34. Meng FL, Guo Z, Huang XJ (2015) Graphene-based hybrids for chemiresistive gas sensors. TrAC – Trends Anal Chem 68:37–47
35. Tang R, Shi Y, Hou Z, Wei L (2017) Carbon nanotube-based chemiresistive sensors. Sensors 17:882
36. Horrillo MC, Martí J, Matatagui D, Santos JP, Sayago I, Gutiérrez J, Martin-Fernandez I, Ivanov P, Grcia I, Cané C (2011) Single-walled carbon nanotube microsensors for nerve agent simulant detection. Sensors Actuators B Chem 157:253–259
37. Wang LC, Tang KT, Kuo CT, Ho CL, Lin SR, Sung Y, Chang CP (2009) A portable electronic nose system with chemiresistor sensors to detect and distinguish chemical warfare agents. 2009 IEEE 3rd Int Conf Nano/Molecular Med Eng Nanomed 2009:302–306.
38. Briglin SM, Burl MC, Freund MS, Lewis NS, Matzger AJ, Ortiz DN, Tokumaru P (2000) Progress in use of carbon-black-polymer composite vapor detector arrays for land mine detection. Detect Remediat Technol Mines Minelike Targets V 4038:530–538
39. Hopkins AR, Lewis NS (2001) Detection and classification characteristics of arrays of carbon black/organic polymer composite chemiresistive vapor detectors for the nerve agent simulants dimethylmethylphosphonate and diisopropylmethylphosponate. Anal Chem 73:884–892
40. Llobet E (2013) Gas sensors using carbon nanomaterials: a review. Sensors Actuators B Chem 179:32–45

41. Wang Y, Zhou Z, Yang Z, Chen X, Xu D, Zhang Y (2009) Gas sensors based on deposited single-walled carbon nanotube networks for DMMP detection. Nanotechnology 20:345502
42. Cattanach K, Kulkarni RD, Kozlov M, Manohar SK (2006) Flexible carbon nanotube sensors for nerve agent simulants. Nanotechnology 17:4123–4128
43. Chuang PK, Wang LC, Kuo CT (2013) Development of a high performance integrated sensor chip with a multi-walled carbon nanotube assisted sensing array. Thin Solid Films 529:205–208
44. Kong L, Wang J, Luo T, Meng F, Chen X, Li M, Liu J (2010) Novel pyrenehexafluoroiso-propanol derivative-decorated single-walled carbon nanotubes for detection of nerve agents by strong hydrogen-bonding interaction. Analyst 135:368–374
45. Kong L, Wang J, Fu X, Zhong Y, Meng F, Luo T, Liu J (2010) p-Hexafluoroisopropanol phenyl covalently functionalized single-walled carbon nanotubes for detection of nerve agents. Carbon 48:1262–1270
46. Fennell J, Hamaguchi H, Yoon B, Swager T (2017) Chemiresistor devices for chemical warfare agent detection based on polymer wrapped single-walled carbon nanotubes. Sensors 17:982
47. Wei L, Shi D, Ye P, Dai Z, Chen H, Chen C, Wang J, Zhang L, Xu D, Wang Z, Zhang Y (2011) Hole doping and surface functionalization of single-walled carbon nanotube chemiresistive sensors for ultrasensitive and highly selective organophosphor vapor detection. Nanotechnology 22:425501
48. Charlie Johnson AT, Staii C, Chen M, Khamis S, Johnson R, Klein ML, Gelperin A (2006) DNA-decorated carbon nanotubes for chemical sensing. Phys Status Solidi Basic Res 243:3252–3256
49. Lee JS, Kwon OS, Park SJ, Park EY, You SA, Yoon H, Jang J (2011) Fabrication of ultrafine metal-oxide- decorated carbon Nano fi bers for DMMP sensor application. ACS Nano 5:7992–8001
50. Robinson JT, Perkins FK, Snow ES, Wei Z, Sheehan PE (2008) Reduced graphene oxide molecular sensors. Nano Lett 8:3137–3140
51. Hu N, Wang Y, Chai J, Gao R, Yang Z, Kong ESW, Zhang Y (2012) Gas sensor based on p-phenylenediamine reduced graphene oxide. Sensors Actuators B Chem 163:107–114
52. Alizadeh T, Soltani LH (2016) Reduced graphene oxide-based gas sensor array for pattern recognition of DMMP vapor. Sensors Actuators B Chem 234:361–370
53. Safavi A, Ahmadi R, Mahyari FA (2014) Simultaneous electrochemical determination of l-cysteine and l-cysteine disulfide at carbon ionic liquid electrode. Amino Acids 46:1079–1085
54. Singh VV, Gupta G, Sharma R, Boopathi M, Pandey P, Ganesan K, Singh B, Tiwari DC, Jain R, Vijayaraghavan R (2009) Detection of chemical warfare agent nitrogen Mustard-1 based on conducting polymer phthalocyanine nanorod modified electrode. Synth Met 159:1960–1967
55. Periyakaruppan A, Arumugam PU, Meyyappan M, Koehne JE (2011) Detection of ricin using a carbon nanofiber based biosensor. Biosens Bioelectron 28:428–433
56. Cinti S, Minotti C, Moscone D, Palleschi G, Arduini F (2017) Fully integrated ready-to-use paper-based electrochemical biosensor to detect nerve agents. Biosens Bioelectron 93:46–51
57. Lee JH, Park JY, Min K, Cha HJ, Choi SS, Yoo YJ (2010) A novel organophosphorus hydrolase-based biosensor using mesoporous carbons and carbon black for the detection of organophosphate nerve agents. Biosens Bioelectron 25:1566–1570
58. Singh VV, Nigam AK, Yadav SS, Tripathi BK, Srivastava A, Boopathi M, Singh B (2013) Graphene oxide as carboelectrocatalyst for in situ electrochemical oxidation and sensing of chemical warfare agent simulant. Sensors Actuators B Chem 188:1218–1224
59. Mahyari M (2016) Electrochemical sensing of thiodiglycol, the main hydrolysis product of sulfur mustard, by using an electrode composed of an ionic liquid, graphene nanosheets and silver nanoparticles. Microchim Acta 183:2395–2401
60. Khan MK, Kerman K, Petryk M, Kraatz HB (2008) Noncovalent modification of carbon nanotubes with ferrocene – amino acid conjugates for electrochemical sensing of chemical warfare agent mimics. Anal Chem 80:2574–2582
61. Suresh S, Gupta AK, Rao VK, Kumar O, Vijayaraghavan R (2010) Amperometric immunosensor for ricin by using on graphite and carbon nanotube paste electrodes. Talanta 81:703–708

62. Wohltjen H, Dessy R (1979) Surface acoustic wave probe for chemical analysis. I. Introduction and instrument description. Anal Chem 51:1458–1464
63. McGill RA, Abraham MH, Grate JW (1994) Choosing polymer coatings for chemical sensors. ChemTech 24:27–37
64. Dominguez DD, Chung R, Nguyen V, Tevault D, McGill RA (1998) Evaluation of SAW chemical sensors for air filter lifetime and performance monitoring. Sensors Actuators B Chem 53:186–190
65. Harsányi G (1995) Polymer films in sensor applications: technology, materials, devices and their characteristics. Technomic Publication Company, Lancaster
66. Wang LC, Tang KT, Kuo CT, Yang SR, Sung Y, Hsu HL, Jehng JM (2009) A high-performance nanocomposite material based on functionalized carbon nanotube and polymer for gas sensing applications. Olfaction Electron Nose Proc 1137:369–372
67. Sayago I, Matatagui D, Fernández MJ, Fontecha JL, Jurewicz I, Garriga R, Muñoz E (2016) Graphene oxide as sensitive layer in love-wave surface acoustic wave sensors for the detection of chemical warfare agent simulants. Talanta 148:393–400
68. Hwang HM, Hwang E, Kim D, Lee H (2016) Mesoporous non-stacked graphene-receptor sensor for detecting nerve agents. Sci Rep 6:1–8
69. Yoon Y, Lee K, Baik C, Yoo H, Min M, Park Y, Lee SM, Lee H (2013) Anti-solvent derived non-stacked reduced graphene oxide for high performance supercapacitors. Adv Mater 25:4437–4444
70. Fifield LS, Grate JW (2010) Hydrogen-bond acidic functionalized carbon nanotubes (CNTs) with covalently-bound hexafluoroisopropanol groups. Carbon 48:2085–2088
71. Hwang E, Hwang HM, Shin Y, Yoon Y, Lee H, Yang J, Bak S, Lee H (2016) Chemically modulated graphene quantum dot for tuning the photoluminescence as novel sensory probe. Sci Rep 6:1–10
72. Pumera M (2011) Graphene in biosensing. Mater Today 14:308–315
73. Xiao X, Tao J, Zhang HZ, Huang CZ, Zhen SJ (2016) Exonuclease III-assisted graphene oxide amplified fluorescence anisotropy strategy for ricin detection. Biosens Bioelectron 85:822–827
74. Li CH, Xiao X, Tao J, Wang DM, Huang CZ, Zhen SJ (2017) A graphene oxide-based strand displacement amplification platform for ricin detection using aptamer as recognition element. Biosens Bioelectron 91:149–154
75. Wang Y, Li Z, Wang J, Li J, Lin Y (2011) Graphene and graphene oxide: biofunctionalization and applications in biotechnology. Trends Biotechnol 29:205–212

Chapter 3
Sensing Volatile Organic Compounds by Phthalocyanines with Metal Centers: Exploring the Mechanism with Measurements and Modelling

Dogan Erbahar, Savas Berber, and Dilek D. Erbahar

Abstract Ab initio density functional theory calculations can be used to study the electronic structure of phthalocyanines (Pcs) with different metal centers and functional groups as well as their interaction with selected organic analytes. Optimum adsorption site of the analytes can be interpreted in terms of the charge distribution of the frontier orbitals of Pc and that the Pc reactivity correlates well with the HOMO-LUMO gap. Calculated analyte-Pc interaction energies provide useful information about the suitability of specific Pc isomers to bind specific molecules, deciding about the suitability for sensing of organic pollutants in aqueous media.

Keywords Organic pollutants · Chemical sensor · Phthalocyanine · Modeling · Ab initio

3.1 Introduction

There is a growing demand for highly sensitive sensors capable of detecting organic pollutants [1]. Among these, phthalocyanines with different metal ion centers called metal phthalocyanines (MPcs) and fluoroalkyloxy substituents attached in different configurations have emerged as promising candidates for sensitive materials.

D. Erbahar (✉)
Faculty of Engineering, Department of Mechanical Engineering, Acibadem, Kadikoy, Dogus University, Istanbul, Turkey
e-mail: derbahar@dogus.edu.tr

S. Berber
Department of Physics, Gebze Technical University, Gebze, Turkey

D. D. Erbahar
TUBITAK Marmara Research Center, Materials Institute, Kocaeli, Turkey

© Springer Nature B.V. 2019
C. Bittencourt et al. (eds.), *Nanoscale Materials for Warfare Agent Detection: Nanoscience for Security*, NATO Science for Peace and Security Series A: Chemistry and Biology, https://doi.org/10.1007/978-94-024-1620-6_3

Phthalocyanines are now well established as sensitive materials for gas sensors [2–5], also they are considered very promising compounds for the use in liquid sensing, too. The initial experimental results confirm high sensitivity towards a wide range of analytes [6–9], but the precise interaction mechanism between the analyte and the sensing molecule is unknown.

Phthalocyanines can coordinate with almost all the metals in the Periodic Table, and these MPc complexes vary greatly in their structures and sensing properties. Existing theoretical investigations of MPcs have focused on the electronic structure of an isolated, single molecule MPc [10–12], the analyte-free film in its β-crystalline form [13] or the molecular interaction of the MPc monomer with a MPc substrate [14]. The other way to tune the structures and sensing properties of phthalocyanine compounds is to introduce different numbers of substituents with different electronic properties onto the peripheral or non-peripheral positions of the phthalocyanine ring. Zhong et al. studied the molecular structures, molecular orbitals and atomic charges of a series of substituted metal free phthalocyanine with tetra and octa configurations on the non-peripheral or peripheral positions of phthalocyanine ring [15]. However, substantial influence of the type of substituent, substitution pattern (tetra, octa), substituent position (nonperipheral and peripheral), and the choices of metal center ion on the electronic structure and sensing mechanism were not generally studied.

Here, we present results of ab initio density functional calculations that provide a better insight into the adsorption bond strength, the preferential adsorption site, the role of particular metal centers replacing the two inner H atoms in Pc in tuning the electronic properties of Pc and selectively discriminating among the analytes, the nature of frontier orbitals, and the fundamental HOMO-LUMO gap. We study specifically fluorinated and non-fluorinated MPcs with different numbers of substituents at peripheral or non-peripheral positions of the phthalocyanine ring, different substitution patterns (tetra, octa) and different metal center ions (Ni, Zn, Cu, Co).

It is found that the optimum adsorption site of the analytes can be interpreted in terms of the charge distribution of the frontier orbitals of Pc and that the Pc reactivity correlates well with the HOMO-LUMO gap. Our computational results are in good agreement with the experimental sensitivity data. Calculated analyte-Pc interaction energies provide useful information about the suitability of specific Pc isomers to bind specific molecules, deciding about the suitability for sensing of organic pollutants in aqueous media.

Computational details are described in Sect. 3.2, and the results are given in Sect. 3.3. After the discussion of the molecular structure in Sect. 3.3.1, the reactive adsorption sites and relative chemical activity of different MPcs are discussed based on the frontier orbitals in Sect. 3.3.2, which are further augmented by our adsorption energy results that are presented in Sect. 3.3.3. Finally, we summarize our results and conclude in Sect. 3.4.

3.2 Computational Details

Geometry optimization and total-energy calculations reported are based on the density functional theory (DFT), as implemented in the SIESTA code [16]. We used the Perdew–Zunger form [17] of the Ceperley–Alder exchange–correlation functional [18] in the local density approximation (LDA) to DFT. The behavior of valence electrons was described by norm-conserving Troullier–Martins pseudopotentials [19] with partial core corrections in the Kleinman–Bylander factorized form [20]. An atomic orbital basis with double-ζ polarization orbitals was used to expand the electronic wavefunctions. Geometry optimization with no symmetry constraints has been performed using a conjugate gradient algorithm and continued until all force components on each atom were less than 0.01 eV/Å.

3.3 Results and Discussion

3.3.1 Molecular Structure

The basic geometry of the phthalocyanine molecule is shown in Fig. 3.1 and the type as well as the location of substitute groups considered in this study are defined in Table 3.1.

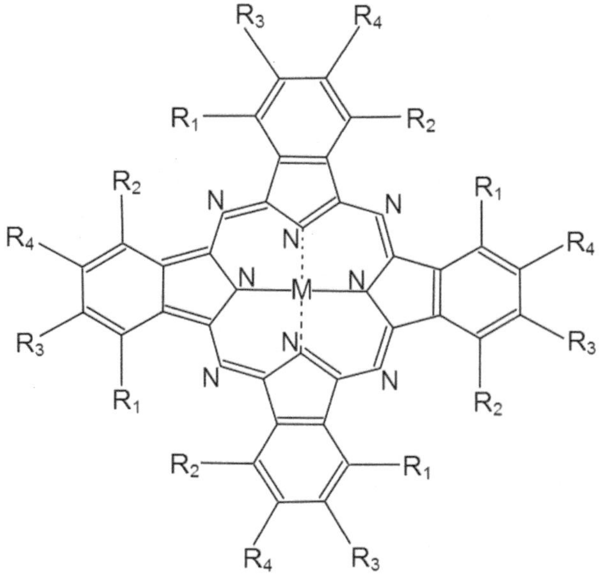

Fig. 3.1 Basic structure of phthalocyanine with a metal (M) center

Table 3.1 Codes used to identify the position of chemical substituents in the basic phthalocyanine structure, shown in Fig. 3.1. The molecules are specified by Code(S)-Code (M), e.g. 2a. The substituent patterns are illustrated in Fig. 3.2

Code(S)	Substituents	Substituent pattern
1	$R_1 = O–CH_2CH_2CH_3$; $R_2 = R_3 = R_4 = H$	tetra, np
2	$R_1 = O–CH_2CF_2CHF_2$; $R_2 = R_3 = R_4 = H$	tetra, np
3	$R_1 = R_2 = R_3 = H$; $R_4 = O–CH_2CH_2CH_3$	tetra, p
4	$R_1 = R_2 = R_3 = H$; $R_4 = O–CH_2CF_2CHF_2$	tetra, p
5	$R_1 = R_2 = H$; $R3 = R_4 = O–CH_2CH_2CH_3$	octa
6	$R_1 = R_2 = H$; $R3 = R_4 = O–$ $CH_2CF_2CHF_2$	octa
Code(M)	Metal	
a	Ni	
b	Zn	
c	Cu	
d	Co	

Different MPc molecules are available varying in the varying the substitute groups R_1, R_2, R_3, R_4 around the benzene ring in the core part of the MPc, where a metal atom is four-coordinated by N atoms as shown in Fig. 3.1.

The first step in classifying Pcs is to distinguish the number of substituents. In octa configuration, there are eight substituent groups that are at the positions of R_1 and R_2 or R_3 and R_4. In tetra configuration there are four substitute groups that could be at any position of R_1, R_2, R_3, R_4.

An octa configuration with peripheral substituent positions is depicted Fig. 3.2a. Another possible geometry for the octa configuration is a non-peripheral structure, which is not considered since the substitute groups hinder each other, and the resulting molecular structure is too complicated for computational modeling.

Tetra compounds are available with substituents in peripheral, p, (Fig. 3.2b) and non-peripheral, np, (Fig. 3.2c) positions. In tetra configuration, the substitution position may be different for each of four benzene rings. Therefore, there are several structural isomers with reduced symmetries. Based on our structure optimization and total energy results, it is found that the structures with C4 symmetry among all possible structure isomers are more stable. Thus, we only consider the structures with C4 symmetry for tetra compounds.

In order to distinguish different structures, each structure is referred by a code name that consists of a number for the substituent pattern, and a letter for the central metal atom type, which are listed in Table 3.1. Note that, in Table 3.1, the even numbers for the substituent pattern correspond to fluorinated systems. The central

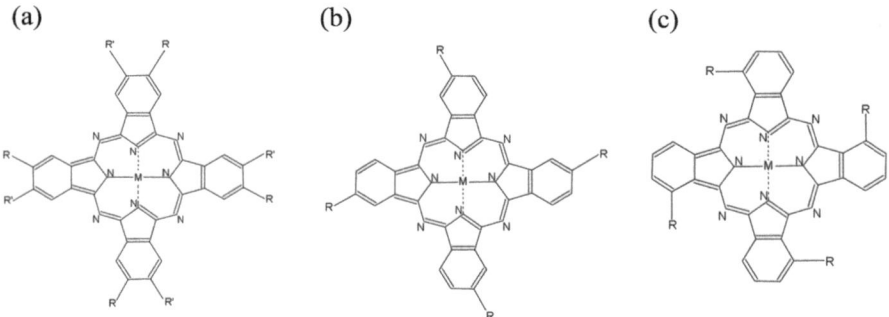

Fig. 3.2 Structural arrangement in phthalocyanine (Pc) molecules. (**a**) peripheral (p) octa Pc, (**b**) peripheral (p) tetra Pc, (**c**) non-peripheral (np) tetra Pc

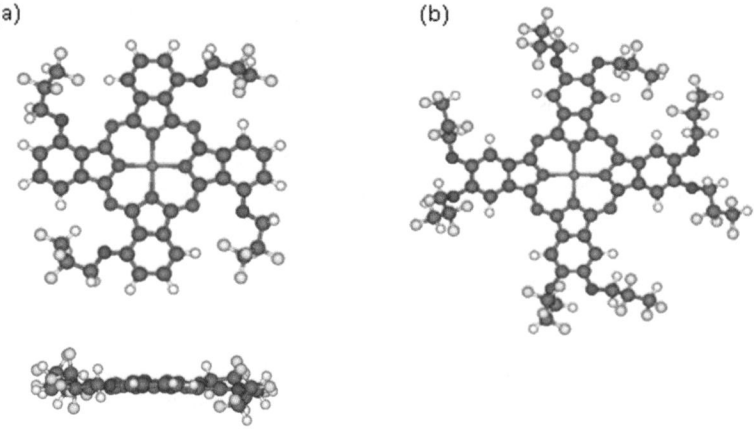

Fig. 3.3 Optimized geometry of (**a**) 2c, top view (top), side view (bottom) and (**b**) 6c phthalocyanines, with the nomenclature given in Table 3.1

atoms in this work are Ni, Zn, Cu, and Co, which are sufficient to compare our computational findings to the experimental results [9].

Calculated atomic structures of CuPc molecules in non-peripheral tetra and peripheral octa geometries are shown in Fig. 3.3a, b respectively. The structures in Fig. 3.3 are obtained starting from the initial geometries with flat profiles. For all investigated MPc's the central metal atom stays in the plane of the Pc core without any puckering. The substituents are seen to have non-flat profiles as seen in Fig. 3.3. Many different initial structures were prepared by varying the rotation of the substitute group around the C-O bond. Although, the structures that are shown in Fig. 3.3 had the lowest total energies, it is found that there are only minute differences between the total energies of competing structures and energetically the best structure.

3.3.2 Active Adsorption Sites and Frontier Molecular Orbitals

Whereas DFT has been designed to correctly reproduce the total charge density and total energy in the electronic ground state, it provides only limited and much less accurate information about individual electronic states. Even though the HOMO-LUMO gap is notoriously underestimated by DFT, still the binding energy trends and the character of Kohn-Sham states provide useful insight into the nature of chemical bonds. In particular, the character of frontier molecular orbitals (MOs) provides valuable information about the molecular reactivity, which usually involves these states [21]. In the following, we will attempt to associate the most reactive sites with the spatial distribution of the frontier MOs.

The key features in the electronic structure of MPcs are briefly discussed in Fig. 3.4 using the spin resolved electronic density of states (DOS) and the frontier orbitals. In Fig. 3.4a, solid lines are used for the majority spin, dashed lines for the minority spin, where the negative of the DOS is shown for the minority spin. The symmetry between the majority and minority spins in Fig. 3.4a is broken near the Fermi level with an exchange splitting of the order of 1 eV. The HOMO state for the majority spin, which is depicted in Fig. 3.4b, is shifted to the LUMO level for the minority spin, which is shown in Fig. 3.4c. The wave functions in Fig. 3.4b, c indicates that the exchange-split states are localized around the metal atom with a strong contribution from d-orbitals.

The spatial distribution of the frontier orbitals in Fig. 3.4b, c point out that the central metal atom should play an important role in the interaction of the CuPc with the analytes. Our results for the other MPcs are in line with the results that are

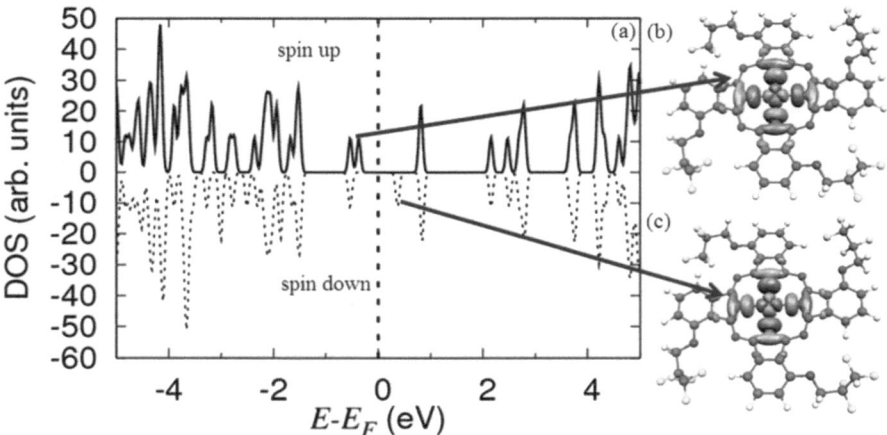

Fig. 3.4 (**a**) Spin-polarized electronic density of states (DOS) near the Fermi level (EF) of tetra-phthalocyanine with a Cu center (2c). Charge distribution of the (**b**) highest occupied and (**c**) lowest unoccupied molecular orbitals. The phase of the HOMO and LUMO wavefunction has been distinguished by gray scale in (**b**) and (**c**)

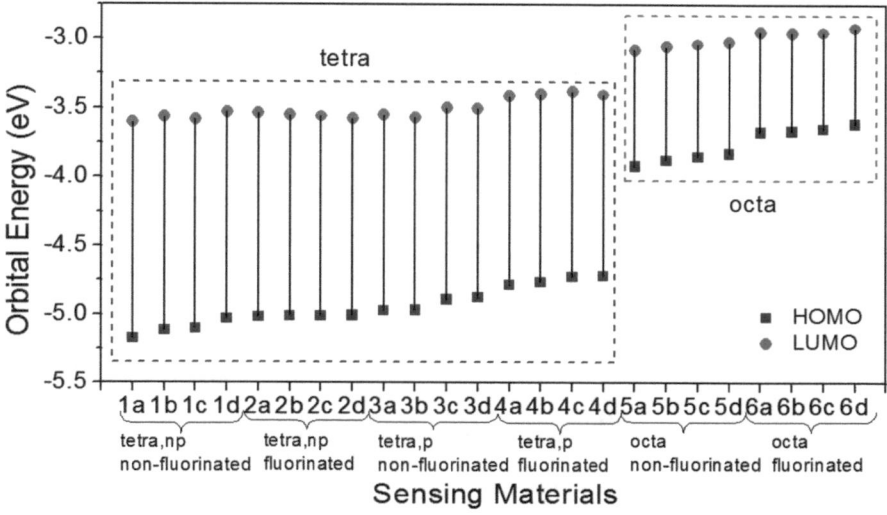

Fig. 3.5 Calculated energies of the highest occupied (HOMO) and lowest unoccupied (LUMO) molecular orbitals of tetra- and octa-phthalocyanine with different metal centers and different functional groups, as specified in Table 3.1

presented in Fig. 3.4. Therefore, for all the MPcs we expect the central atom to be the main adsorption site for the analytes.

Next, we use the HOMO-LUMO energy gap as an indicator of the chemical reactivity of the MPc molecules. The HOMO and LUMO energy levels are shown in Fig. 3.5, where the levels are with respect to the vacuum level. The chemical reactivity of a sensing material directly correlates with the sensitivity values measured in the experiments.

The energy levels in Fig. 3.5 are grouped in terms of the adsorption position, where for example 1a, 1b, 1c and 1d form a group. The central metal atom is varied within the same group. For example, 1a has Ni at the center while 1b has Zn. In each group, the energy gap narrows moving from Ni, Zn, Cu to Co. The most striking feature is the large difference between the HOMO-LUMO gaps of tetra and octa configurations. Interpreting the HOMO-LUMO gap as an indicator for the reactivity, octa geometries should be more reactive, therefore show better sensing ability. Comparing different groups, fluorinated systems have narrower energy gaps, which indicate that fluorinated substitute groups have higher electron affinity. Therefore, the sensor characteristics of fluorinated systems are expected to be better than that of non-fluorinated systems. The comparison of our results in the same group reveals that Cu and Co central atoms should be chosen for far better sensing properties. Note that since the dependence of sensitivity on the central metal atom type is the same for all groups, our conclusions about the preferred metal for a sensor application is valid conclusion for all substituents, and atomic configurations.

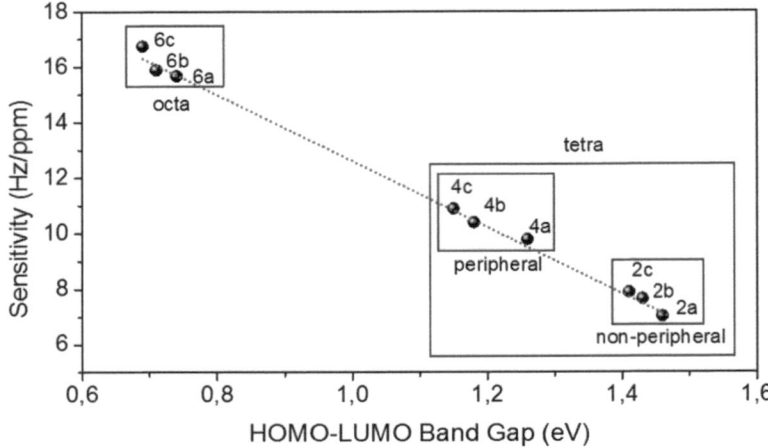

Fig. 3.6 A comparison of the experimental sensitivity values taken from Ref. [9] of various tetra- and octa-phthalocyanines with the calculated HOMO-LUMO gaps. Red line is a guide to the eye

The energy gap values of peripheral tetra compounds are smaller than that of non-peripheral tetra compounds, as shown in Fig. 3.5, which indicates that the substituents at the peripheral positions have a more significant influence on the electronic structure than those at non peripheral positions. Therefore, we expect the peripheral configuration to exhibit better sensor characteristics, which is indeed the case as shown in Fig. 3.6. The experimental data used in Fig. 3.6 is taken from Ref. [9].

3.3.3 Interaction Between Organic Molecules and Phthalocyanine

In the following, the adsorption energies are used to determine the relativity sensitivity of sensor materials. Higher adsorption energies translate into better sensitivities in sensor applications. In Fig. 3.7, experimental sensitivity results for structure 7c, taken from Ref. [9], are compared to our calculated adsorption energies, which are directly related to the sensitivity.

The adsorption energy is defined as the difference between the total energies before and after the adsorption. The total energy before the adsorption is taken as the summation of the total energies of an isolated analyte molecule and an isolated MPc. According to our definition of the adsorption energy, higher values indicate energetically more favorable adsorption geometries. Therefore, the magnitude of the adsorption energy directly correlates with the experimental sensitivity values. In agreement with the experimental results, the Pc-based sensors are better for

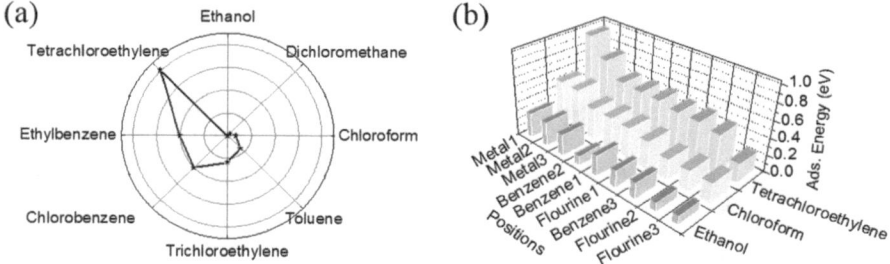

Fig. 3.7 (**a**) Observed polar plot sensitivities showing the overall affinity and preferences of the sensitive material 6c upon exposure to different organic compounds (data from Ref. [9]). (**b**) Calculated binding energy of chloroform, ethanol and tetrachloroethylene on different adsorption sites of tetra-phthalocyanine. The Pc nomenclature is given in Table 3.1

tetrachloroethylene, and less-efficient for chloroform. The calculated adsorption energies for ethanol, presented in Fig. 3.7b, are the lowest. Therefore, the MPcs are not efficient in sensing ethanol. Since our preliminary results indicate that the adsorption energy for ethanol is very close to the value for water molecules, we believe sensing ethanol in aqueous environment is nearly impossible. Although we show only the results for 7c compound in Fig. 3.7, we performed structure optimization and total energy calculations for all the systems and obtained adsorption energies. The general trend in Fig. 3.7b, and the conclusions we reached are robust for all the systems we consider in this work. Comparing our adsorption energy based results in Fig. 3.7 with our HOMO-LUMO energy gap values in Fig. 3.5, we find that the qualitative conclusions that we reached based on the HOMO-LUMO gap values are in good agreement with the more quantitative analysis.

In order to gain a better understanding for the microscopic origin of these findings, we next concentrate on the exact adsorption positions and orientations of the analytes. We distinguish three major adsorption regions: central metal atom, benzene ring and substitute groups. For each adsorption site, several orientations of the analytes were considered to obtain the optimum adsorption geometries after a full structure relaxation in conjugate-gradient algorithm. For example, in Fig. 3.7b, there are three adsorption energies for the metal site, namely metal1, metal2, and metal3. The number of different orientations at each site is selected until all possibilities are exhausted. As we deduced from the frontier orbitals, the adsorption energies in Fig. 3.7b also point out that the metal site is the dominant adsorption location.

Selected optimized geometries are depicted in Fig. 3.8 as ball-and-stick models to explain the main results for the exact atomic structures of analytes on MPc sensing molecules. Since tetrachloroethylene is a planar molecule, it differs from the other analytes in Fig. 3.8, and presents stronger orientation dependence in adsorption energies. Energetically most stable atomic structure of the tetrachloroethylene, which is depicted in Fig. 3.8a, is parallel to the Pc plane which is the result of increased interaction area in the parallel configurations. Although the metal-analyte

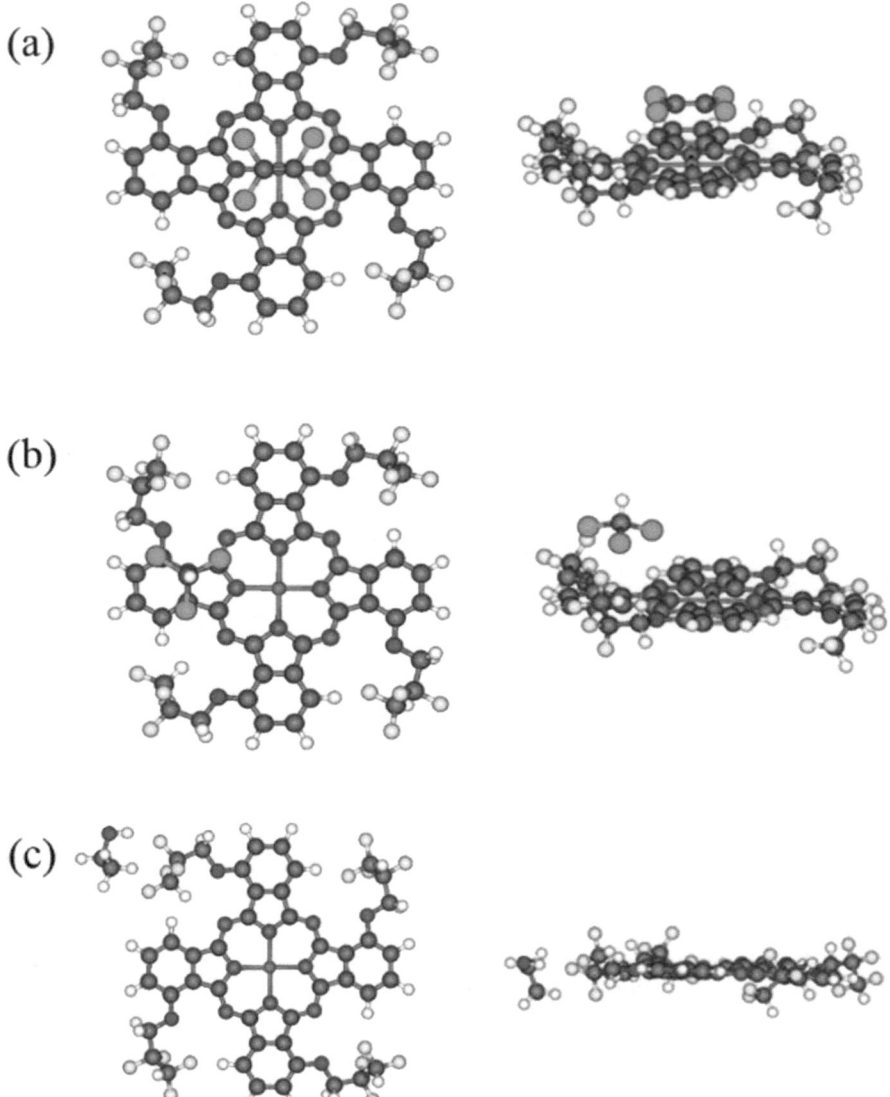

Fig. 3.8 Preferential adsorption of (**a**) tetrachloroethylene at the metal site, (**b**) chloroform at the benzene site, and (**c**) ethanol at the fluorine site

interaction dominates the energetics, a weak bonding between the analyte and the Pc contributes enough to alter the optimum orientation. We observe the same tendency for other analytes: when adsorbed on the central atom, the analyte prefers the orientation that maximizes the interaction area with the MPc.

If the analyte adsorbs on the adsorption sites other than the metal site, the analyte of course chooses to maximize the interaction area. However, the optimum adsorption orientation may also be influenced by the anisotropy of the analyte. For example, it is not clear a priori if the H in chloroform should point to the MPc, or point away. Therefore, in our calculations, we distinguish these two distinct cases, where either halogens face toward the Pc or the hydrogen points to the Pc molecule. As shown in Fig. 3.8b, we observe the tendency of halogens to be at the interface with the MPc. Moreover, in aqueous environment, the H atom in the analyte should prefer making H-bonds with water molecules, and thus should point away from the MPc.

The optimum atomic structure of an ethanol adsorbed on a substitute group is shown in Fig. 3.8c. As mentioned above, since ethanol shows the smallest adsorption energy it could move around the substituent easier than other analytes. We notice that the bonding is less directional, and ethanol can form energetically almost degenerate structures. Therefore, the real adsorption geometry may not as well determine as the optimum structures at other adsorption sites.

As mentioned above, we observe that the HOMO-LUMO energy gap is a good indicator for determining the best sensing material. Of course, a sensor may show different sensitivities for different analytes. Following our findings, we believe that the analytes with smaller energy gaps are more reactive, and our sensors should be more sensitive to those analytes. Calculated HOMO-LUMO energy gaps reveal that tetrachloroethylene has the smallest gap among the analytes that we considered here. Therefore, tetrachloroethylene is expected to be sensed at higher sensitivities than other analytes on the basis of the energy gap values. This expectation is in good agreement with both our calculated adsorption energies and experimentally observed sensitivities [9].

3.4 Summary and Conclusions

We perfomed ab initio structural optimization, total energy, and electronic structure calculations of the metal phthalocyanines (MPcs) that are candidates for next generation aqueous sensors. In addition, we have determined the interaction energies of several analytes with the MPcs. We find that the optimum adsorption site of the analytes can be interpreted in terms of the charge distribution of the frontier orbitals of Pc and that the Pc reactivity correlates well with the HOMO-LUMO gap. We find that the central metal atom plays an important role in increasing the sensitivity of Pc sensors, and the chemical composition as well as the configuration of the substitute groups influences the sensitivity. Our computational results are in good agreement with the experimental sensitivity data. Calculated analyte-Pc interaction energies provide useful information about the suitability of specific Pc isomers to bind specific molecules, deciding about the suitability for sensing of organic pollutants in aqueous media.

References

1. James D, Scott SM, Ali Z, O'Hare WT (2005) Chemical sensors for electronic nose systems. Microchim Acta 149(1):1–17. https://doi.org/10.1007/s00604-004-0291-6
2. Zhou R, Josse F, Gopel W, Ozturk ZZ, Bekaroglu O (1996) Phthalocyanines as sensitive materials for chemical sensors. Appl Organomet Chem 10(8):557–577. https://doi.org/10.1002/(Sici)1099-0739(199610)10:8<557::Aid-Aoc521>3.3.Co;2-V
3. Basova TV, Tasaltin C, Gurek AG, Ebeoglu MA, Ozturk ZZ, Ahsen V (2003) Mesomorphic phthalocyanine as chemically sensitive coatings for chemical sensors. Sensor Actuators B-Chem 96(1–2):70–75. https://doi.org/10.1016/S0925-4005(03)00487-8
4. Mumyakmaz B, Ozmen A, Ebeoglu MA, Tasaltin C (2008) Predicting gas concentrations of ternary gas mixtures for a predefined 3D sample space. Sensor Actuators B-Chem 128(2):594–602. https://doi.org/10.1016/j.snb.2007.07.062
5. Ozmen A, Tekce F, Ebeoglu MA, Tasaltin C, Ozturk ZZ (2006) Finding the composition of gas mixtures by a phthalocyanine-coated QCM sensor array and an artificial neural network. Sensor Actuators B-Chem 115(1):450–454. https://doi.org/10.1016/j.snb.2005.10.007
6. Giancane G, Guascito MR, Malitesta C, Mazzotta E, Picca RA, Valli L (2009) QCM sensors for aqueous phenols based on active layers constituted by tetrapyrrolic macrocycle Langmuir films. J Porphyrins Phthalocyanines 13(11):1129–1139. https://doi.org/10.1142/S1088424609001467
7. Harbeck M, Erbahar DD, Gurol I, Musluoglu E, Ahsen V, Ozturk ZZ (2010) Phthalocyanines as sensitive coatings for QCM sensors operating in liquids for the detection of organic compounds. Sensor Actuators B-Chem 150(1):346–354. https://doi.org/10.1016/j.snb.2010.06.062
8. Erbahar DD, Gurol I, Ahsen V, Ozturk ZZ, Musluoglu E, Harbeck M (2011) Explosives detection in sea water with phthalocyanine quartz crystal microbalance sensors. Sens Lett 9(2):745–748. https://doi.org/10.1166/sl.2011.1607
9. Harbeck M, Erbahar DD, Gurol I, Musluoglu E, Ahsen V, Ozturk ZZ (2011) Phthalocyanines as sensitive coatings for QCM sensors: comparison of gas and liquid sensing properties. Sensor Actuators B-Chem 155(1):298–303. https://doi.org/10.1016/j.snb.2010.12.038
10. Day PN, Wang ZQ, Pachter R (1998) Calculation of the structure and absorption spectra of phthalocyanines in the gas-phase and in solution. Theochem J Mol Struct 455(1):33–50. https://doi.org/10.1016/S0166-1280(98)00238-3
11. Ishikawa N, Maurice D, HeadGordon M (1996) An ab initio study of excited states of the phthalocyanine magnesium complex and its cation radical. Chem Phys Lett 260(1–2):178–185. https://doi.org/10.1016/0009-2614(96)00828-7
12. Nguyen KA, Pachter R (2001) Ground state electronic structures and spectra of zinc complexes of porphyrin, tetraazaporphyrin, tetrabenzoporphyrin, and phthalocyanine: a density functional theory study. J Chem Phys 114(24):10757–10767. https://doi.org/10.1063/1.1370064
13. Lozzi L, Santucci S, La Rosa S, Delley B, Picozzi S (2004) Electronic structure of crystalline copper phthalocyanine. J Chem Phys 121(4):1883–1889. https://doi.org/10.1063/1.1766295
14. Yamaguchi T (1997) Electronic states of copper phthalocyanine adsorbed on Si(001)2x1 surface. J Phys Soc Jpn 66(3):749–756. https://doi.org/10.1143/Jpsj.66.749
15. Zhong AM, Zhang YX, Bian YZ (2010) Structures and spectroscopic properties of non-peripherally and peripherally substituted metal-free phthalocyanines: a substitution effect study based on density functional theory calculations. J Mol Graph Model 29(3):470–480. https://doi.org/10.1016/j.jmgm.2010.09.003
16. Soler JM, Artacho E, Gale JD, Garcia A, Junquera J, Ordejon P, Sanchez-Portal D (2002). Pii S0953-8984(02)30737-9) The SIESTA method for ab initio order-N materials simulation. J Phys-Condens Matter 14(11):2745–2779. https://doi.org/10.1088/0953-8984/14/11/302
17. Perdew JP, Zunger A (1981) Self-interaction correction to density-functional approximations for many-electron systems. Phys Rev B 23(10):5048–5079. https://doi.org/10.1103/PhysRevB.23.5048

18. Ceperley DM, Alder BJ (1980) Ground-state of the electron-gas by a stochastic method. Phys Rev Lett 45(7):566–569. https://doi.org/10.1103/PhysRevLett.45.566
19. Troullier N, Martins JL (1991) Efficient pseudopotentials for plane-wave calculations. 2. Operators for fast iterative diagonalization. Phys Rev B 43(11):8861–8869. https://doi.org/10.1103/PhysRevB.43.8861
20. Kleinman L, Bylander DM (1982) Efficacious form for model pseudopotentials. Phys Rev Lett 48(20):1425–1428. https://doi.org/10.1103/PhysRevLett.48.1425
21. Woodward RB, Hoffmann R (1969) The conservation of orbital symmetry. Angewandte Chemie International Edition in English 8(11):781–853. https://doi.org/10.1002/anie.196907811

Chapter 4
Chemical Sensors for VOC Detection in Indoor Air: Focus on Formaldehyde

Marc Debliquy, Arnaud Krumpmann, Driss Lahem, Xiaohui Tang, and Jean-Pierre Raskin

Abstract This chapter will present a brief overview of the current sensors for VOC detection, in particular formaldehyde which has become one of the most problematic gases in indoor air. Many sensing technologies were exploited for this purpose but, in this chapter, we will focus on the impedimetric sensors. These sensors consist in a sensitive layer deposited on an insulating substrate fitted with a pair of electrodes. The detection is based on the change of conductivity of the sensitive layer due to surface interactions with the target gas provoking an electron transfer. This kind of sensor acts as a simple variable resistance and is often called chemiresistor. By principle, these sensors are simple, easy to integrate in classical electronics and cheap. Considering the nature of the sensitive coating, we can distinguish several families: metal oxide sensors, semiconductor polymer sensors or based on graphene. All 3 types of sensors will be described in this chapter.

Keywords Sensors · Volatile organic compounds · Metal oxides

4.1 Introduction

Volatile Organic Compounds in indoor air have become these last years the subject of a big concern. The main emission sources are: furniture, paintings, varnishes, wood protection, construction materials, etc. In particular, formaldehyde is now considered by the authorities as one of the priority pollutants because of its

M. Debliquy (✉) · A. Krumpmann
Material Science Department, Université de Mons (UMONS), Mons, Belgium
e-mail: Marc.Debliquy@umons.ac.be

D. Lahem
Materials Science Department, Materia Nova, Mons, Belgium

X. Tang · J.-P. Raskin
Institute of Information and Communication Technologies, Electronics and Applied Mathematics (ICTEAM), Université catholique de Louvain (UCL), Louvain-la-Neuve, Belgium

© Springer Nature B.V. 2019
C. Bittencourt et al. (eds.), *Nanoscale Materials for Warfare Agent Detection: Nanoscience for Security*, NATO Science for Peace and Security Series A: Chemistry and Biology, https://doi.org/10.1007/978-94-024-1620-6_4

carcinogenic [1] character and because of the multiplication of sources in our close environment. The world health organization (WHO) guideline for indoor air formaldehyde concentration is 80 ppb (0.1 mg/m^3) [2]. Methods based on air sample collection and lab measurements are available and are the most used methods [3]. They are accurate and reliable but these methods are expensive, are not real time and some statistics issues are raised because of that. That is why, it is important to detect and measure formaldehyde in real time, in situ and with a low cost equipment since this data is essential for preventive actions (choice of the materials/furniture treatment, ventilation or indoor air purification). Devices based on chemical sensors are a good solution as these systems can be very sensitive, low cost and easily integrated in common electronics to build portable systems [4]. A chemical sensor consists in a sensitive coating that interacts with the target molecule (absorption or adsorption) resulting in a change of its physical properties (color, refractive index, density, conductivity, permittivity, etc....). This change is then measured (transduction mechanism) and transformed in an exploitable electrical signal. Of course, the performances of the sensor (sensitivity, selectivity, response time, recovery time, lifetime, etc) critically depend on the sensitive material and the challenge is to determine the suitable materials. Different transduction mechanisms were investigated: electrochemical sensors [5–7], quartz crystal microbalances (QCM) [8–11], surface acoustic wave (SAW) sensors [12–14], optical fiber sensors [15–18], microcapacitors [19] semiconductor sensors [20–25].

In this chapter, we will focus on electrochemical systems exploiting organic or inorganic semiconductor materials. These sensors are based on the monitoring of the conductivity changes of a semiconducting sensitive coating due to a reversible doping by the adsorbed gases. The sensor acts just like a variable resistance and is often called chemoresistive sensor or chemoresistor.

This chapter will shortly present an overview of chemoresistive sensors for formaldehyde detection based on metal oxides on molecularly imprinted conducting polymers and on graphene.

4.2 Working Principles

The detection principle of chemoresistors is based on the conductivity change of those materials when they are in contact with certain gases [26–29]. The general scheme of such a sensor is presented in Fig. 4.1 [30].

When a gas is adsorbed onto the surface, if the interaction is strong enough (meaning chemisorption), an electron transfer between the semiconductor and the adsorbed species takes place resulting in an increase (decrease) of the charge carrier concentration in the semiconductor and as a consequence an increase (decrease) of the conductivity. As there is an equilibrium between the gas concentration in the atmosphere and the adsorbed quantity, a direct relation is expected between the conductivity change and the gas concentration. This phenomenon can be observed with organic or inorganic semiconductors or even with hybrid materials.

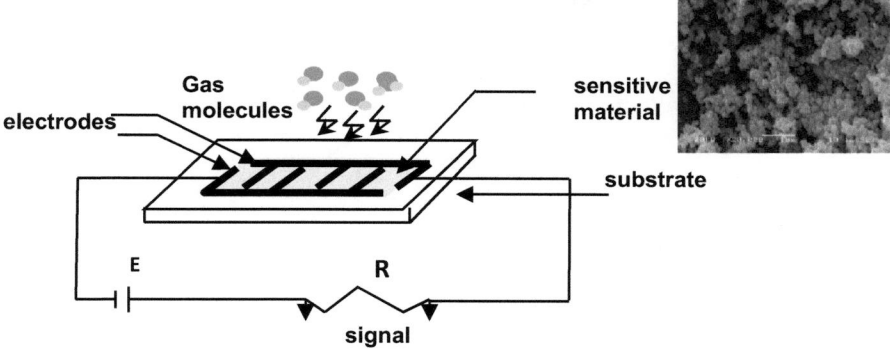

Fig. 4.1 Scheme of a semiconductor sensor

The sensor acts just like a variable resistance and is often called chemoresistive sensor or chemoresistor.

As the detection principle deals with chemisorption and surface reactions with the gas, an activation energy is generally needed. Otherwise, the response to the gas and the recovery are too slow or even no response is observed. That is why a heating element is usually added on the substrate (very often in the form of a coil). Gas sensors working at room temperature is the quest for the Holy Grail.

Athough organic materials are potentially very attractive because they can be more easily modified than inorganic materials and so "tailor" the performances, they encountered considerably less success than the inorganic semiconductors, in particular metal oxides which encountered a commercial success. The main reasons are the bad long-term stability and it is often impossible to use them at temperatures at which gas-solid interactions proceed rapidly and reversibly.

Seiyama, demonstrated with a ZnO thin film that gas sensing is possible with simple electrical devices [31]. He studied a simple chemoresistive device sensitive to propane based on ZnO thin films operating at 485 °C. Taguchi fabricated and patented the first chemoresistive gas sensor device for practical applications using tin dioxide (SnO_2) as the sensitive material [32]. This paved the way to intense researches to extend the principle to numerous applications of the semiconductor gas sensors.

Because of their simplicity, low cost, small size and ability to be integrated into electronic devices, chemical sensors have been the object of an extensive work these last two decades as they have a big potential in all kinds of applications: industrial emission control, household security, vehicle emission control and environmental monitoring, agricultural, biomedical, etc., [33–38].

The detection mechanism for metal oxide sensors (often called MOS) is represented in Fig. 4.2 [27, 28, 30, 39].

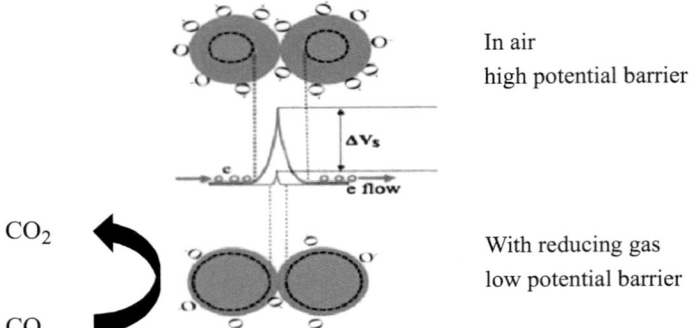

In air
high potential barrier

CO_2

CO

With reducing gas
low potential barrier

Fig. 4.2 Detection mechanism for metal oxide sensors

For an n-type semiconductor, in air, oxygen will be adsorbed at the surface of the oxide. An electron transfer occurs from the oxide to the oxygen leading to the formation of oxygen ions O^{2-}, O^- or O_2^-.

The nature of the ion depends on the temperature. The consequence is that the surface of the crystals is depleted, a potential barrier appears and the conductivity is decreased. Very often, the sensitive films are constituted of grains in contact and the potential barrier will modulate the transfer of electrons between the grains and so the global conductivity.

So for n-type semiconductors, the global conductance can be expressed as: [40, 41]

$$G = k \cdot \sigma_b \cdot \exp\left(\frac{-eV_s}{k_bT}\right) \tag{4.1}$$

with

G, the conductance of the sensitive layer;
k, geometrical factor (including thickness of the sensitive layer, length, width of the electrodes);
σ_b, bulk conductivity;
V_s, potential at the surface.

This formula is established assuming the electron transfer limitation by the surface potential.

Adsorption of the gases can modify this potential leading to conductivity changes. For instance, reduction of the oxygen pressure in the atmosphere and so reduction of the adsorbed oxygen surface concentration leads to an increase of the conductivity.

In general, for n-type semiconductors, the contact with reducing gases (CO, H_2, CH_4, ...) will increase the conductivity. Indeed, reducing gases can react with adsorbed oxygen releasing then the trapped electrons, which increases the charge carrier concentration. For oxidizing gases (NO_2, Cl_2, O_3, ...), their adsorption will cause an increase of the barrier because of the trapping of electrons.

For p-type semiconductors, the effects are opposite.

The nature of the oxide is a key factor for the choice of the sensitive layer for a given gas but the defects at the surface of the material play an important role as they are adsorption centers for gases. The control of the surface is a major concern for gas sensing. The most studied metal oxides for gas sensing are SnO_2, ZnO, In_2O_3, WO_3, TiO_2 as n-type semiconductors and NiO, Cr_2O_3, CuO as p-type semiconductors and composites [28].

The main drawbacks of these sensors are: the lack of selectivity due to the fact that the interaction between the semiconductor and the target gas is not always specific and the need to heat the sensitive layer to rather high temperatures leading to increased power consumption. The current developments strive to reduce these drawbacks by studying a lot of other materials and by exploiting the opportunities provided by the new nanoscale technologies. Nanotechnologies nowadays, enable the manipulation of matter at the molecular or atomic level and allow then to control the morphology of the sensitive materials. It is quite clear that grain-size reduction at nanometric scale and tailoring of the crystallography can enhance the detection properties of metal oxides. A bigger sensitivity, a lower operating temperature and a better selectivity are expected.

Depending on the application of interest and availability of fabrication methods, different surface morphology and configurations of the metal-oxides have been achieved; including single crystals, thin films, thick films and one dimensional (1-D) nanostructures [42]. Among all these, 1-D nanostructures have recently attracted much attention because of their potential applications in gas sensors [43–47]. 1-D nanostructures are particularly suited to this application because of their high surface-to-volume ratio as well as their good chemical and thermal stabilities.

To enhance the sensing characteristics of the metal oxide gas sensors, it is very common to modify the surface by adding suitable promoters (metal particles, foreign metal oxide, ions) on the metal oxide layer [48, 49]. For instance, Pt promotes the gas-sensing reaction by the spill-over mechanism (massively exploited for heterogenous catalysis). Pt clusters catalyze the dissociation of the gases favouring then the reactions with the adsorbed oxygen species. Figure 4.3 shows the mechanism with hydrogen. The result is an increase of the sensitivity to hydrogen and a lowering of the working temperature.

Organic semiconductors have been studied for three decades for gas sensing. Among the organic semiconductors, owing to their exceptional optical, electronic and electrochemical properties [50–52], metallophthalocyanines have attracted considerable interest. In particular, these organic molecules, because of the presence of delocalized π electrons, are sensitive to oxidizing or reducing gases at ppm concentrations. Indeed, the adsorption of gases presenting a redox character can involve an electron transfer during the interaction with these molecules leading to a change of the charge carrier number and as a consequence a change of the conductivity. This electron transfer can also lead to a change of the optical spectrum. As the interaction with the gases is reversible, phthalocyanines were studied as sensitive element for conductive and optical gas sensors [53–58]. Another important characteristic for the use in gas detection is the remarkable chemical and thermal stability of the phthalocyanine derivatives.

Fig. 4.3 Spill over
mechanism with hydrogen
enhancing the oxidation rate

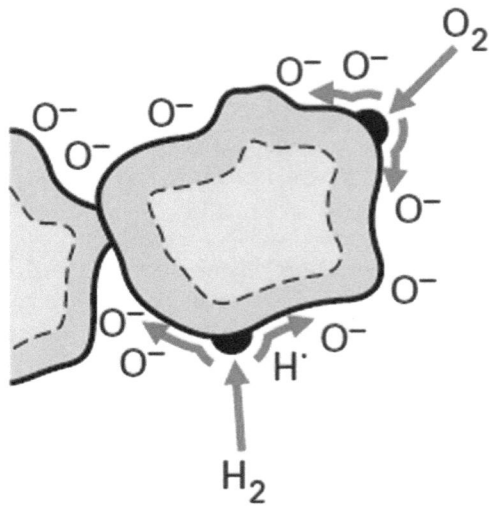

The following equations illustrate the reaction with an oxidizing gas (like NO_2) of metal phthalocyanine (MPc).

$$MPc + A_{Ox} \rightarrow \left(MPc^+, A_{Ox}{}^-\right) \tag{4.2}$$

$$\left(MPc^+, A_{Ox}{}^-\right) \rightarrow MPc^+ + A_{Ox} \tag{4.3}$$

$$MPc^+ + A_{Ox}{}^- \rightarrow MPc + A_{Ox}{}^- + h^+ \tag{4.4}$$

MPc usually shows a p-type character in air as it is in contact with oxidizing gases of the atmosphere.

An important difference with the metal oxides is that gases can diffuse in the lattice of the organic materials. The effect is not limited to the surface of the crystals and a reversible charge transfer can happen in the bulk. The diffusion can be slow limiting then the response speed. That is why, these materials are usually exploited as thin films deposited by various means (vacuum evaporation, Langmuir-Blodgett) [53, 54, 57]. These molecules were essentially studied for the detection of NO_2, H_2S, O_3 and NH_3.

The other important family of sensitive organic materials is the family of the conducting polymers. As detailed in paragraph 4, these polymers show a sensitivity to a lot of gases and in particular volatile organic compounds. These materials and derivatives were used in electronic noses for the recognition of odors.

For conducting polymers, because of the incorporation of the target molecules in the bulk not only the charge carrier concentration can vary but also the mobility.

The main advantage of the organic materials is that they can operate at room temperature.

To overcome the shortcomings of these materials and to take benefits of synergetic effects, lots of works have been carried out for the study of hybrid structures mixing organic and inorganic compounds.

More recently, carbon based materials attracted the attention of the researchers for the use in gas sensing. Carbon nanotubes (CNT) were extensively studied for various applications [59–61]. CNT are not per se sensitive or selective to gases. Therefore, these structures need to be functionalized. This word means that the surface of the CNT is modified by grafting specific molecules that have an affinity for the gas to detect. Doing so, the electronic structure of the CNT is locally modified by the grafted molecule. When the gas is adsorbed on this molecule, the electronic conduction in the CNT is modified. Another possibility is to create heterojunctions with another material (for instance a metal oxide or metal cluster) forming a composite with strong synergetic effects. This strategy led to the development of very sensitive gas sensors working at room temperature with good reversibility and short response times [62, 63].

In the same vein, graphene, considered as one of the most promising 2D materials because of its unique properties, was studied for gas sensing. This 2D material has indeed all the carbon atoms on the surface and presents then a huge specific surface area. Like CNT, graphene cannot operate properly alone and needs to be functionalized [64]. Some developments of graphene based gas sensors are presented in paragraph 5.

4.3 VOC Sensors Based on Metal Oxides

Semiconductor metal oxide gas sensors, well known for their high sensitivity, have been considered as promising candidates for monitoring the harmful VOCs due to their advantages such as high sensitivity, simple manufacture technique, low cost, stability and rapid response and recovery time [23, 65, 66].

For the detection of formaldehyde, HCHO, various metal oxides have been studied including particularly the n-type semiconducting oxides SnO_2, ZnO, WO_3 and In_2O_3 [67–71]. P-type metal oxide semiconductors have also been studied as sensing materials for HCHO detection [72, 73]. As an important p-type semiconductor, nickel oxide (NiO) has been extensively investigated due to its catalytic properties for the oxidation of formaldehyde [74–76].

In one of our previous works [76], we showed that the synthesis route for the NiO, and it is a general problem for metal oxide sensors, is of primary importance on the results. Figure 4.4 shows the morphology of the sensitive material of NiO synthesized following different routes and Fig. 4.5 shows their responses to formaldehyde.

The response S in % is defined as:

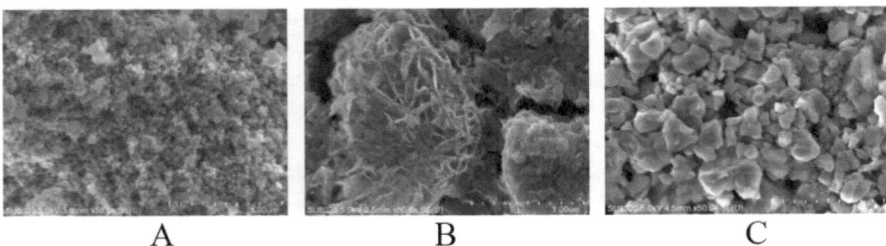

Fig. 4.4 FE-SEM image showing nanostructures NiO obtained from precursor prepared by (**a**) sol-gel from nickel chloride, (**b**) precipitation from nickel sulfate and (**c**) precipitation from nickel malonate [76]

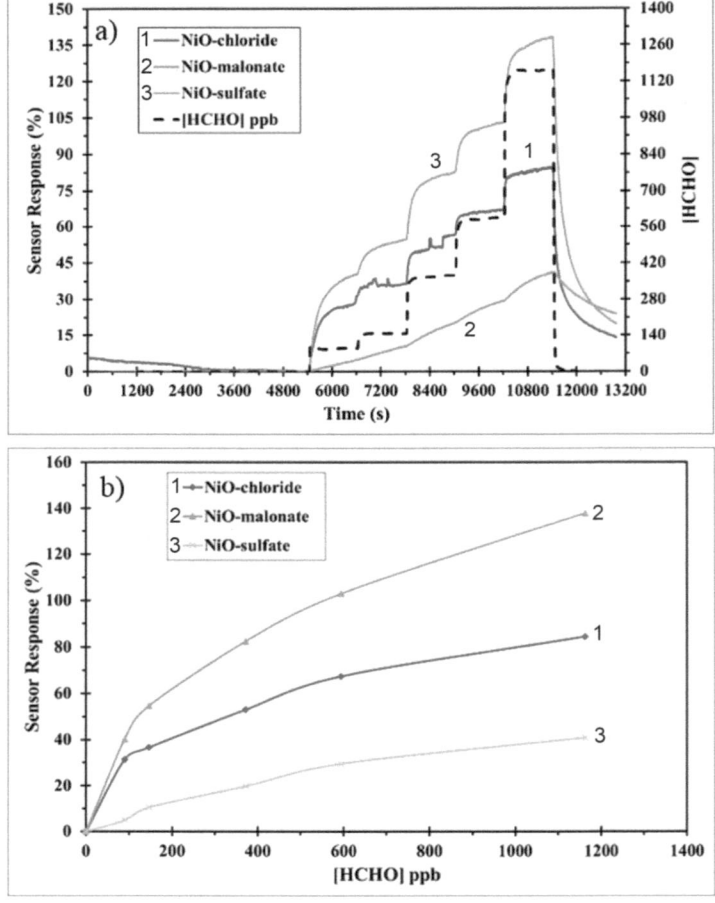

Fig. 4.5 (**a**) Dynamic response of different NiO sensors exposed to formaldehyde at 200 °C. (**b**) Response versus concentration of different NiO sensors

$$S = \frac{R_{gas} - R_{air}}{R_{air}} \times 100 \qquad (4.5)$$

Where, R_{gas} and R_{air} are the resistance of sensor after exposure to formaldehyde and clean air, respectively.

However, one of the critical issues currently limiting the wide use of these oxides is their lack of selectivity towards formaldehyde [77], and the need to heat the sensitive layer to rather high temperatures usually around 200–400 °C, leading to increased power consumption.

To overcome the problem of cross-sensitivity and enhance the HCHO sensitivity, various ways have been suggested in the literature, which include the addition of a noble catalytic metal (such as platinum and gold) and the use of composite metal oxides. Indeed, various sensitive materials such as doped-SnO_2 [78–80], doped-ZnO [81, 82], doped-CuO [73], CdO-mixed In_2O_3 [83], $LaFe_{1-x}Zn_xO_3$ [84], $La_{1-x}Pb_xFeO_3$ [77] and Cd-doped TiO_2-SnO_2 [85] have been reported to be selective and more sensitive to HCHO at ppm or sub-ppm levels.

In order to reduce the power consumption, the area of the sensitive layer has been miniaturized using micro-machined hotplates, MEMS technology [86]. An example is shown in Figs. 4.1 and 4.6 (home made design in University of Louvain-la-Neuve).

In the last decade, MEMS-based formaldehyde sensing devices have been proposed. The NiO thin film has been used as the formaldehyde sensing layer in references [80, 87, 88] and Pd-doped SnO_2 in reference [79].

On the other hand, the development of HCHO sensing materials operating at room-temperature sensors has been reported. S. Lin et al. have prepared a room-temperature formaldehyde gas sensor using the TiO_2 nanotube array [89]. L. Peng et al. [90] have demonstrated that applying UV light irradiation on the zinc oxide nanorods is an effective approach to achieve good response to formaldehyde at room temperature. H. Mu et al. [91] have reported that ZnO/graphene sensors showed high response to formaldehyde at room-temperature.

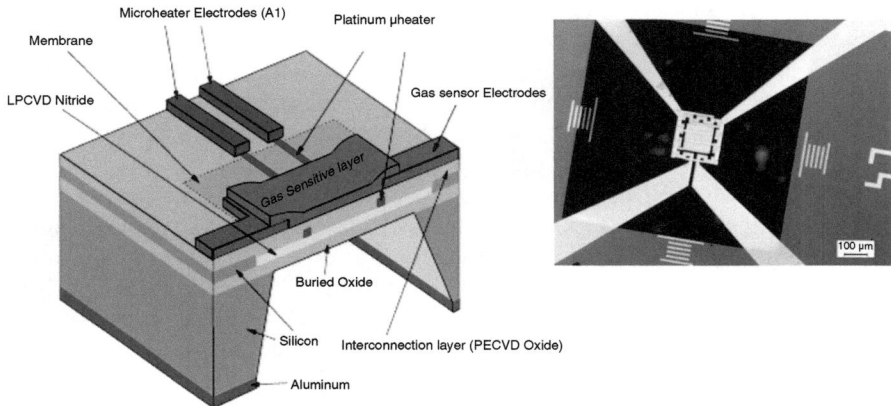

Fig. 4.6 Structure of a micro hotplate on a silicon substrate with embedded platinum heater

4.4   Sensors Based on Conducting Polymers

A large number of polymers have been investigated as sorbents for building sensing coatings [92]. Among the various polymers, the utilization of conducting polymers as sensitive layers in gas sensors is considered as a promising approach because they offer the possibility to work at room temperature in contrast with the metal oxides. Polypyrrole was one of the first polymers used in gas sensor for detection of alcohols and other organic vapors [93–95]. The second most studied material is polyaniline which has demonstrated a good sensing ability at room temperature; good environmental stability, high electrical conductivity, good reversibility, reproducibility and good performance [96–99]. However, this kind of sensors still suffers from a lack of selectivity. This lack of selectivity was very often compensated by using arrays of sensors combined in an electronic nose [100, 101].

Molecularly imprinted polymers (MIP) are polymers that are synthesized incorporating the target molecule we want to detect as a template. Functional monomers form a complex around the template and are linked afterwards to form a polymer constituted of a series of "cages" trapping the template. Once the synthesis is complete, the template molecule is extracted, leaving a molecular cavity imprinted in the polymer matrix that allows the polymer to selectively recognize the target molecule [102, 103]. The goal was to mimick what is done in living organisms. These elements are cheap, easy to synthesize and can be adapted to any kind of surface. Very soon, the potential of these materials for chemical sensing (in gas or liquid phase) was envisaged for all kinds of applications [104–110]. MIP coatings were then studied as sorbents for gas sensing based on QCM [111, 112], on optical fibers [18, 113–115]. Using conducting polymers for building MIP's hopefully allows to get semiconductor gas sensors working at room temperature with a good selectivity.

Sensors based on polypyrrole deposited by electropolymerization were shown to be able to detect acetaldehyde in the range [1–20 ppm].

In reference [116], the sensor is a chemoresistor and it consists in a MIP sensitive layer, polypyrrole (PPy) deposited on a pair of interdigitated gold electrodes laying on a SiO_2/Si substrate. MIP Polypyrrole can be synthetized by various methods but, as the sensitive layer has to be deposited on metal electrodes, electropolymerization seems convenient to insure adhesion. For comparison, non-imprinted PPy films called NIP were prepared under the same conditions but without template. The behaviours of MIP and NIP are compared by impedance measurements and mass adsorption measurements with a quartz crystal microbalance.

The sensors consist in gold interdigitated electrodes deposited on a SiO_2/Si substrate (Fig. 4.1). The chips are bonded in a 2 pin-TO header (the third pin is the ground). The sensitive film will be directly grown by electropolymerization on the electrodes. The width and the finger spacing of the electrodes is 2 μm, a 5-μm-thick MIP film can cover the metallic electrodes with a continuous film (Fig. 4.7).

Fig. 4.7 Picture of the fabricated sensors: (**a**) electrodes before deposition (**b**) general view (**c**) electrodes covered with MIP PPy film (**d**) SEM picture

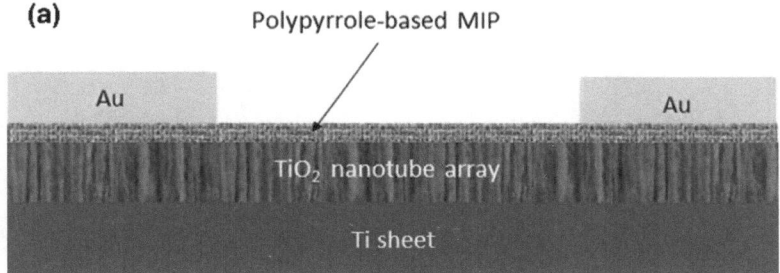

Fig. 4.8 Scheme of the sensor MIP on TiO$_2$ nanotube array

The advantages of this method are a good adhesion and the fact that the sensitive layer is deposited only on the useful area avoiding further photolithography steps.

The obtained results show a limited sensitivity of this simple sensor. Similar results can be obtained with formaldehyde. One of the reasons is that the structure of the coating is rather compact and the specific surface area is small. So, to improve the sensitivity, the MIP film was grown on a porous template.

In reference [117], the detailed study of a sensor based on a polypyrrole film grown on TiO$_2$ nanotubes is given (Fig. 4.8).

The MIP consists in a mixture of 3-carboxylic acide pyrrole and pyrrole grown by electropolymerization on TiO$_2$ nanotubes (NTA).

The nanotubes were produced by anodization of a Titanium foil in an ethylene glycol bath. Figure 4.9 shows the structures of the NTA and the MIP film. The MIP film has a porous fishnet-like structure with 20 nm diameter wires. This structure is compared to the structure obtained on flat gold surface keeping the same electropolymerization conditions (Fig. 4.9d).

The sensor shows interesting sensitivity at room temperature for formaldehyde with high selectivity towards other organic gases and humidity. This is due to the specific interaction with formaldehyde in the cavities like sketched in Figs. 4.10 and 4.11.

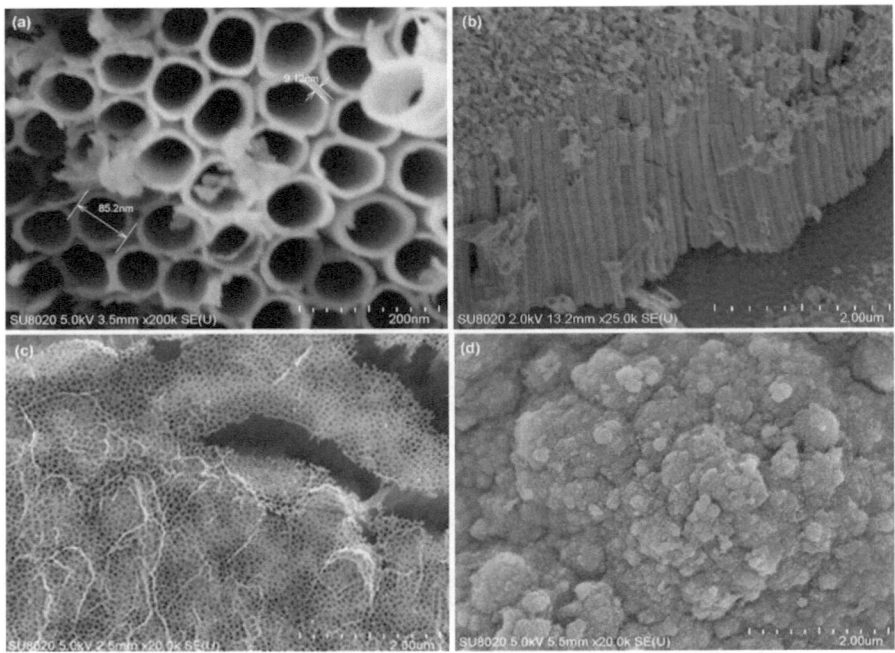

Fig. 4.9 Scanning electron microscope images of (**a**) top view of TiO$_2$ nanotube array; (**b**) cross-section view of TiO$_2$ nanotube array; (**c**) thin layer of molecularly imprinted polypyrrole synthesized on TiO$_2$ nanotube array; (**d**) thick polypyrrole film on flat substrate

Fig. 4.10 Dependence between conductance responses and HCHO concentrations for the polypyrrole-based MIP/TiO$_2$-NTA sensor

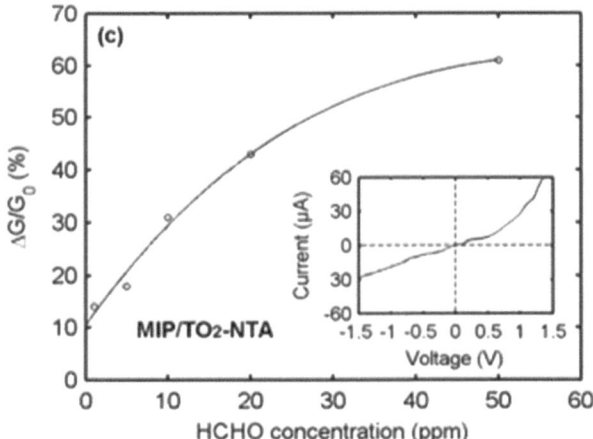

Fig. 4.11 Chemical scheme illustration for the cavities on the polypyrrole-based MIP layer to detect formaldehyde

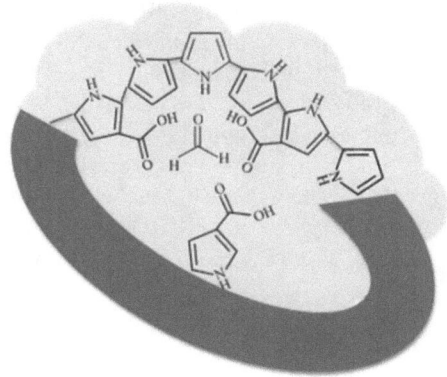

4.5 Functionalized Graphene Sensor for Formaldehyde Detection

Since the first graphene sensor was demonstrated by Novoselov's group in 2007 [118], the research on graphene-based gas sensors has sharply increased these past 10 years. Graphene-based materials have been exploited as sensing layers for detecting various types of gases, such as NH_3, NO_2, H_2, CO_2, CO, CH_4, SO_2, H_2S, and O_2 [119–125]. However, the study of graphene-based sensors for volatile organic compounds (VOCs) detection is relatively immature (only focus on acetone and ethanol) [126]. Very few works on graphene-based sensor for formaldehyde (HCHO) detection were reported [127].

Single layer graphene has largest surface-to-volume ratio, possessing the potential ability to detect a single molecule. Moreover, the high carrier mobility of graphene inherently ensures low electrical noise and low power consumption in graphene sensors [128]. However, pristine graphene is chemically inert due to free of dangling bonds [129] and then weakly adsorbs gas molecules, while defective graphene has stronger adsorption ability (meaning higher sensitivity). This high sensitivity is related to defects or traps in graphene where the gas molecules can be readily grafted. The sensing principle roots on a change of graphene's resistance when the gas molecules are grafted on the sensor surface. Nonetheless, similar resistance changes could be induced under exposure to different gases [130], meaning a lack of selectivity.

The graphene functionalization is a crucial step to improve the sensitivity and selectivity of graphene sensors. The functionalization can be classified as covalent and non-covalent pathways. The former includes hydrogenation [131], chlorination [132], oxidation [133], fluorination [134] and grafting of functional groups [135]. Unfortunately, covalent functionalization can lead to many unwanted damages for graphene structure, and then degrade electronic properties [136]. Non-covalent functionalization refers to physical modification of graphene by DNAs [137, 138] proteins [139], peptides [140], and nanoparticles [141]. It has the major advantage of fully preserving the graphene lattice and thus the electrical characterization.

Organic molecules play an important role in the graphene functionalization [142, 143]. Here we comparatively exploit two categories of graphene sensors, one functionalized with TFQ (2,3,5,6,-Tetrafluorohydroquinone) molecules and the other one with amine (NH_2-terminated monolayer) groups for HCHO detection. Our results show that TFQ molecules with non-covalent bonds act as specific links between graphene and HCHO, simultaneously preserving the original properties of graphene. The TFQ molecules not only increases the sensor's response but also its selectivity.

The fabrication of the graphene sensors consists of electrode patterning, graphene transfer, and sensor functionalization. The insert in Fig. 4.1a shows a fabricated graphene sensor. A pair of Au electrodes is patterned on a SiO_2/Si substrate by optical lithography and lift-off process. Then chemical vapor deposition (CVD) grown graphene is transfered on the substrate by using PMMA resist. Two kinds of organic molecules are used for the graphene functionalization. One is TFQ molecules (Alfa Aesar, 0.3 wt% in acetone), which is dropped on the graphene sensor. This functionalization is non-covalent, which could be performed through π-π stacking or hydrophilic interactions [144]. Therefore, graphene should retain its original sp^2 conjugated structure. The other is amine groups, which are covalently bond on the graphene sensor by using the well-known diazonium salts addition reaction. The detail can be found in [145]. Milowska et al. performed theoretical studies of transport properties of graphene functionalized with amine groups [146]. Their results predict that chemisorption of amine groups leads to breaking one of π-bonds and sp^2 to sp^3 re-hybridization of C-C bonds.

The fabricated graphene sensor is wire-bonded to 24-pin dual-in-line package (Fig. 4.12a), which is placed in a sealed chamber. The graphene sensor is assembled with a CMOS chip platform through a cable. The platform is linked to a computer by a USB interface as shown in Fig. 4.12b. The platform has electrical connectors for bias and resistance measurements. Hereafter, we refer to the resistance measurements carried out by a DC bias of 0.5 V at atmospheric pressure with a temperature of 22 °C and a relative humidity of 50%. Air and HCHO are used as carrier and target gas, respectively. The sensor resistance response is defined as $\Delta R/R_0 = (R-R_0)/R_0$, where R_0 and R are the sensor resistance before and after exposure to the target gas, respectively.

Figure 4.13 shows Raman spectra of single-layer (SL) graphene functionalized by amine groups and TFQ molecules. We observe the defect-activated peak (D peak located at ~ 1356 cm^{-1}), the broadening and position downshift of the G and 2D peaks in amine-functionalized graphene. The D peak can only be observed when the crystal symmetry is broken by point defects [147]. The increased number of defects results in the broadening of the G and 2D peaks. The addition of new chemical bonds can relax residual short-range strain, leading to the position downshift of both peaks. These observations confirm that the amine functionalization is covalent. The Raman result is consistent with the theoretical prediction of Milowska. The absence of the D peak is evident in the case of TFQ organic molecules which are non-covalently

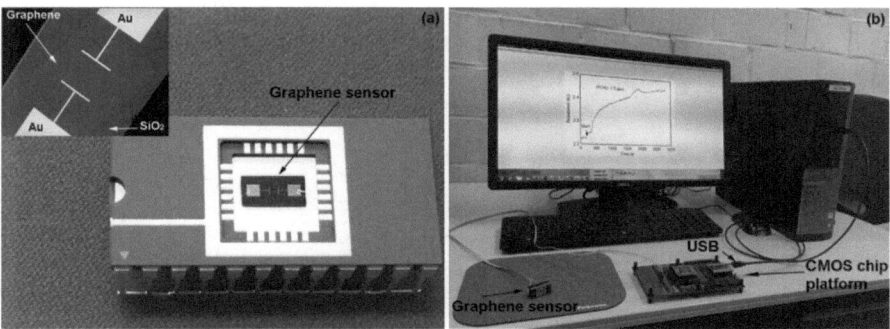

Fig. 4.12 (**a**) TFQ-functionalized graphene sensor, where CVD graphene is transferred on a pair of Au electrodes lying on a SiO$_2$/Si substrate. (**b**) Sensor measurement setup, a graphene sensor wire-bonded to 24-pin dual-in-line package is assembled on a CMOS chip platform, which is linked with a computer by a USB interface. From reference [145] with permission (IoP publishing 2017)

Fig. 4.13 Raman spectra for graphene functionalized by amine groups and TFQ molecules. The Raman spectrum of single layer (SL) graphene before functionalization is also displayed at the bottom. From reference [145] with permission (IoP publishing 2017)

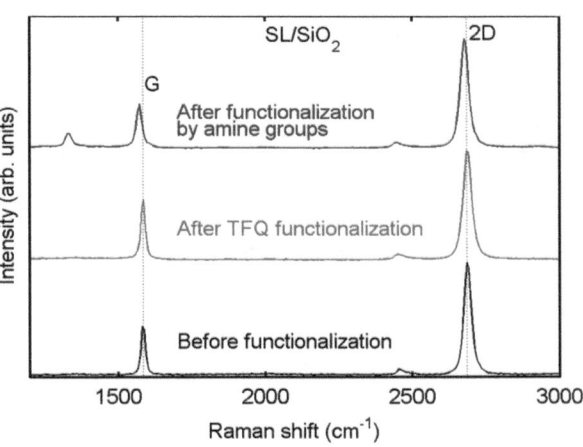

adsorbed on graphene by π-π bonds or hydrophilic interactions. The G peak, more sensitive to charge doping [148], remains at the same position. These facts indicate that the TFQ functionalization induces smaller doping and less defects/traps than amine.

Figure 4.14a shows the resistance response of the TFQ-functionalized graphene sensor for different HCHO concentrations. Figure 4.14b shows the resistance behaviour for a TFQ-functionalized graphene sensor and a bare graphene sensor, respectively. The former exhibits a resistance response of 10% for a HCHO concentration of 1.5 ppm, while the latter gives a negligible resistance response even if it is exposed to a high HCHO concentration of 50 ppm. The density functional theory predicts that HCHO molecule is an electron donor, specifically, the adsorption of a HCHO molecule donates 0.021–0.039 electrons to graphene [149]. Our previous experiment results [150] indicated that the as-transferred graphene is p-type rather

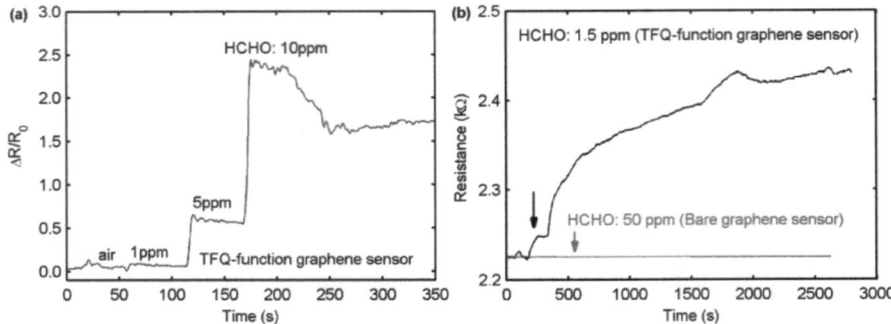

Fig. 4.14 Resistance behaviour of typical graphene sensors at 20 °C and at relative humidity of 50%: (**a**) TFQ-functionalized sensor for HCHO concentrations in a range from 1 to 10 ppm. (**b**) TFQ-functionalized sensor for 1.5 ppm HCHO, and bare graphene sensor for 50 ppm HCHO. From reference [145] with permission (IoP publishing 2017)

than pristine, thereby its Fermi level is upshifted. The electrons transfer from the adsorbed HCHO molecules to the as-transferred graphene becomes energetically unfavourable. Besides, the binding energy between HCHO molecule and bare graphene is small, because of a large distance between them [151]. Therefore, the absorption of HCHO molecules on bare graphene and the charge transfer process do not take place. As long as the graphene sensor is functionalized by TFQ molecules, a drastic increase of its resistance is observed as shown in Fig. 4.3b. The possible mechanism is that the hydroxyl groups in TFQ interact with HCHO molecules to form weak and reversible intermediate products. These intermediate products are closer to graphene, which would make the charge transfer easier. The electrons donated by HCHO are transferred to graphene through the intermediate products, thereby depleting holes in the p-type graphene and increasing the graphene resistance. As a result, the TFQ functionalization improves the graphene sensitivity to HCHO molecules.

Figure 4.15a, b show the resistance responses of the TFQ-functionalized and amine-functionalized graphene sensors for the detection of acetone-saturated and ethanol-saturated vapors, respectively. The linkages between amine groups and graphene are based on covalent bonds, creating defects or traps in graphene. When the amine-functionalized sensor is exposed to acetone or ethanol vapor, the non-specific gas molecules are unselectively grafted defects or traps on graphene. This causes the resistance variation, more specifically, 10% and 3% for the acetone-saturated and the ethanol-saturated vapors, respectively. Rumyantsev's team found that different organic vapors can induce similar resistance changes in defective graphene [152]. However, the TFQ molecules are physically adsorbed on graphene without creating defects or traps. They can specifically react with HCHO molecules to form changed intermediate products and transfer electrons to graphene. There-fore, the TFQ-functionalized graphene sensor has a good selectivity to HCHO compared with the amine-functionalized sensor.

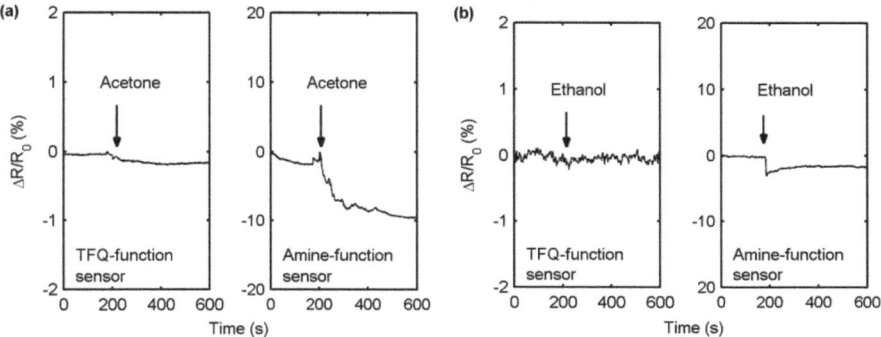

Fig. 4.15 Resistance responses of the graphene sensors functionalized with TFQ molecules and amine groups: (**a**) for acetone-saturated vapour and for (**b**) ethanol-saturated vapour at 20 °C and at relative humidity of 50%. From reference [145] with permission (IoP publishing 2017)

It is worth noting that the recovering time of the TFQ-functionalized graphene sensor is slow (minutes) and this problem can be overcome, for example, by using ultra-violet induced signal recovery [153] or low-temperature heating [144]. In addition, the life time of the TFQ-functionalized graphene sensor is shorter than that of the covalent-functionalized sensor. Nevertheless, non-covalent functionalization TFQ can be easily cleaned by acetone. Our sensor can be re-functionalization and reused.

4.6 Conclusions

This chapter presented a brief overview of the current chemical sensors for VOC detection, in particular formaldehyde which has become one of the most problematic gases in indoor air. Among all possible technologies, we focussed on resistive sensors, called chemoresistors. By principle, these sensors are simple, easy to integrate in classical electronics and cheap. Many sensitive materials were studied and still there is room for improvement of the performances. We can be sure that progress in nanotechnology and knowledge of the sensing mechanism will allow to develop more sensitive, selective and reliable sensors based on this simple technology.

References

1. IARC, [Formaldehyde, 2-Butoxyethanol and 1-tert-Butoxypropan-2-ol. (2006) IARC monographs on the evaluation of carcinogenic risks to humans, vol. 88], World Health Organization, Lyon, 39–325
2. World Health Organization (2010) Regional Office for Europe, "WHO guidelines for indoor air quality: selected pollutants", Geneva, ISBN: 9789289002134
3. Vairavamurthy A, Roberts JM, Newman L (1992) Methods for determination of low molecular weight compounds in the atmosphere: a review. Atmos Environ 26A:1965–1993

4. Chung PR, Tzeng CT et al (2013) Formaldehyde gas sensors: a review. Sensors 13:4468–4484
5. Fleet B, Gunasingham H (1992) Electrochemical sensors for monitoring environmental pollutants. Talanta 39:1449–1457
6. Sato T, Plashnitsa VV, Utiyama M, Miura (2010) N Potentiometric YSZ-based sensor using NiO sensing electrode aiming at detection of volatile organic compounds (VOCs) in air environment. Electrochem Commun 12:524–526
7. Mead MI, Popoola OAM et al (2013) The use of electrochemical sensors for monitoring urban air quality in low-cost, high-density networks. Atmos Environ 70:186–203
8. Si P, Mortensen J, Komolov A et al (2007) Polymer coated quartz crystal microbalance sensors for detection of volatile organic compounds in gas mixtures. Anal Chim Acta 597:223–230
9. Shafiq Islam AKM, Ismail Z et al (2005) Transient parameters of a coated quartz crystal microbalance sensor for the detection of volatile organic compounds (VOCs). Sensors Actuators B109:238–243
10. Khot LR, Panigrahi S, Lin D (2011) Development and evaluation of piezoelectric-polymer thin film sensors for low concentration detection of volatile organic compounds related to food safety applications. Sensors Actuators B Chem 153:1–10
11. Fan X, Du B (2012) Selective detection of trace p-xylene by polymer-coated QCM sensors. Sensors Actuators B166–167:753–760
12. Clifford KH, Lindgren RE et al (2003) Development of a surface acoustic wave sensor for in-situ monitoring of volatile organic compounds. Sensors 3:236–247
13. Fang M, Vetelino K, Rothery M et al (1999) Detection of organic chemicals by SAW sensor array. Sensors Actuators B56:155–157
14. Fernández MJ, Fontecha JL, Sayago I et al (2007) Discrimination of volatile compounds through an electronic nose based on ZnO SAW sensors. Sensors Actuators B127:277–283
15. Wolfbeis OS (2002) Fiber-optic chemical sensors and biosensors. Anal Chem 74:2663–2678
16. Elosua C, Matias IR, Bariain C et al (2006) Volatile organic compound optical fiber sensors: a review. Sensors 6:1440–1465
17. Yoon J, Chae SK, Kim JM (2007) Colorimetric sensors for volatile organic compounds (VOCs) based on conjugated polymer-embedded electrospun fibers. J Am Chem Soc 129:3038–3039
18. González-Vila Á, Debliquy M, Lahem D et al (2017) Molecularly imprinted electropoly-merization on a metal-coated optical fiber for gas sensing applications. Sensors Actuators B244:1145–1151
19. Patel SV, Mlsna TE et al (2003) Chemicapacitive microsensors for volatile organic compound detection. Sensors Actuators B96:541–553
20. Lee DS, Jung JK, Lim J. W et al (2001) Recognition of volatile organic compounds using SnO2 sensor array and pattern recognition analysis. Sensors Actuators B77: 228–236
21. Zhang WM, Hu JS et al (2007) Detection of VOCs and their concentrations by a single SnO2sensor using kinetic information. Sensors Actuators B123:454–460
22. Mishra RK, Sahay PP (2012) Synthesis characterization and alcohol sensing property of Zn-doped SnO2 nanoparticles. Ceram Int 38:2295–2304
23. Zeng W, Tian-Mo L (2010) Gas-sensing properties of SnO2–TiO2-based sensor for volatile organic compound gas and its sensing mechanism. Phys B Condens Matter 405:1345–1348
24. Lahem D, Lontio FR et al (2016) Formaldehyde gas sensor based on nanostructured nickel oxide and the microstructure effects on its response. In: IC-MAST2015 IOP Conf. Series: materials science and engineering 108
25. Zhang YM, Lin YT et al (2014) A high sensitivity gas sensor for formaldehyde based on silver doped lanthanum ferrite. Sensors Actuators B190:171–176
26. Neri G (2015) First fifty years of chemoresistive gas sensors. Chemosensors 3:1–20
27. Moseley PT, Norris J, Williams DE (1991) Techniques and mechanisms in gas sensing. In: Adam Hilger

28. Korotcenkov G, Cho BK (2017) Metal oxide composites in conductometric gas sensors: achievements and challenges. Sensors Actuators B 244:182–210
29. Kanan SM, El-Kadri OM et al (2009) Semiconducting metal oxide based sensors for selective gas pollutant detection. Sensors 9:8158–8196
30. Decroly A, Krumpmann A, Debliquy M et al (2016) Nanostructured TiO2 Layers for Photovoltaic and Gas Sensing Applications, INTECH Book "Green Nanotechnology". ISBN 978-953-51-4692-6
31. Seiyama T, Kato A (1962) A new detector for gaseous components using semiconductor thin film. Anal Chem 34:1502–1503
32. Seiyama T (1988) Chemical sensors-current status and future outlook. In: Seiyama T (ed) Chemical Sensor Technology, vol 1. Elsevier, Amsterdam
33. Yamazoe N (1991) New approaches for improving semiconductor gas sensors. Sensors Actuators B Chem 5:7–19
34. Shimizu Y, Egashira M (1999) Basic aspects and challenges of semiconductor gas sensors. MRS Bull 24:18–24
35. Yamazoe N (2005) Toward innovations of gas sensor technology. Sensors Actuators B 108:2–14
36. Gurlo A, Bârsan N, Weimar U (2006) In: Fierro JLG (ed) Gas sensors based on semiconductiong metal oxides. In metal oxides: chemistry and applications. CRC Press, Boca Raton, p 683
37. Aleixandre M, Gerboles M (2012) Review of small commercial sensors for indicative monitoring of ambient gas. Chem Eng Trans 30:169–174
38. Bârsan N, Hübner M, Weimar U (2011) Conduction mechanisms in SnO2 based polycrystalline thick film gas sensors exposed to CO and H2 in different oxygen backgrounds. Sensors Actuators B 157:510–517
39. Bârsan N, Tomescu A (1995) Calibration Procedure for SnO2-based Gas Sensors. Thin Solid Films 259:91–95
40. Niebling G, Schlachter A (1995) Qualitative and quantitative gas analysis with non-linear interdigital sensor arrays and artificial neural networks. Sensors and Actuators B26–27:289
41. Yamaura H, Tamaki J, Moriya K et al (1997) Highly selective CO sensor using indium oxide doubly promoted by cobalt oxide and gold. J Electrochem Soc 144
42. Mochida T, Kikuchi K, Kondo T, Ueno H, Matsuura Y (1995) Highly sensitive and selective H2S gas sensor from r.f. sputtered SnO2 thin film. Sensors Actuators B 25:433–437
43. Tricoli A, Righettoni M, Pratsinis SE (2009) Minimal cross-sensitivity to humidity during ethanol detection by SnO2-TiO2 solid solutions. Nanotechnology 20:315502
44. Cederquist A, Gibbons E, Meitzler A (1976) Characterization of Zirconia and Titania Engine Exhaust Gas Sensors for air/fuel feedback control systems. SAR Tech Pap. https://doi.org/10.4271/7602
45. Kolmakov A, Moskovits M (2004) Chemical sensing and catalysis by one-dimensional metal-oxide nanostructures. Annu Rev Mater Res 34:151–180
46. Arafat MM, Dinan B et al (2012) Gas sensors based on one dimensional nanostructured metal-oxides: a review. Sensors 12:7207–7258
47. Thong LV, Hoa ND et al (2010) On-chip fabrication of SnO2-nanowire gas sensor: the effect of growth time on sensor performance. Sensors Actuators B Chem 146:361–367
48. Huang MH, Mao S, Feick H et al (2001) Room-temperature ultraviolet nanowire nanolasers. Science 292:1897–1899
49. Yang Z, Li LM, Wan Q et al (2008) High-performance ethanol sensing based on an aligned assembly of ZnO nanorods. Sensors Actuators B Chem 135:57–60
50. Wan Q, Li QH, Chen YJ et al (2004) Fabrication and ethanol sensing characteristics of ZnO nanowire gas sensors. Appl Phys Lett 84:3654–3656
51. Jones T, Bott B, Thorpe S (1989) Fast response metal phthalocyanine-based gas sensors. Sensors Actuators B 17:467–474
52. Simon J, André JJ (1985) Molecular semiconductors. Springer, Berlin/Heidelberg

53. Wright JD (1991) Gas adsorption on phthalocyanines and its effects on electrical properties. Prog Surf Sci 31:1–60
54. Mukhopadhyay S, Hogarth CA (1994) Gas sensing properties of phthalocyanine Langmuir–Blodgett films. Adv Mater 6:162–164
55. Capone S, Mongelli S et al (1999) Gas sensitivity measurements on NO2 sensors based on Copper(II) tetrakis(n-butylaminocarbonyl) phthalocyanine LB films. Langmuir 15:1748–1753
56. Simon J, Bouvet M, Bassoul P (1994) The encyclopedia of advanced materials. Pergamon, Oxford, pp 1680–1692
57. Rodriguez-Mendez ML, Aroca R, Desaja JA (1993) Electrochromic and gas adsorption properties of Langmuir-Blodgett films of lutetium bisphthalocyanine complexes. Chem Mater 5(7):933–937
58. Weiss R, Fischer J (2006) Lanthanide phthalocyanine complexes. The porphyrin handbook, 1st ed., vol. 16, Kadish K, Smith KM, Guilard R (eds); Academic Press Inc., New York, pp 171–246
59. Paolo Bondavallia P, Leganeux P, Pribat D (2009) Carbon nanotubes based transistors as gas sensors: state of the art and critical review. Sensors Actuators B 140:304–318
60. Espinosa EH, Ionescu R, Chambon B et al (2007) Hybrid metal oxide and multiwall carbon nanotube films for low temperature gas sensing. Sensors Actuators B 127:137–142
61. Helbling T, Pohle R, Durrer L et al (2008) Sensing NO2 with individual suspended single-walled carbon nanotubes. Sensors Actuators B 132:491–497
62. Bittencourt C, Felten A, Espinosa EH et al (2006) Evaporation of WO3 on carbon nanotube films: a new hybrid film. Smart Mater Struct 15:1555–1560
63. Ionescu R, Espinosa EH et al (2006) Oxygen functionalisation of MWNT and their use as gas sensitive thick-film layers. Sensors Actuators B 113:36–46
64. Mu H, Zhang Z et al (2014) High sensitive formaldehyde graphene gas sensor modified by atomic layer deposition zinc oxide films. Appl Phys Lett 105:033107
65. Zhua BL, Xie CS, Wu J et al (2006) Influence of Sb, In and Bi dopants on the response of ZnO thick films to VOC's. Mater Chem Phys 96:459–465
66. Elmi I, Zampolli S et al (2008) Development of ultra-low-power consumption MOX sensors with ppb-level VOC detection capabilities for emerging applications. Sensors Actuators B Chem 135:342–351
67. Daza L, Dassy S, Delmon B (1993) Chemical sensors based on SnO2 and WO3 for the detection of formaldehyde: cooperative effects. Sensors Actuators B Chem 10:99–105
68. Lee DS, Jung JK et al (2001) Recognition of volatile organic compounds using SnO2 sensor array and pattern recognition analysis. Sensors Actuators B 77:228–236
69. Zhu BL, Xie CS, Wang WY (2004) Improvement in gas sensitivity of ZnO thick film to volatile organic compounds (VOCs) by adding TiO2. Mater Lett 58:624–629
70. Srivastava AK (2003) Detection of volatile organic compounds (VOCs) using SnO2 gas-sensor array and artificial neural network. Sensors Actuators B 96:24–37
71. Lai X, Wang D et al (2010) Ordered arrays of bead-chain-like In2O3 nanorods and their enhanced sensing performance for formaldehyde. Chem Mater 22:3033–3042
72. Castro-Hurtado I, Herrán J et al (2011) Studies of influence of structural properties and thickness of NiO thin films on formaldehyde detection. Thin Solid Films 520:947–952
73. Gou X, Wang G et al (2008) Chemical synthesis, characterisation and gas sensing performance of copper oxide nanoribbons. J Mater Chem 18:965–969
74. Dirksen JA, Duval K, Ring TA (2001) NiO thin film formaldehyde gas sensor. Sensors Actuators B Chem 80:106–115
75. Lee CY, Chiang CM, Wang YH et al (2007) A self-heating gas sensor with integrated NiO thin film for formaldehyde detection. Sensors Actuators B Chem 122:503–510
76. Lahem D, Lontio FR et al (2016) Formaldehyde gas sensor based on nanostructured nickel oxide and the microstructure effects on its response. In: Proceedings IC-MAST2015 IOP Conf. Series: materials science and engineering 108

77. Zhang L, Hu JF et al (2005) Formaldehyde sensing characteristics of perovskite La0.68Pb0.32FeO3 nano-materials. Physica B 370:259–263
78. Wang J, Liu L, Cong SY et al (2008) An enrichment method to detect low concentration formaldehyde. Sensors Actuators B Chem 134:1010–1015
79. Wang J, Zhang P et al (2009) Silicon-based micro-gas sensors for detecting formaldehyde. Sensors Actuators B Chem 136:399–404
80. Lv P, Tang ZA et al (2008) Study on a microgas sensor with SnO2–NiO sensitive film for indoor formaldehyde detection. Sensors Actuators B Chem 132:74–80
81. Han N, Tian Y, Wu X et al (2009) Improving humidity selectivity in formaldehyde gas sensing by a two-sensor array made of Ga-doped ZnO. Sensors Actuators B Chem 138:228–235
82. Han N, Chai L, Wang Q et al (2010) Evaluating the doping effect of Fe, Ti and Sn on gas sensing property of ZnO. Sensors Actuators B Chem 147:525–530
83. Chen T, Zhou Z, Wang Y (2008) Effects of calcining temperature on the phase structure and the formaldehyde gas sensing properties of CdO-mixed In2O3. Sensors Actuators B Chem 135:219–223
84. Huang S, Qin H, Song P et al (2007) The formaldehyde sensitivity of LaFe1−xZnxO3-based gas sensor. J Mater Sci 42:9973–9977
85. Zeng W, Tianmo Liu T et al (2009) Selective detection of formaldehyde gas using a Cd-doped TiO2-SnO2 sensor. Sensors 9:9029–9038
86. Wollenstein J, Plaza JA et al (2003) A novel single chip thin film metal oxide array. Sensors Actuators B Chem 93:350–355
87. Lee CY, Chiang CM, Chou PC et al (2005) A novel microfabricated formaldehyde gas sensor with NiO thin film, sensors for industry conference p, pp 1–5
88. Wang YH, Lee CY et al (2008) Enhanced sensing characteristics in MEMS-based formaldehyde gas sensors. Microsyst Technol 14:995–1000
89. Lin S, Li D et al (2011) A selective room temperature formaldehyde gas sensor using TiO2 nanotube arrays. Sensors Actuators B 156:505–509
90. Peng L, Zhao Q et al (2009) Ultraviolet-assisted gas sensing: a potential formaldehyde detection approach at room temperature based on zinc oxide nanorods. Sensors Actuators B 136:80–85
91. Mu H, Zhang Z et al (2014) Highly sensitive formaldehyde graphene gas sensor modified by atomic layer deposition zinc oxide films. Appl Phys Lett 105:033107
92. Adhikari B, Majumdar S (2004) Polymers in sensor applications. Prog Polym Sci 29:699–766
93. Bartlett PN, Archer PB et al (1989) Conducting polymer gas sensors Part I: fabrication and characterization. Sensors Actuators B Chem 19:125–140
94. Bartlett PN, Sk L-C (1989) Conducting polymer gas sensors Part II: response of polypyrrole to methanol vapour. Sensors Actuators B Chem 19:141–150
95. Bartlett PN, Sk L-C (1989) Conducting polymer gas sensors Part III: results for four different polymers and five different vapours. Sensors Actuators B Chem 20:287–292
96. Agbor NE, Petty MC et al (1995) Polyaniline thin films for gas sensing. Sensors Actuators B Chem 28:173–179
97. Anitha G, Subramanian E (2005) Recognition and exposition of intermolecular inter-action between CH2Cl2 and CHCl3 by conducting polyaniline materials. Sensors Actuators B Chem 107:605–615
98. Li ZF, Blum FD, Bertino MF et al (2008) One-step fabrication of a polyaniline nanofiber vapor sensor. Sensors Actuators B Chem 134:31–35
99. Antwi-Boampong S, Bel Bruno JJ (2013) Detection of formaldehyde vapor using conductive polymer films. Sensors Actuators B Chem 182:300–306
100. Lange U, Roznyatovskaya NV, Mirsky VM (2008) Conducting polymers in chemical sensors and arrays. Anal Chim Acta 614:1–26
101. Guadarrama A, Rodrıguez-Méndez ML et al (2001) Electronic nose based on conducting polymers for the quality control of the olive oil aroma: discrimination of quality, variety of olive and geographic origin. Anal Chim Acta 432:283–292

102. Haupt K, Linares AV et al (2011) In: Haupt K (ed) Molecularly imprinted polymers. Springer, Berlin Heidelberg, pp 1–28
103. Vasapollo G, Del Sole R et al (2011) Molecularly imprinted polymers: present and future prospective. Int J Mol Sci 12:5908–5945
104. Kröger S, Tumer APF et al (1999) Imprinted polymerbased sensor system for herbicides using differential-pulse voltammetry on screen-printed electrodes. Anal Chem 71:3698–3702
105. Luo C, Liu M, Mo Y et al (2001) Thickness-shear mode acoustic sensor for atrazine using molecularly imprinted polymer as recognition element. Anal Chim Acta 428:143–148
106. Tan Y, Yin J, Liang C et al (2001) A study of a new TSM bio-mimetic sensor using a molecularly imprinted polymer coating and its application for the determination of nicotine in human serum and urine. Bioelectrochemistry 53:141–148
107. Manbohi A, Shamaeli E, Alizadeh N (2014) Nanostructured conducting molecularly imprinted polypyrrole film as a selective sorbent for benzoate ion and its application in spectrophotometric analysis of beverage samples. Food Chem 155:186–191
108. Justino CIL, Freitas AC et al (2015) Recent developments in recognition elements for chemical sensors and biosensors. TrAC—Trends Anal Chem 68:2–17
109. Whitcombe MJ, Kirsch N, Nicholls IA (2014) Molecular imprinting science andtechnology: a survey of the literature for the years 2004–2011. J Mol Recognit 27:297–401
110. Sharma PS, D'Souza F, Kutner W (2012) Molecular imprinting for selective chemical sensing of hazardous compounds and drugs of abuse. TrAC—TrendsAnal Chem 34:59–76
111. Hirayama K, Sakai Y, Kameoka K et al (2002) Preparation of a sensor device with specific recognition sites for acetaldehyde by molecular imprinting technique. Sensors Actuators B 86:20–25
112. Ihdene Z, Mekki A et al (2014) Quartz crystal microbalance VOCs sensor based on dip coated polyaniline emeraldine salt thin films. Sensors Actuators B 203:647–654
113. Wu N, Feng L et al (2009) An optical reflected device using a molecularly imprinted polymer film sensor. Anal Chim Acta 653:103–108
114. Lépinay S, Ianoul A, Albert J (2014) Molecular imprinted polymer-coated optical fiber sensor for the identification of low molecular weight molecules. Talanta 128:401–407
115. Cennamo N, Donà A, Pallavicini P et al (2015) Sensitive detection of 2,4,6-trinitrotoluene by tridimensional monitoring of molecularly imprinted polymer with optical fiber and five-branched gold nanostars. Sens Actuators B 208:291–298
116. Debliquy M, Dony N et al (2016) Acetaldehyde chemical sensor based on molecularly imprinted polypyrrole. Procedia Eng 168:569–573
117. Tang X, Raskin JP, Lahem D (2017) A formaldehyde sensor based on molecularly-imprinted polymer on a TiO_2, nanotube array. Sensors 17:675
118. Schedin F, Geim AK et al (2007) Detection of individual gas molecules absorbed on graphene. Nat Mater 6(9):652–655
119. Chen CW, Hung SC et al (2011) Oxygen sensors made by monolayer graphene under room temperature. Appl Phys Lett 99(24):243502
120. Fowler JD, Matthew JA, Tung VC et al (2009) Practical chemical sensors from chemically derived graphene. ACS Nano 3(2):301–306
121. Juree H, Lee S et al (2015) A highly sensitive hydrogen sensor with gas selectivity using a PMMA membrane-coated Pd nanoparticle/single-layer graphene hybrid." ACS Appl Mater Interfaces, 150209062057005
122. Pandey PA, Wilson NR, Covington JA (2013) Pd-doped reduced graphene oxide sensing films for H2 detection. Sensors Actuators B Chem 183:478–487
123. Lu G, Ocola LE, Chen J (2009) Reduced graphene oxide for room temperature gas sensors. Nanotechnology 20:445502
124. Hu N, Yang Z. Wang Y et al (2014) Ultrafast and sensitive room temperature NH_3 gas sensors based on chemically reduced graphene oxide. Nanotechnology 25, no. 2:025502
125. Le H, Zhang Z et al (2015) Multifunctional graphene sensors for magnetic and hydrogen detection. ACS Appl Mater Interfaces 7(18):9581–9588

126. Wang T, Da H, Zhi Y et al (2016) A review on graphene-based gas/vapor sensors with unique properties and potential applications. Nano-Micro Lett 8(2):95–119
127. Tahe A, Soltani LH (2013) Graphene/poly(methylMethacrylate) chemiresistor sensor for formaldehyde odor sensing. J Hazard Mater 248–249:401–406
128. Schedin F, Geim AK, Morozov SV et al (2007) Detection of individual gas molecules absorbed on graphene. Nat Mater 6(9):652–655
129. Wangyang F, Jiang L, van Geest EP et al (2017) Sensing at the surface of graphene field-effect transistors. Adv Mater 29(6):1603610
130. Sergey R, Liu G, Shur MS, Potyrailo RA, Balandin AA (2012) Selective gas sensing with a single pristine graphene transistor. Nano Lett 12(5):2294–2298
131. Elias DC,. Nair RR, Mohiuddin TMG et al (2009) Control of graphene's properties by reversible hydrogenation: evidence for graphane. Science 323(5914):610–613
132. Bo L, Zhou L et al (2011) Photochemical chlorination of graphene. ACS Nano 5:5957–5961
133. Dimiev AM, Tour JM (2014) Mechanism of graphene oxide formation. ACS Nano 8(3):3060–3068
134. Wei F, Long P et al (2016) Two-dimensional fluorinated graphene: synthesis, structures, properties and applications. Adv Sci 3:1500413
135. Hang Z, Bekyarova E et al (2011) Aryl functionalization as a route to band gap engineering in single layer graphene devices. Nano Lett 11:4047–4051
136. Milowska KZ, Majewski JA (2013) Stability and electronic structure of covalently functionalized graphene layers: covalently functionalized graphene layers. Phys Status Solidi B 250:1474–1477
137. Lin C-T, Loan PTK et al (2013) Label-free electrical detection of DNA hybridization on graphene using hall effect measurements: revisiting the sensing mechanism. Adv Funct Mater 23:2301–2307
138. Ye L, Goldsmith BR, Kybert NJ et al (2010) DNA-decorated graphene chemical sensors. Appl Phys Lett 97:083107
139. Lerner MB, Matsunaga F et al (2014) Scalable production of highly sensitive nanosensors based on graphene functionalized with a designed G protein-coupled receptor. Nano Lett 14:2709–2714
140. Lingyan F, Wu L et al (2012) Detection of a prognostic indicator in early-stage cancer using functionalized graphene-based peptide sensors. Adv Mater 24:125–131
141. Bochen Z, Uddin MA et al (2016) Temperature dependent carrier mobility in graphene: effect of Pd nanoparticle functionalization and hydrogenation. Appl Phys Lett 108:093102
142. Xiaochen D, Fu D et al (2009) Doping single-layer graphene with aromatic molecules. Small 5:1422–1426
143. Zhu Y, Yufeng H et al (2016) A graphene-based affinity nanosensor for detection of low-charge and low-molecular-weight molecules. Nanoscale 8:5815–5819
144. Liu J, Liu Z, Barrow CJ et al (2015) Molecularly engineered graphene surfaces for sensing applications: a review. Anal Chim Acta 859:1–19
145. Tang X, Mager N et al (2017) Defect-free functionalized graphene sensor for formaldehyde detection. Nanotechnology 28:055501
146. Milowska KZ, Majewski JA (2014) Graphene-based sensors: theoretical study. J Phys Chem C 118:17395–17401
147. Ferrari AC (2007) Raman spectroscopy of graphene and graphite: disorder, electron–phonon coupling, doping and nonadiabatic effects. Solid State Commun 143:47–57
148. Yaping D, Lu Y, Kybert NJ et al (2009) Intrinsic response of graphene vapor sensors. Nano Lett 9:1472–1475
149. Ruoxi W, Zhang D et al (2006) Boron-doped carbon nanotubes serving as a novel chemical sensor for formaldehyde. J Phys Chem B 110:18267–18271
150. Reckinger N, Tang X et al (2016) Oxidation-assisted graphene heteroepitaxy on copper foil. Nanoscale 8:18751–18759

151. Mei C, Zhao YP (2009) Adsorption of formaldehyde molecule on the intrinsic and Al-doped graphene: a first principle study. Comput Mater Sci 46:1085–1090
152. Rumyantsev S, Liu G, Shur MS et al (2012) Selective gas sensing with a single pristine graphene transistor. Nano Lett 12:2294–2298
153. Vineet D, Surwade SP et al (2010) All-organic vapor sensor using inkjet-printed reduced graphene oxide. Angew Chem Int Ed 49:2154–2157

Chapter 5
Gas Sensing Using Monolayer MoS$_2$

Ruben Canton-Vitoria, Nikos Tagmatarchis, Yuman Sayed-Ahmad-Baraza, Chris Ewels, Dominik Winterauer, Tim Batten, Adam Brunton, and Sebastian Nufer

Abstract In this chapter we explore the possibilities of using MoS$_2$ for chemical gas sensing. We first introduce monolayer MoS$_2$ and discuss the different reconstructed phases that can be produced in terms of their atomic and electronic structure. We show how these properties can vary drastically from their bulk counterparts, and how MoS$_2$ can be taken as a useful model for other transition metal dichalcogenides (TMDs). We next explore different routes to tune the material properties through chemical functionalisation, before summarising the current literature on gas sensing using MoS$_2$. We discuss the possibilities for Raman

Author contributed equally with all other contributors.Ruben Canton-Vitoria and Yuman Sayed-Ahmad-Baraza

R. Canton-Vitoria · N. Tagmatarchis
Theoretical and Physical Chemistry Institute, National Hellenic Research Foundation, Athens, Greece

Y. Sayed-Ahmad-Baraza · C. Ewels
Institut des Materiaux Jean Rouxel (IMN), UMR6502 CNRS, Universite de Nantes, Nantes, France

D. Winterauer
Institut des Materiaux Jean Rouxel (IMN), UMR6502 CNRS, Universite de Nantes, Nantes, France

Renishaw plc, UK, Wotton-under-Edge

T. Batten
Renishaw plc, UK, Wotton-under-Edge

A. Brunton
M-Solv Ltd. Oxford, United Kingdom

S. Nufer (✉)
M-Solv Ltd. Oxford, United Kingdom

University of Sussex, Brighton, United Kingdom
e-mail: sebastian.nufer@m-solv.com

© Springer Nature B.V. 2019
C. Bittencourt et al. (eds.), *Nanoscale Materials for Warfare Agent Detection: Nanoscience for Security*, NATO Science for Peace and Security Series A: Chemistry and Biology, https://doi.org/10.1007/978-94-024-1620-6_5

71

spectroscopy of MoS_2 as a highly selective route to gas sensing, and the future possibilities for this as-yet largely unexplored application for this important family of monolayer TMDs.

Keywords Gas sensing · MoS_2 · Transition metal dichalcogenides

5.1   Introduction

Gas sensing has received an increasing amount of attention in recent years. Its relevance has increased in a number of different applications such as industrial production, the automotive industry, the medical sector, environmental monitoring and notably in security [1].

Nanomaterials are emerging into gas sensor technology as they provide a high surface to volume ratio which favours gas adsorption and increases the sensitivity [2]. There are different approaches on the exploitation of the properties of the nanomaterials in a gas sensor structure [3]. In the electrochemical approach, the conductance of the material is changing due to adsorption of a gas. To directly read out the conductance the nanomaterial is paired with interdigitated electrodes, which are deposited on to the material [4]. Putting the sensitive material in a field-effect transistor the source-drain current is measured [5] or the threshold voltage is monitored [6]. Another electrochemical route is measuring the quantum capacitance, which is sensitive to the chemical interaction of the material [7].

There are other methods to measure low concentrations of gases. A colorimetric detection where the colour of the exposed material is analysed [8]. The frequency of surface acoustic waves on the nanomaterial is sensitive to chemical interaction with gases [9]. There are optical methods to measure concentrations of gases using the properties of nanomaterials like surface Plasmon [10] or Raman spectroscopy [11]. Mostly graphene and carbon nanotubes have been incorporated in these configurations for gas sensors so far. It is most likely that the nanomaterials now emerging will also use these kinds of structures to sense the smallest traces of gases in their field of operation. In the expanding field of gas sensing technology MoS_2 is a promising new 2D nanomaterial.

5.2   Structure and Properties of MoS_2

The intense research in recent years in graphene has shown how reduced dimensionality can drastically change the properties of a material. The study of graphene has progressed very rapidly and the knowledge gained and the techniques and methodologies developed for these studies can be extended to other 2D materials [12, 13].

Among the range of different 2D materials, the family of transition metal dichalcogenides (TMDs), and specially MoS$_2$ has attracted a lot of attention. Due to its properties, MoS$_2$ is very promising for different applications in sensing, nanoelectronics, optoelectronics, spintronics, catalysis, energy storage and biomedicine [12, 14–16]. The electronic properties of single or few layers MoS$_2$ are sensitive to internal or external factors [17, 18], which combined with the fact that MoS$_2$ can exist as different polymorphs with different properties [19] that can be chemically functionalised [20, 21] makes MoS$_2$ a great material candidate for sensing applications.

5.2.1 Structure

MoS$_2$ is a member of the family of materials called layered TMDs. The general stoichiometric formula of TMDs is MX$_2$, where M is transition metal (from groups IV to X), and X is a chalcogenide. TMDs formed from metal of groups IV to VII and the chalcogens S, Se, Te are predominantly layered [15]. Bulk MoS$_2$ is composed of 2D layers stacked by weak Van der Waals forces, which allows its exfoliation [12, 15]. A single layer is composed of a plane of hexagonally arranged Mo atoms sandwiched between two planes of hexagonally arranged S atoms, held together by predominantly covalent Mo-S bonds [15].

MoS$_2$ can be present as different polymorphs [15, 19]. A nomenclature, based on Ramsdell notation [22, 23], is used for distinguishing the different polymorphs, which consists in a number representing the number of MoS$_2$ layers in a unit cell and a letter representing the symmetry (H for hexagonal, T for trigonal, and R for rhombohedral) [15, 24]. This notation is commonly used even in the case of single layer MoS$_2$, where the number 1 followed by the symmetry letter is used [15, 19]. This can cause confusion as the single layer material is named as a bulk material with 1 layer in the unit cell.

For a single layer, two different phases exist regarding the coordination symmetry of the metallic atom (Fig. 5.1a): trigonal prismatic coordination, with AbA stacking of the 3 atomic layers and D$_{3h}$ symmetry (1H phase); or octahedral coordination, with AbC stacking and D$_{3d}$ symmetry (ideal 1T phase) [15, 19]. The trigonal prismatic coordination is the most stable and the one found in natural MoS$_2$. Metastable octahedrically coordinated phases can be produced in specific conditions and tend to present structural distortions [15, 19]. This issue will be addressed in more detail later in the text.

When considering the stacking of single MoS$_2$ layers, different polytypes are possible depending on the stacking sequence. Regarding the trigonal prismatic coordination, two bulk polytypes have been identified: 2H and 3R (Fig. 5.1b). The 2H has a stacking sequence of AbA BaB and is the polytype commonly found in natural MoS$_2$, while the 3R has a stacking sequence of AbA CaC BcB (where capital letters represents S atoms and low case represents Mo) [15, 24]. The way on how we stack single layers can change the symmetry of the material, which can drastically

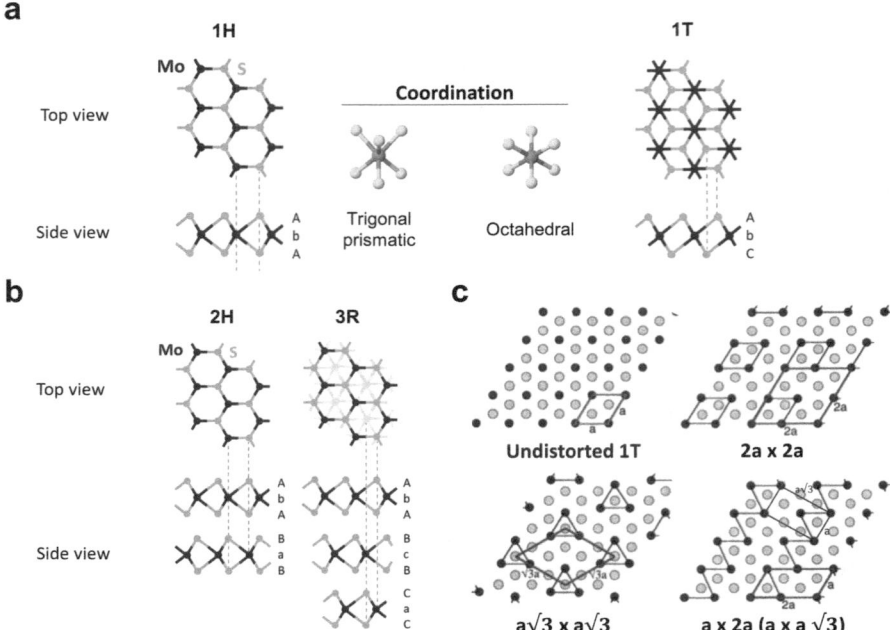

Fig. 5.1 Atomic structure of (**a**) monolayer MoS$_2$ polymorphs 1H and 1T showing the different metallic coordination for each one, and (**b**) the bulk polytypes with trigonal prismatic coordinated MoS$_2$ 2H and 3R. The stacking of layers within a unit cell is shown. (**c**) Structure of the ideal undistorted 1T phase and the three types of superlattices reported for the distorted phases. The unit cells showing the superlattice type are shown, and the proposed clustering for Mo atoms (dark circles) is represented with lines connecting the atoms. Only one of the proposed clustering compatible with a 2a × 2a superlattice is shown. (Panels **a** and **b** adapted from [142] published by the Royal Society of Chemistry, Panel **c** adapted with permission from [74]. Colors code for atoms in the original reference. Copyright 1999 American Chemical Society)

change some properties [12, 25–28]. MoS$_2$ with octahedral coordination can also form bulk phases. For example, the intercalation of 2H-MoS$_2$ with butyl lithium is reported to transform to a 1T phase with only one layer per unit cell [29, 30]. Nevertheless, the atoms within a layer are generally distorted from the ideal 1T phase forming superlattices.

5.2.2 Properties of 1H/2H-MoS$_2$

Monolayer and few-layer MoS$_2$ have very interesting properties that could be useful for sensing applications [12, 17, 31]. They can be obtained, in general, by mechanical or liquid exfoliation from bulk MoS$_2$, or by synthesis via chemical vapour deposition (CVD) [15, 31]. Some liquid exfoliation methods (such as Li intercalation) can produce a phase change from the 2H to 1T phase [15, 19, 31].

MoS₂ possesses very good mechanical properties appropriated for flexible nanodevices. Studies made on 5–25 layers [32] or monolayer MoS₂ [33, 34] show that it has high mechanical strength. The studies made on monolayer MoS₂ show a 2D elastic modulus of ~170 to 180 N/m (only around half of that of graphene) and a Young's modulus of ~270 GPa [33, 34], which is slightly higher than stainless steel [33]. Thin-film transistors developed with CVD grown MoS₂ exhibited high mechanical flexibility with no degradation of their electrical characteristics upon mechanical bending [35], which suggests that MoS₂ could be a good candidate for flexible nanodevices.

The remarkable electrical and optical properties of MoS₂ arise from its electronic structure [12, 25, 36]. Bulk 2H-MoS₂ has an indirect band gap semiconductor that transforms to a direct band gap at the monolayer regime. This can be explained looking at the band structure of MoS₂ (Fig. 5.2). In the bulk case, the valence band maximum (VBM) is located at Γ point and the conduction band minimum (CBM) is located halfway between the Γ-K line leading to the indirect gap. When we reduce the number of layers the energy of the indirect gap is increased, and MoS₂ finally becomes a direct band gap at K point when we reach the monolayer regime. The gap at K point is not much affected by the number of layers. This evolution is not

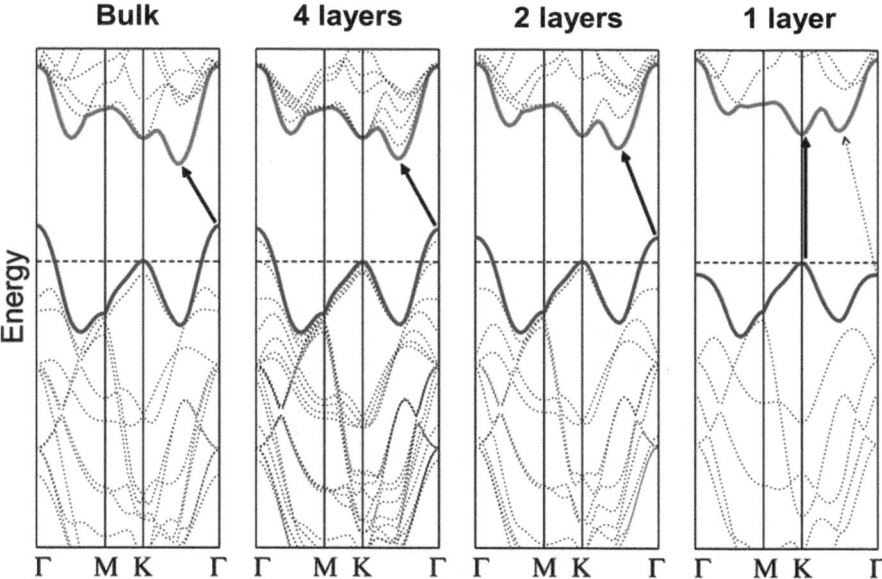

Fig. 5.2 Calculated band structures for MoS₂ with different number of layers at DFT-GGA level. The solid arrow shows the evolution of the band gap as we reduce the number of layers. Multilayer MoS₂ have an indirect gap while monolayer MoS₂ shows a direct gap at K point. (Adapted with permission from [37]. Copyright 2010 American Chemical Society)

only due to confinement effects, but also from the interaction of the pz orbitals from S atoms in neighboring layers, that participate in the indirect gap [25, 37–40]. The VBM at the K point in monolayer MoS_2 is split due to spin-orbit coupling [41–46].

The consequences of the band structure can be observed in the optical properties of MoS_2. Three main peaks appear in the absorption/reflectance spectra of MoS_2 in the range of $\sim$1.8–3 eV: two peaks denoted as A and B in the low energy part corresponding to excitonic transitions at the K point from the spin split bands, and a third one at higher energy, labelled as C, associated with higher-lying interband transitions around Γ point [47–51]. For thin layers of MoS_2 peaks A and B show little change with the number of layers [37, 40]. Photoluminiscence (PL) properties are greatly sensitive to the number of layers. Due to the indirect to direct gap transition, PL, which is negligible in the bulk case, emerges when approaching the monolayer regime [37, 40]. Mak et al. performed PL studies on free-standing samples of few-layers MoS_2 and found a dramatic enhancement of the quantum yield (QY) of more than 10^4 for the monolayer compared to the bulk case, with a QY of 4×10^{-3} for monolayer MoS_2 [40]. The PL spectra consisted in a single peak at 1.90 eV for the monolayer attributed to the direct optical band gap, while for few-layers they found 3 peaks: two peaks (A and B) related with the direct gap, separated by 150 meV, and a third peak (I) attributed to the indirect band gap, which progressively shifts towards the bulk indirect gap value of 1.29 eV as the number of layers increases. The two peaks A and B are attributed to the 2 excitons arising from the splitting of the valence band at K point due to interlayer coupling and spin-orbit coupling. Splendiani et al. also found a great enhancement of the PL QY when going from bulk to monolayer for MoS_2 samples deposited on Si/SiO_2; but, in this case, the two emission peaks attributed to the A and B excitonic transitions are still present in monolayer MoS_2 at 677 nm (1.83 eV) and 627 nm (1.98 eV) [37]. Adsorption of gases on monolayer MoS_2 has been shown to quantitatively enhance the PL due to charge transfer between MoS_2 and the physisorbed molecules [51]. This modulation of the PL has been shown to be reversible, and could be relevant for gas sensing applications. The combination of the optical direct band gap and the semiconducting properties of MoS_2 makes this material very promising for optoelectronic applications such as photodetectors or light emitting devices [12, 31].

The effect of the number of layers can also be observed in the Raman spectra. The difference in Raman shift between the two characteristic modes E^1_{2g} and A_{1g} varies with the number of layers and can provide a convenient method for determining layer thickness [52, 53].

Due to a high exciton binding energy in monolayer MoS_2, the electronic band gap is larger than the optical band gap [25, 31, 36, 43, 44, 49, 54]. The electronic band gap for monolayer MoS_2 has been calculated to be $\sim$2.80 eV [44, 55, 56], while a smaller value of 2.16 eV has been experimentally obtained by scanning tunnelling spectroscopy (STS) for monolayer MoS_2 on graphite [57]. It has been proposed that the substrate could affect the value of the electronic band gap measured [55]. Ryou et al. [55] studied theoretically how the band gap can be modulated with the substrate and the nearby dielectric environment, finding that the band gap tends to reduce with increasing the environmental dielectric constant. This approach could be used for band gap tuning in Field-Effect Transistors (FETs).

Monolayer MoS_2 has been explored as an interesting candidate for the development of new ultrathin electronic devices, due to the possibilities offered by its band gap [12, 31]. The room-temperature carrier mobility in monolayer MoS_2 has been predicted to be limited to ~410 cm^2 V^{-1} s^{-1} by optical phonon scattering [58]. Nevertheless, obtaining good real-life mobilities of devices fabricated with MoS_2 is an issue [12]. Early experiments reported mobilities of ~0.5–3 cm^2 V^{-1} s^{-1} [31, 59]. These mobilities can be enhanced by dielectric engineering employing high-κ dielectric materials [12, 31, 58, 60]. Most MoS_2 FET devices are n-type, but p-type or ambipolar can also be fabricated [12, 31]. As example of the performance achievable by a monolayer MoS_2 FET device, the n-type FET realized by Radisavljevic et al. using the high-κ dielectric material HfO_2 showed high current on/off ratio exceeding 10^8, room-temperature mobility of 200 cm^2 V^{-1} s^{-1}, and a subthreshold swing of 74 mV per decade [61].

Some interesting properties arise from the symmetry of MoS_2. Few-layer MoS_2 having an even number of layers have inversion centre, while an odd number of layers lacks it [12]. Monolayer MoS_2 may have applications in spintronics and valleytronics, due to the combination of spin-orbit-coupling and the absence of inversion symmetry, which produce a spin splitting of the valence band and spin-valley coupling [12, 15, 17, 36]. The valence band splitting induced by spin-orbit-coupling has been observed in experiments[41–43] and calculations [36, 41, 42, 44–46] with values around 150 meV. The lack of inversion symmetry in monolayer and odd-layered MoS_2 also cause them to present piezoelectricity, in contrast to even-layered and bulk MoS_2 [27, 28].

Strain can be used to modulate the band gap and electronic properties of MoS_2 [12, 18, 33, 36, 62–64]. For example a decrease in the optical gap and direct to indirect transition has been observed under uniaxial tensile strain [63]. The properties of MoS_2 can also be tuned by other strategies as gating with electric fields, alloying, combining with other materials to form heterostructures, changing the dimensions of the nanosheets, introducing defects or doping [12, 17, 18].

The edges of MoS_2 sheets can have an influence in the properties of the material, and can be important for gas sensing as molecules could adsorb there [12, 65]. It has been shown that edges play an important role in catalytic reactions [66, 67], in comparison with the basal plane. The morphology of the edge depends on different factors such as synthesis conditions [68–71] or particle size [72]. Different edge morphologies could present different properties. For example, metallic character or magnetism has been predicted for specific edge morphologies [69, 73].

5.3 1T-MoS₂

Different MoS_2 phases with Mo atom presenting octahedral coordination can be produced from the 2H/1H phase through alkali metal intercalation

[15, 19, 74–83] or electron beam irradiation [84]. This opens the possibility of tuning the properties of MoS_2 through phase engineering [19]. The atomic structure and properties of the materials produced by the different methods have been investigated with controversy during last decades [74, 75, 78]. Different experimental studies show that, generally, the structure of these materials corresponds to distorted 1T phases presenting different Mo-Mo clusterization [74–79, 81–83]. Nevertheless, the presence of the ideal 1T phase has also been reported [77, 78, 80, 84]. Despite the structural variability found for the octahedral coordinated MoS_2 phases, many times they are called in the literature as $1T-MoS_2$. Sometimes, the distorted 1T phases are called 1T', 1T"... but different structures can appear under the same name in different publications (e.g., 1T" [78, 85]). In order to avoid confusion, all the octahedral coordinated phases (distorted or not distorted) will be referred as O_h-MoS_2 regardless the specific atomic structure, within this discussion.

DFT studies on single-layer MoS_2 [78, 85, 86] have predicted that the ideal 1T phase is unstable in the neutral state, and that the 1H structure is the most stable one. Nevertheless, these studies show that different distorted 1T phases based on Mo-Mo clustering could exist as metastable states.

Three types of superstructures have been proposed (Fig. 5.1c): 2a × 2a, a × 2a (or alternatively a × a$\sqrt{3}$) and a$\sqrt{3}$ × a$\sqrt{3}$ [19, 87]. The type of structure formed depends on the preparation conditions. An a × 2a superstructure based on parallel Mo zig-zag chains have been found in restacked MoS_2, previously exfoliated from Li intercalated MoS_2 [74, 75]. This structure can be alternatively described as an orthorhombic a × a$\sqrt{3}$ cell. When this restacked MoS_2 is modified with Br_2 a change from the zig-zag chains structure to a a$\sqrt{3}$ × a$\sqrt{3}$ occurs [75]. This superstructure is also found in the layers of O_h-MoS_2 obtained from the oxidation of $K_x(H_2O)_yMoS_2$ [81, 82]. Different types of 2a × 2a superstructures have also been reported or predicted for O_h-MoS_2 at different conditions [30, 76, 78, 86, 87].

The phase transformation from 2H to O_h-MoS_2, and the type of superstructure formed seems to be related with the amount of charge present in the MoS_2 layers. Alkali intercalated MoS_2 materials are usually the precursors for the production of O_h-MoS_2. The intercalated alkali metals could transfer negative charge to the MoS_2 layers, which could be the driving force for the phase transformation. The charge transferred would alter the electron counting of the d orbital of Mo, and different superstructures would be stabilised depending on the degree of charge transferred [19, 75, 76, 86]. Wypych et al. performed STM [82] and electron diffraction studies [83] of the materials produced from the oxidation of $K_x(H_2O)_yMoS_2$ with different reaction times (i.e. different degree of oxidation). The superstructure a × 2a composed of Mo zig-zag chains was found in the incompletely oxidised sample, while the a$\sqrt{3}$ × a$\sqrt{3}$ was preferentially found after longer oxidation times. This lead them to propose the following scheme for the structural changes in the whole process [82]: K_zMoS_2, z ≈ 0.7 (2a × 2a) → $K_x(H_2O)_yMoS_2$, x ≈ 0.3 (a × a$\sqrt{3}$) → $K_x(H_2O)_yMoS_2$, x < 0.3 (a × 2a) → "1T-MoS_2" (a$\sqrt{3}$ × a$\sqrt{3}$). Similar results have been found for restacked layers prepared from the exfoliation of Li intercalated MoS_2. Heising and Kanatzidis reported that these restacked layers displaying the a × 2a superstructure retain some negative charge, and similarly convert to the (a$\sqrt{3}$ × a$\sqrt{3}$) after chemical modification with Br$_2$ [75]. They have also proposed reaction schemes illustrating the different phases obtained after different treatments of Li intercalated MoS$_2$ (LiMoS$_2$). The changes produced during the lithiation process have been studied by *in situ* HRTEM of the electrochemical intercalation of lamellar 2H-MoS$_2$ nanosheets [76]. It was found that a phase transition from 2H to O$_h$-LiMoS$_2$ occurs, with the formation of different superlattices. Different structures are proposed for this superlattices: Mo zig-zag chains (a × 2a/a × a$\sqrt{3}$), ribbon-chain superstructure (a × a$\sqrt{3}$) and diamond-chain superstructure (2a × 2a). Interestingly these structures disappeared and reduced to the "1T" structure after few days, and were considered to be transient states. The process of lithiation has also been studied theoretically for the bulk [88] and monolayer regime [86]. In both cases distorted 1T structures with Mo diamond-chain superstructure (2a × 2a) are found at high Li concentrations as the most stable phases. In addition, it has been studied how the relative energy between the 2H/1H phase and the O$_h$-MoS$_2$ phases change with Li concentration (and number of electrons injected, in case of the monolayer), and an inversion of the relative stability, favouring the O$_h$-MoS$_2$ over the 2H/1H is obtained when a specific threshold in Li concentration or number of electrons is reached. In the case of the monolayer regime [86], in the absence of Li atoms and charge, a metastable structure based on Mo zig-zag chains (a × 2a superstructure) similar to that found in the experimental studies of restacked MoS$_2$ is predicted.

Recently, STEM studies have been performed in single-layer MoS$_2$ [77, 78, 84]. In samples prepared through exfoliation of Li intercalated MoS$_2$, domains of the 1H, ideal 1T, and the Mo-zig-zag chain superstructure coexist [77, 78]. In addition, a 2a × 2a superstructure, not based on Mo diamonds, is found to coexist in a very low extension with the other phases [78]. It has been observed that electron-beam irradiation in STEM experiments can produce phase transitions. Transitions from the Mo zig-zag phase to the 1T [77] or the 1H phase [78] in the chemically exfoliated samples, have been reported. In addition, the direct transformation 1H to 1T due to the electron-beam irradiation in CVD grown MoS$_2$ single-layer doped with Re (0.6%) has been reported by Lin et al. [84]. They have also investigated the atomic mechanism of different phase transformations induced by the electron beam. Interestingly, they can locally control the 1H → 1T transformation. As 1H and 1T phase have, respectively, semiconducting and metallic character, this can be very useful for the design of nanoelectronic devices. They have used this technique to produce different nanodevice prototypes architectures: Schottky diode, nanoscale transistor, quantum lead and quantum dots.

O$_h$-MoS$_2$ phases have been found to be metastable, and covert back to the 2H/1H with temperature or aging [30, 30, 75, 79, 80, 82]. DSC measurements in restacked O$_h$-MoS$_2$ produced by exfoliation of Li intercalated MoS$_2$ suggest that this transition could occur at ∼90 or ∼98 °C depending on the starting phases involved [75],but higher temperatures have been as well proposed depending also on the type of material [30, 80, 82].

O$_h$-MoS$_2$ has been found to be electrically conducting [75, 80, 81], with some debate regarding the actual phase responsible of the phenomenon [75]. Theoretical studies have predicted that the ideal 1T phase should be metallic, while a very small direct gap is opened for the Mo zig-zag chains superstructure [78, 85, 86].

In order to be able to compare the structure and stability of some of the superstructures found across the literature, we have performed DFT calculations on a selected set of structures: 1H, ideal 1T and distorted 1T phases with different Mo-Mo clustering: Mo zig-zag chains (a × 2a), Mo trimers (2a × 2a) and Mo diamonds (2a × 2a).

DFT calculations on the relative stability and the structural parameters of different phases can be found in the literature. Kan et al. [86] compared the 1H, ideal 1T, and Mo zig-zag chains (a × 2a) structures, while in a different study, Chou et al. [78] included a 2a × 2a superstructure based on trimers of Mo atoms in the comparison. In addition, a different 2a × 2a superstructure based on diamond shape clustering of Mo atoms was predicted by Calandra [85], and was compared to all structures mentioned before, except the Mo trimers 2a × 2a superstructure. If we want to be able to compare the stability and parameters of these 5 phases, they need to be calculated using the same computational method and conditions.

In the present work, DFT calculations were carried out for the different phases at the local density approximation (LDA) level using the code AIMPRO [89–91]. The spin-averaged charge density is fitted to plane waves with an energy cut-off of 100 Ha, while Kohn-Sham wave functions were constructed using localised Cartesian Gaussian orbital functions (50 for Mo, $l \leq 3$, and 28 for S, $l \leq 2$). Relativistic pseudopotentials generated by Hartwigsen, Goedecker and Hutter [92] were used. In order to aid convergence of the self-consistency cycle, a finite electron temperature of 0.04 eV was used. Energies are taken to be self-consistent when the error drops below 10^{-5} Ha. Primitive cells were constructed for the different phases, and the structures were optimised allowing lattice parameters and atom positions to change until the convergence criteria is reached (maximum atomic position change in a given iteration $<10^{-4}$ Bohr radius and/or energy change $<10^{-5}$ Ha). For 1H and 1T the cell is constrained to be hexagonal (a = b, $\alpha = 120°$), while for the distorted phases all lattice parameters are allowed to change independently. The cells consisted in one MoS_2 unit for 1H and 1T phases, 2 MoS_2 units (1 × 2 supercell) for the Mo zig-zag chains (a × 2a) superstructure, and 4 MoS_2 units (2 × 2 supercell) for the two 2a × 2a superstructures (and the 2 × 2 supercell of 1T phase), where gamma-centred k-point meshes generated under the Monkhorst-Pack scheme [93] of 24 × 24 × 1, 24 × 12 × 1 and 12 × 12 × 1 were used respectively, in order to obtain a consistent k-point density across phases. The interlayer cell parameter was fixed at c = 3.108 nm, in order to avoid interaction between layers.

The 1H phase is found to be the most stable one, while the ideal 1T phase is found to be the less stable one, with an energy difference per MoS_2 unit of 0.84 eV. From the distorted 1T phases, the Mo zig-zag chains superstructure (a × 2a) is found to be the most stable (0.59 eV per MoS_2 unit relative to 1H) followed very close by the two 2a × 2a phases with an energy difference per MoS_2 unit with respect to 1H of 0.67 and 0.68 eV for the Mo trimers and Mo diamonds superstructures respectively. These results are in agreement with previous calculations from the literature: 0.82 [68], 0.83 [75], 0.85 [76] eV for the ideal 1T phase, 0.64 [75] eV for Mo diamonds,

0.63 [68] eV for Mo trimers, and 0.51 [76], 0.54 [75] eV, 0.55 [68] eV for the Mo zig-zag chains superstructures.

We also performed an optimisation for the 1T phase in a 2 × 2 supercell. If the symmetry of the starting structure is broken by slightly displacing one S atom (~1.6 pm in every coordinate), and all lattice parameters are allowed to vary independently, the structure is found to distort to the Mo trimers superstructure during the run. This suggests that the ideal 1T phase could be an unstable phase in neutral state, in agreement with previous studies [68, 75, 76].

The Mo diamonds superstructure found in first place was highly symmetric with the four Mo-Mo bond distances of the perimeter of the diamond being approximately equal (287 pm). This structure has an inversion centre located at the geometric centre of the diamond, in contrast to the other 2a × 2a superstructure formed by trimers. This is important because, as we have already commented, some properties depend on the presence or absence of an inversion centre. We carried out some calculations in order to see if the structure changes when we slightly distort it. If a displacement in the order of 10.6 pm at each "y" and "z" coordinate for one S atom lying on top of the diamond, or in the order of 5.3 pm for one Mo atom located at the corner of the diamond, the Mo diamonds structure is recovered after the optimisation runs, although slightly less symmetric. The previous four equivalent Mo-Mo bonds forming the diamond, now appear as a pair of 285 pm and another pair of 288 pm (structures labelled as "S." and "AS." respectively in Table 5.1). The energy difference between the more and less symmetric diamonds structures is negligible (just 0.14 meV per MoS$_2$ unit in favour of the less symmetric one). If a higher distortion is introduced in the starting geometry (displacement of ~10.6 pm at each "y" and "z" coordinates of the same Mo atom) the structure transforms into the more stable Mo trimers superstructure. This suggests that the diamonds superstructure could be also an unstable structure, or a metastable structure that could transform easily to the trimers one. Further studies need to be done in order to determine the thermal and kinetic stability of this phase. The structural parameters and relative energies of the different structures found are summarised in Table 5.1. The result that the 1H phase is the most stable one, can explain the experimentally observed phase changes of O$_h$-MoS$_2$ towards the 1H/2H phase with heating or aging. In addition, the fact that the ideal 1T phase is usually found, in theoretical studies, as an unstable phase can also explain why usually O$_h$-MoS$_2$ is found formed by distorted phases. Nevertheless, the ideal 1T phase has also been reported, and further studies considering other effects (such as charge, adsorbates, defects . . .) should be done in order to explain its presence.

We have shown that several distorted O$_h$-MoS$_2$ phases can exist as metastable phases. Among the distortions studied in this work, the Mo zig-zag chains (a × 2a) superstructure is the most stable one, and should be the distorted phase preferentially formed. Interestingly, this superstructure is usually found in Li exfoliated MoS$_2$, or in MoS$_2$ obtained from the incomplete oxidation of K$_x$(H$_2$O)$_y$MoS$_2$. Nevertheless, although this superstructure has been found to be metastable by different DFT studies [68, 75, 76], the investigations made on restacked O$_h$-MoS$_2$ from Li

Table 5.1 (following page) Relative energies with respect to 1H phase and structural parameters calculated for different phases of MoS_2. Top and side views of the structures obtained are illustrated (top), with the unit cells represented by dashed lines. Two alternative cells are considered for the Mo zig-zag superstructure (a × 2a and a × a√3). The different type of superlattices are represented as a × 2a (a × a√3) and 2a × 2a, while the two actual lattice constants for each cell are labelled as a_0 and b_0. Proposed clustering of Mo atoms is highlighted with thick lines on the illustrations, and the corresponding associated bond distances are represented in bold in the table. For the Mo diamonds structures, the structural differences are visually imperceptible and only one structure is shown ("asym."). The energy difference between the two diamonds structures is <0.5 meV, and not considered in the table

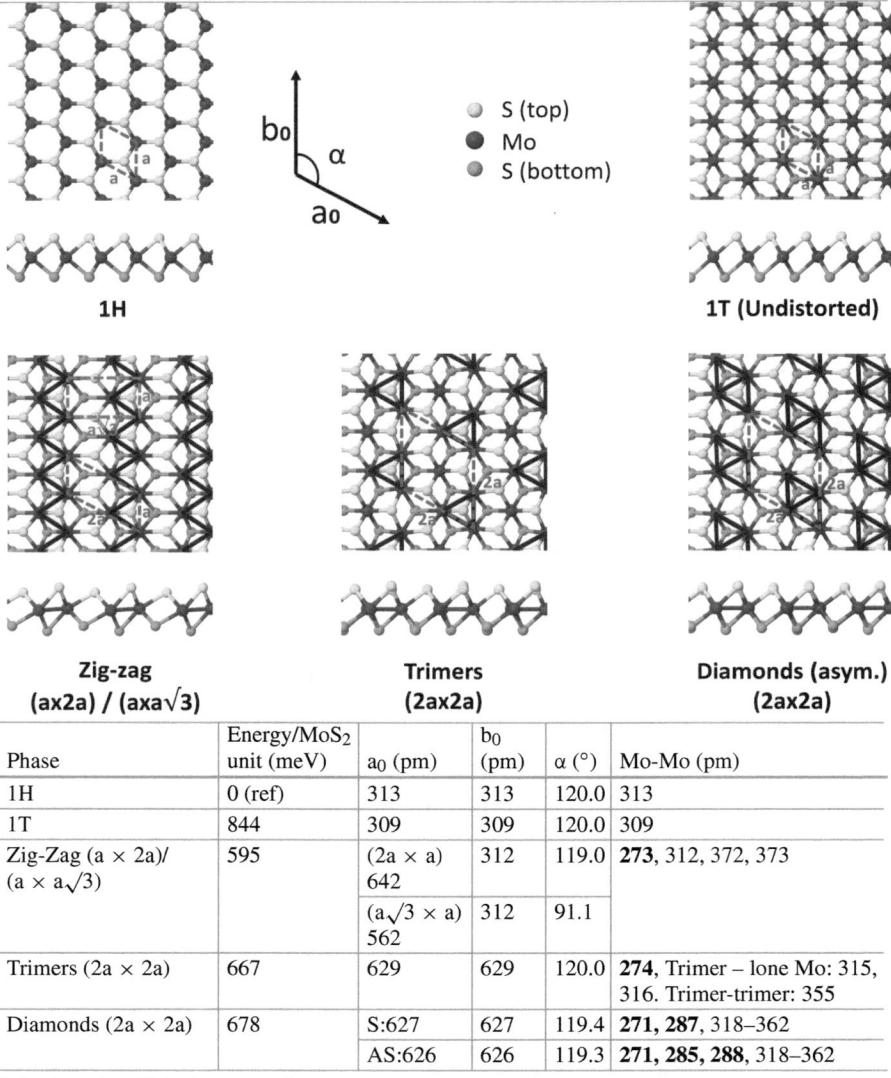

Phase	Energy/MoS$_2$ unit (meV)	a_0 (pm)	b_0 (pm)	α (°)	Mo-Mo (pm)
1H	0 (ref)	313	313	120.0	313
1T	844	309	309	120.0	309
Zig-Zag (a × 2a)/ (a × a√3)	595	(2a × a) 642	312	119.0	**273**, 312, 372, 373
		(a√3 × a) 562	312	91.1	
Trimers (2a × 2a)	667	629	629	120.0	**274**, Trimer – lone Mo: 315, 316. Trimer-trimer: 355
Diamonds (2a × 2a)	678	S:627	627	119.4	**271, 287**, 318–362
		AS:626	626	119.3	**271, 285, 288**, 318–362

S Symmetric, *AS* Anti-symmetric

exfoliated MoS$_2$ [65] or O$_h$-MoS$_2$ obtained from the oxidation of K$_x$(H$_2$O)$_y$MoS$_2$ [72, 73], have related this phase with the presence of negative charge, and a superstructure consisting on Mo trimers (a$\sqrt{3}$ × a$\sqrt{3}$) is preferentially found when these samples are oxidised. This Mo trimers phase is different from the Mo trimers (2a × 2a) superstructure studied here. While the Mo trimers phase based in a 2a × 2a superlattice leave some Mo atoms not participating in any cluster (Table 5.1), in the phase based in an a$\sqrt{3}$xa$\sqrt{3}$ superlattice all Mo atoms participate in a trimer (Fig. 5.1c), and hence, it is expected to be more stable than the 2a × 2a one. This phase should be considered in future studies for a proper comparison of the different superstructures in the neutral state, as it could be the most stable one at this charge state.

We have also shown that different structures compatible with a 2a × 2a superstructure are possible. These phases are higher in energy than the Mo zig-zag chains phase and are expected to appear in much less proportions. In addition, the 2a × 2a superstructures are very close in energy, with an energy difference per MoS$_2$ unit of 12 meV between the most stable Mo trimers and the less stable Mo diamonds superstructures. The Mo diamonds superstructure can be understood as two trimers sharing and edge and forming the diamond shape. If we pull one of the Mo atoms which does not share an edge, along the long axis of the diamond, the two trimers forming the diamond become non-equivalent, until one of the trimer is broken and, eventually, we end up with the Mo trimers superstructure, where this Mo atom is equidistant with the surrounding trimers. The transformation from the distorted Mo diamonds superstructure to the Mo trimers that we found in our simulations is reminiscent of this process. The distortion introduced to the Mo diamonds superstructure produced an increase of the energy per MoS$_2$ unit of only 48 meV. This, accompanied by the fact that the 2 phases are very close in energy suggests that the potential energy surface for these transformations could be relatively flat, and they could interconvert easily at room temperature.

We have to note that in these calculations we are not taking into account spin-orbit-coupling which is known to split the valence band in MoS$_2$ around 150 meV. In addition, the results obtained are only valid at 0K, and vibrational states and entropic factors should be considered in order to correctly predict the relative stability of the phases at room temperature. In order to know how long the different metastable structures could exist at working conditions before they transform to more stable phases, kinetic factors should be taken into account. Energy barriers for the different transformations could be calculated using methods such as nudge elastic band (NEB).

It has been shown that MoS$_2$ has a wide phase variability. As these phases have different properties, being able to control their stability will open new possibilities in the design of electronic nanodevices. In addition, as the phase stability is dependent on the physical and chemical conditions, those changes could potentially be used for sensing applications.

5.4   Chemical Functionalization of MoS$_2$

To be incorporated in a sensor structure the MoS$_2$ can be functionalized to enhance sensitivity and selectivity. Basic functionalization can be done using non-covalent and covalent functionalization. Here we show the covalent functionalization without changing the basal plane structure using 1,2-dithiolanes.

Exfoliation of MoS$_2$ in liquid media is the method of choice to obtain macroscopic quantities of flakes suitable not only for spectroscopic and microscopy characterization but most importantly for performing reactions en route for the development of novel hybrid materials. Furthermore, depending on the exfoliation methodology applied, it is possible to acquire semiconducting or metallic MoS$_2$ [94–98]. Herein, we employed the chlorosulfonic-assisted strategy to exfoliate MoS$_2$ of the 2H semiconducting polytype.[99] Notably, the exfoliated MoS$_2$ possess edge defects with strong affinity for sulfur-based moieties. In this frame, covalent functionalization of such exfoliated MoS$_2$ with 1,2-dithiolanes results on hybrid materials with interesting properties.[100] A representative protocol, shown in Fig. 5.3, concerns the reaction of 1,2-dithiolane-based derivatives **1a** carrying a BOC-protective terminal amine and **2a** carrying a pyrene chromophore, with exfoliated MoS$_2$ for 48 h at 50–80 °C in DMF, furnishing materials **2a** and **2b**, respectively (Fig. 5.1). Moreover, acidic treatment of **2a** yields **2c** by liberating the terminal amine in the form of the corresponding ammonium salt. A benefit of **2c** is related with solubility achieved in polar protic solvents, namely, despite solubility of **2a** and **2b** is NMP and DMF, the ammonium functionalized MoS$_2$ **2c** possesses enhanced solubility in MeOH.

Markedly, the presence of free –NH$_2$ units in **2c** allows sensitive sensing and capture of species carrying -COOH. In fact, the amount of free –NH$_2$ units in **2c** is confirmed by the Kaiser test and quantitative calculated to be 103 μmol g^{-1}. As far as **2a** and **2b** concerns, the amount of organic units decorating MoS$_2$ is

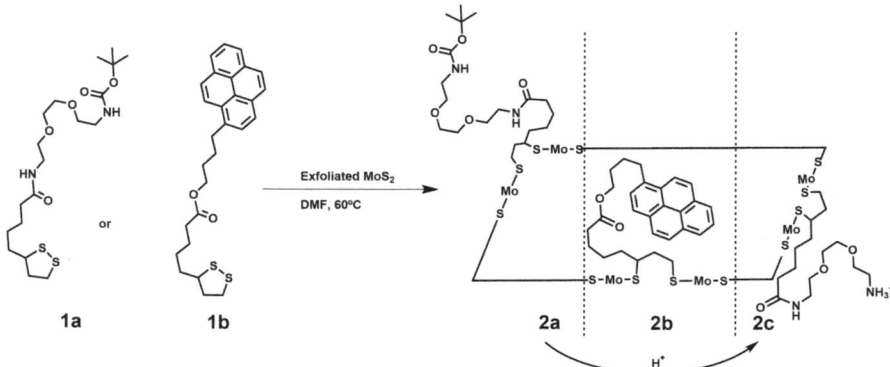

Fig. 5.3 Reaction route for obtaining nanohybrids 2a–c from exfoliated semiconducting MoS$_2$

estimated by thermogravimetric analysis (TGA). The mass loss for **2a** and **2b** in the temperature range 200–500 °C under an inert atmosphere is found to be 5.7% and 2.2%, respectively (Fig. 5.4), and corresponds to the thermal decomposition of the organic species attached onto MoS$_2$. The latter is calculated as 1 per every 50 and 160 MoS$_2$ units in **2a** and **2b**, respectively. Considering that the covalent functionalization in **2a** and **2b** takes place at edge sites, the small functionalization degree is reasonable.

The success of MoS$_2$ exfoliation was monitored by Raman spectroscopy upon 514 nm excitation. There are two characteristic modes discernable in both exfoliated MoS$_2$ and functionalized MoS$_2$-pyrene **2b** material. The energy difference between A$_{1g}$ mode located at around 408 cm^{-1} and the E$^1_{2g}$ mode at around 382 cm^{-1} is usually used for evaluating the average number of layers in exfoliated MoS$_2$.[52, 101] Hence, since for exfoliated MoS$_2$ the energy difference between A$_{1g}$ and E$^1_{2g}$ is calculated to be 24 cm^{-1} (Fig. 5.5), it is reasonable to claim the presence of 3–4 layers in average. Furthermore, considering that functionalization of MoS$_2$ with n-doping materials results on decrease of the relative intensity for the A$_{1g}$ mode, while the opposite phenomenon, namely enhanced intensity, is observed when MoS$_2$ is functionalized with p-doping materials,[102–107] the reduction in the intensity of

Fig. 5.4 Thermographs for exfoliated MoS$_2$ (gray) as compared with **2a** (black) and **2b** (dotted), obtained under nitrogen atmosphere

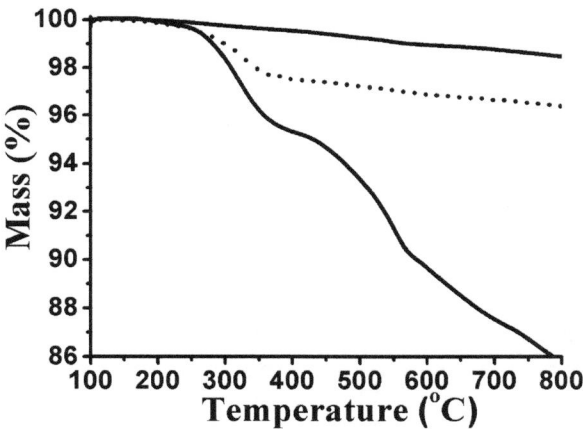

Fig. 5.5 Raman spectra for exfoliated MoS$_2$ (grey) as compared with MoS$_2$-pyrene **2b** (black), upon excitation at 514 cm^{-1}

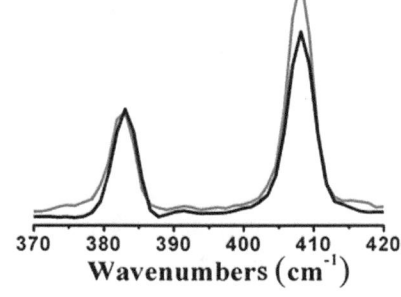

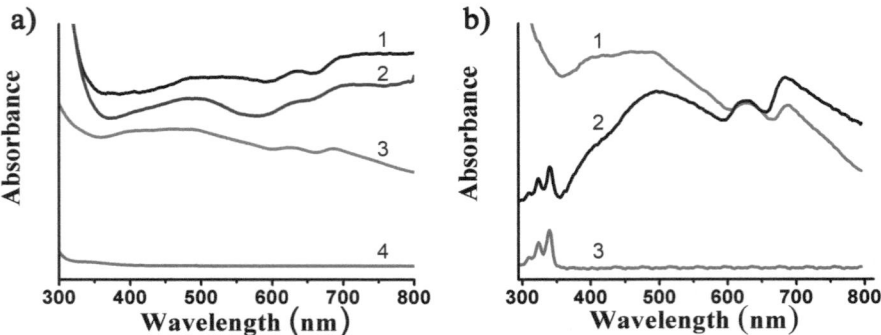

Fig. 5.6 UV-Vis spectra for exfoliated MoS₂ (curve 3) as compared with (a) functionalized nanohybrid **2c** (curve 2), dithiolane-based derivative **1a** (curve 4) and functionalized nanohybrid **2a** (curve 1) in DMF, and (b) 1,2-dithiolane-based pyrene derivative **1b** (curve 3) as compared with functionalized MoS₂-pyrene nanohybrid **2b** (curve 2) in MeOH

A_{1g} in **2b** is attributed to n-doping, via electron-transfer from pyrene. The latter is further justified with the aid of photoluminescence assays – see below.

The semiconducting nature of MoS₂ in hybrids **2a–2c** is a clear advantage, as opposed to the corresponding metallic polytype, when its novel electronic and optical properties are employed targeting sensing applications. Focusing in the UV-Vis absorption spectra for **2a** and **2c**, the characteristic absorption bands owed to MoS₂ are clearly identified at 401, 494, 626 and 681 nm (Fig. 5.6). On the other hand, for the pyrene-modified **2b**, on top of the absorptions due to MoS₂, additional bands attributed to pyrene centered at 309, 323 and 340 nm are present. However, those absorption bands in **2c** are simple superimpositions of the bands due to free pyrene **1b** and exfoliated MoS₂, hence implying the absence of meaningful electronic interactions between the two species at the ground state.

Focusing on possible electronic interactions between pyrene and MoS₂ within **2c** at the excited state, photoluminescence assays are carried out. In more detail, the strong and characteristic emission of free pyrene **1b** centred at 375, 396 and 417 nm upon excitation at 390 nm is quenched in **2c** (Fig. 5.7a). The photoluminescence suppression of pyrene in **2c** is indicative of charge and/or energy transfer as the decay mechanism of the singlet excited state of pyrene. In order to study the dynamics of the photoluminescence, time resolved assays were performed. In this frame, time-correlated-single-photon-counting measurements were employed to acquire the photoluminescence lifetime profiles for **2c**. The fluorescence decay profile for free pyrene **1b** was measured and correlated with that owed to **2c**. Exciting pyrene at 375 nm, the decay of emission at 396 nm shows a monoexponenetial behavior with lifetime 3.79 ns. For hybrid **2b**, the fluorescence decay profile is biexponential (Fig. 5.7b), with a slower lifetime 2.43 ns resembling the one of free pyrene in **1b** and a faster one with lifetime 260 ps attributed to the singlet excited state deactivation of pyrene via MoS₂.

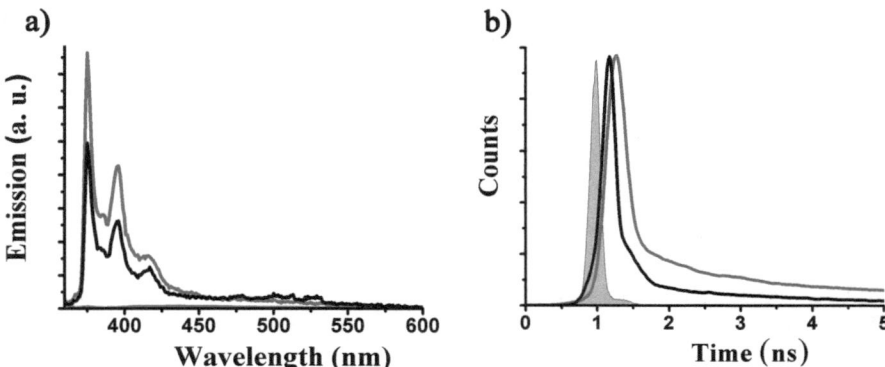

Fig. 5.7 (**a**) Photoluminescence spectra of free pyrene **1b** (top) as compared to nanohybrid **2b** (bottom) upon excitation at 340 nm in MeOH for samples possessing equal absorbance at the excitation wavelength. (**b**) Decay profiles at 396 nm of free pyrene **1b** (red) as compared to nanohybrid **2b** (left), upon 375 nm excitation

Covalent functionalization allows to tune the properties of the MoS$_2$ without changing its structure. Being able to tune the bandgap of the 2D material is a desired characteristic in order to incorporate the material into gas sensing devices to make it more sensitive and selective.

5.5   Gas Sensing Using MoS$_2$

MoS$_2$ is interesting for gas sensing application due to its high surface to volume ratio and because of its tuneable bandgap which means it can be incorporated into a field-effect transistor (FET) structure [108]. Another advantage is the selective surface reactivity enabling the differentiation between different gases [109]. MoS$_2$ has been used mostly in the electrochemical route to detect targeted analytes. Within the conductance regime the FET structure, the direct measurement using interdigitated electrodes and thin film transistors have been investigated.

The FET structure consists of a Si substrate used as a back gate on which a SiO$_2$ layer is used as insulating material. On top the MoS$_2$ forms the channel on which source and drain electrodes have been deposited using photolithography. The MoS$_2$ used is deposited either with the mechanical exfoliation technique or with CVD. Li et al. [110] used the exfoliation method to deposit single and up to 4 layers of MoS$_2$ into a FET structure. They found the single layer to be unstable but got a sensor response measuring the change of the source drain current of the transistor using 2 layers when 0.3 ppm NO is applied. Late et al. [111] found that multilayer MoS$_2$ is more sensitive than bilayer structures when NH$_3$, NO$_2$ or water vapour is applied. They changed gate voltages and exposed the device to light to change its sensing properties showing that the device can be additionally tuned to increase the limit of detection. Yoon et al. [112] investigated in the Schottky diode device where they used platinum as their contact material and exfoliated MoS_2 in single and up to 4 layer structures. The gate voltage dependent source drain current was measured indicating n-type behaviour of the MoS_2 towards NH_3 and NO_2. Cho et al. [113] also exposed their FET with exfoliated MoS_2 to the same gases and reached a limit of detection of 1.2 ppm for NO_2 using graphene as their source and drain contacts. They transferred their device on to a flexible substrate without having any loss in performance making it applicable in wearable technology. Liu et al. [114] formed a FET structure using CVD grown MoS_2 and formed a Schottky junction using gold as their contact. They reached a limit of detection of 20 ppb for NO_2 and 1 ppm for NH_3 measuring the source drain current. Beside they studied the Schottky contact modulation due to adsorption of the gases in the transistor structure.

The other popular route to go is the direct conductance measurement using interdigitated electrodes. Perkins et al. [109] contacted exfoliated MoS_2 and exposed it to several gases among them precursor of warfare agents and showed the enhanced selectivity compared to carbon nanotube devices. Their limit of detection for certain materials was as low as 12 ppb. Lee et al.[115] formed a conductive gas sensor using CVD grown MoS_2 layer on which gold was evaporated as contacts. Exposure to NH_3 showed a limit of detection of 0.3 ppm showing that CVD based device can compete with the devices based on exfoliation. Cho et al. [116] investigated the charge transfer of molecules in gas form with CVD grown MoS_2 in a sensor structure. Using photoluminescence measured during the measurement of NH_3 and NO_2 showed that adsorption mostly occurs at the surface of the material. Further simulations have been done to strengthen the results found in the experiments. They also showed the increased sensitivity of the device at elevated temperatures as well as the faster sensor recovery when heated.

He et al. [117] fabricated a thin film transistor (TFT) on PET using reduced graphene oxide as source and drain material on which MoS_2 is spin coated out of solution. They functionalised the MoS_2 transistor using platinum nanoparticles and exposed the device to NO_2 with a detection limit of 0.5 ppm with the functionalised device. The device underwent bending tests and showed excellent stability characteristics exceeding the performance of graphene based flexible TFTs. Liu et al. [118] introduced an exfoliated boron nitride (BN) passivation layer on to their exfoliated MoS_2 TFT to increase the stability over time. They have compared their sensing properties to non-passivated devices using polar and non-polar solvents. The introduction of a BN layer did not prevent the gas detection, the response was different to the pristine devices though.

This overview shows that MoS_2 has been incorporated in a lot of electrochemical configurations showing a low limit of detection for a wide range of gases and an increased selectivity compared to other nanomaterials. Beside the electrochemical structure to detect gases there are many different other as shown in the introduction. One promising route for MoS_2 is the optical route, which we discuss in more detail in the next section.

5.6   Gas Sensing Using Raman Spectroscopy

Raman gas sensing, especially remote Raman gas sensing, attracted large attention in the late 1960s as lasers became widely used as Raman excitation sources [119, 120]. It emerged alongside other laser based probing techniques referred to as LIDAR [121, 122]. Main targets were atmospheric constituents [120] and air pollutants such as NO_2 [120, 122, 123].

Semiconducting gas sensors emerged around the same time [124–127]. The lowering of semiconductor manufacturing cost heralded their dominance in today's gas sensing applications. The use Raman spectroscopy and other optically based gas sensing techniques continued, although almost exclusively in applications, where direct access to the target gases is impossible, such as remote sensing [119–122], enclosed gases in geological samples [128], combustion chambers [129] and power plant gas turbines [130]. More recent developments in Raman gas sensing employed hollow-core photonic bandgap fibres [131].

With the advent of 2D materials such as single molecular layers of MoS_2 [132] and graphene [133] as active sensor materials, research in semiconductor gas sensors got a new boost. Allotropes of carbon such as reduced graphene oxide [134, 135] and 2D networks of carbon nanotubes [136, 137] have been successfully used as gas sensor materials, with detection sensitivities down to the single molecule level [138]. Raman spectroscopy plays a vital role in graphene characterisation [139], and has also been demonstrated as a potential detection technique of graphene adsorbed molecules [11].

Resistivity based sensors with monolayer MoS_2 were developed [109, 116, 140, 141], using Raman spectroscopy to characterise the MoS_2 monolayer [140, 141] and demonstrating the sensitivity of the MoS_2 Raman spectrum to its ambient atmosphere [140] or solvent [106].

The change in Raman spectrum is usually characteristic to the type molecule adsorbed [106] and could in some cases be used to compensate for lacking specificity.

5.7   Conclusions

Nanomaterials show promising characteristics for gas sensing applications. Their high surface area to volume ratio allows quick responses and low limits of detection. The 2D architecture of materials such as MoS_2 naturally adapts itself to gas sensing with high relative surface area and large surface capture area for absorbed gas species.

Sensor structures often use the electrochemical route to monitor the environment. Selectivity is a crucial characteristic of the sensor and can be tuned by functionalising the materials. In the current chapter we have shown that monolayers of MoS_2 are particularly suitable for sensing applications. The different phases of MoS_2, namely H, T and T', are well understood. All three phases are present in the metallic, whereas in the semiconducting phase only the H phase is present. DFT calculations show that the stability under covalent functionalization with 1,2-dithiolanes onto monolayers of MoS_2 is higher at the edge than in the basal plane.

Covalent functionalisation of MoS_2 with 1,2-ditholates was proved experimentally. The MoS_2 properties are changed without modifying the structure of the basal plane. The covalent functionalisation using electron donor groups derivates of 1,2-dithiolanes allows to tune the bandgap and are shown in the chapter, proofing the feasibility of the functionalization of the MoS_2. This provides a useful route to allow tuning of the sensitivity to specific target molecules, without modifying the underlying architecture of an overlapping matrix of MoS_2 sheets for example. The pristine material has been incorporated into various device structures using its electrochemical properties. The understanding of the MoS_2 and the ability to functionalize the material can lead to better performances of the MoS_2 based gas sensing devices in terms of limit of detection and selectivity.

In-situ Raman spectroscopy appears a promising route to new types of sensing device where the well understood spectrum of the MoS_2 can be used to detect small traces of concentrations in the operation regime.

Acknowledgments This project has received funding from the European Union's Horizon 2020 research and innovation programme under the Marie Sklodowska-Curie grant agreement N° 642742.

References

1. Liu X, Cheng S, Liu H et al (2012) A survey on gas sensing technology. Sensors (Basel) 12:9635–9665
2. Jimenez-Cadena G, Riu J, Rius FX (2007) Gas sensors based on nanostructured materials. Analyst 132:1083–1099
3. Varghese SS, Varghese SH, Swaminathan S et al (2015) Two-dimensional materials for sensing: graphene and beyond. Electronics 4:651–687
4. Yuan W, Shi G (2013) Graphene-based gas sensors. J Mater Chem A 1:10078–10091
5. Dan Y, Lu Y, Kybert NJ et al (2009) Intrinsic response of graphene vapor sensors. Nano Lett 9:1472–1475
6. Chikkadi K, Muoth M, Hierold C (2013) Hysteresis-free, suspended pristine carbon nanotube gas sensors. Solid-State Sensors, Actuators and Microsystems (TRANSDUCERS & EUROSENSORS XXVII), 2013 Transducers & Eurosensors XXVII: The 17th International conference on. pp 1637–1640
7. Xia J, Chen F, Li J, Tao N (2009) Measurement of the quantum capacitance of graphene. Nat Nanotechnol 4:505–509. https://doi.org/10.1038/nnano.2009.177
8. Wang X, Sun X, Hu PA et al (2013) Colorimetric sensor based on self-assembled Polydiacetylene/graphene-stacked composite film for vapor-phase volatile organic compounds. Adv Funct Mater 23:6044–6050
9. Arsat R, Breedon M, Shafiei M et al (2009) Graphene-like nano-sheets for surface acoustic wave gas sensor applications. Chem Phys Lett 467:344–347
10. Cittadini M, Bersani M, Perrozzi F et al (2014) Graphene oxide coupled with gold nanoparticles for localized surface plasmon resonance based gas sensor. Carbon 69:452–459

11. Liu M, Chen W (2013) Graphene nanosheets-supported Ag nanoparticles for ultrasensitive detection of TNT by surface-enhanced Raman spectroscopy. Biosens Bioelectron 46:68–73
12. Ganatra R, Zhang Q (2014) Few-layer MoS$_2$: a promising layered semiconductor. ACS Nano 8:4074–4099
13. Jiang J-W (2015) Graphene versus MoS$_2$: a short review. Front Phys 10:287–302
14. Zhang G, Liu H, Qu J, Li J (2016) Two-dimensional layered MoS$_2$: rational design, properties and electrochemical applications. Energy Environ Sci 9:1190–1209
15. Chhowalla M, Shin HS, Eda G et al (2013) The chemistry of two-dimensional layered transition metal dichalcogenide nanosheets. Nat Chem 5:263–275
16. Chen Y, Tan C, Zhang H, Wang L (2015) Two-dimensional graphene analogues for biomedical applications. Chem Soc Rev 44:2681–2701
17. Heine T (2014) Transition metal chalcogenides: ultrathin inorganic materials with tunable electronic properties. Acc Chem Res 48:65–72
18. Wang H, Yuan H, Hong SS et al (2015) Physical and chemical tuning of two-dimensional transition metal dichalcogenides. Chem Soc Rev 44:2664–2680
19. Voiry D, Mohite A, Chhowalla M (2015) Phase engineering of transition metal dichalcogenides. Chem Soc Rev 44:2702–2712
20. Chen X, McDonald AR (2016) Functionalization of two-dimensional transition-metal Dichalcogenides. Adv Mater 28:5738–5746
21. Presolski S, Pumera M (2016) Covalent functionalization of MoS$_2$. Mater Today 19:140–145
22. Rouxel J (1986) Reactivity and phase transitions in transition metal dichalcogenides intercalation chemistry. J Chim Phys 83:841–850
23. Ramsdell LS (1947) Studies on silicon carbide. Am Mineral 32:64–82
24. Kolobov AV, Tominaga J (2016) Bulk TMDCs: review of structure and properties. 2-dimensional transition-metal dichalcogenides. Springer, Cham, pp 29–77
25. Zhao W, Ribeiro RM, Eda G (2014) Electronic structure and optical signatures of semiconducting transition metal dichalcogenide nanosheets. Acc Chem Res 48:91–99
26. Jiang T, Liu H, Huang D et al (2014) Valley and band structure engineering of folded MoS$_2$ bilayers. Nat Nanotechnol 9:825–829
27. Zhu H, Wang Y, Xiao J et al (2015) Observation of piezoelectricity in free-standing monolayer MoS$_2$. Nat Nanotechnol 10:151–155
28. Wu W, Wang L, Li Y et al (2014) Piezoelectricity of single-atomic-layer MoS$_2$ for energy conversion and piezotronics. Nature 514:470
29. Py M, Haering R (1983) Structural destabilization induced by lithium intercalation in MoS$_2$ and related compounds. Can J Phys 61:76–84
30. Dungey KE, Curtis MD, Penner-Hahn JE (1998) Structural characterization and thermal stability of MoS$_2$ intercalation compounds. Chem Mater 10:2152–2161
31. Wang QH, Kalantar-Zadeh K, Kis A et al (2012) Electronics and optoelectronics of two-dimensional transition metal dichalcogenides. Nat Nanotechnol 7:699–712
32. Castellanos-Gomez A, Poot M, Steele GA et al (2012) Elastic properties of freely suspended MoS$_2$ nanosheets. Adv Mater 24:772–775
33. Bertolazzi S, Brivio J, Kis A (2011) Stretching and breaking of ultrathin MoS$_2$. ACS Nano 5:9703–9709
34. Liu K, Yan Q, Chen M et al (2014) Elastic properties of chemical-vapor-deposited monolayer MoS$_2$, WS$_2$, and their bilayer heterostructures. Nano Lett 14:5097–5103
35. Pu J, Yomogida Y, Liu K-K et al (2012) Highly flexible MoS$_2$ thin-film transistors with ion gel dielectrics. Nano Lett 12:4013–4017
36. Liu G-B, Xiao D, Yao Y et al (2015) Electronic structures and theoretical modelling of two-dimensional group-VIB transition metal dichalcogenides. Chem Soc Rev 44:2643–2663
37. Splendiani A, Sun L, Zhang Y et al (2010) Emerging photoluminescence in monolayer MoS$_2$. Nano Lett 10:1271–1275
38. Kuc A, Zibouche N, Heine T (2011) Influence of quantum confinement on the electronic structure of the transition metal sulfide TS$_2$. Phys Rev B 83:245213
39. Molina-Sánchez A, Sangalli D, Hummer K et al (2013) Effect of spin-orbit interaction on the optical spectra of single-layer, double-layer, and bulk MoS$_2$. Phys Rev B 88:045412

40. Mak KF, Lee C, Hone J et al (2010) Atomically thin MoS_2 a new direct-gap semiconductor. Phys Rev Lett 105:136805
41. Alidoust N, Bian G, Xu S-Y, et al (2013) Observation of monolayer valence band spin-orbit effect and induced quantum well states (QWS) in MoX_2. arXiv preprint arXiv:1312.7631
42. Dou X, Ding K, Jiang D et al (2016) Probing spin-orbit coupling and interlayer coupling in atomically thin molybdenum disulfide using hydrostatic pressure. ACS Nano 10:1619–1624
43. Klots A, Newaz A, Wang B et al (2014) Probing excitonic states in suspended two-dimensional semiconductors by photocurrent spectroscopy. Sci Rep 4:6608
44. Qiu DY, Felipe H, Louie SG (2013) Optical spectrum of MoS_2: many-body effects and diversity of exciton states. Phys Rev Lett 111:216805
45. Zhu Z, Cheng Y, Schwingenschlögl U (2011) Giant spin-orbit-induced spin splitting in two-dimensional transition-metal dichalcogenide semiconductors. Phys Rev B 84:153402
46. Zibouche N, Kuc A, Musfeldt J, Heine T (2014) Transition-metal dichalcogenides for spintronic applications. Ann Phys 526:395–401
47. Li Y, Chernikov A, Zhang X et al (2014) Measurement of the optical dielectric function of monolayer transition-metal dichalcogenides: MoS_2, $MoSe_2$, WS_2, and WSe_2. Phys Rev B 90:205422
48. Rigosi AF, Hill HM, Li Y et al (2015) Probing interlayer interactions in transition metal dichalcogenide heterostructures by optical spectroscopy: MoS_2/WS_2 and $MoSe_2/WSe_2$. Nano Lett 15:5033–5038
49. Hill HM, Rigosi AF, Roquelet C et al (2015) Observation of excitonic Rydberg states in monolayer MoS_2 and WS_2 by photoluminescence excitation spectroscopy. Nano Lett 15:2992–2997
50. Kozawa D, Kumar R, Carvalho A, et al. (2014) Photocarrier relaxation in two-dimensional semiconductors. arXiv preprint arXiv:1402.0286
51. Tongay S, Zhou J, Ataca C et al (2013) Broad-range modulation of light emission in two-dimensional semiconductors by molecular physisorption gating. Nano Lett 13:2831–2836
52. Li H, Zhang Q, Yap CCR et al (2012) From bulk to monolayer MoS_2: evolution of Raman scattering. Adv Funct Mater 22:1385–1390
53. Lee C, Yan H, Brus LE et al (2010) Anomalous lattice vibrations of single-and few-layer MoS_2. ACS Nano 4:2695–2700
54. Berkelbach TC, Hybertsen MS, Reichman DR (2013) Theory of neutral and charged excitons in monolayer transition metal dichalcogenides. Phys Rev B 88:045318
55. Ryou J, Kim Y-S, Santosh K, Cho K (2016) Monolayer MoS_2 bandgap modulation by dielectric environments and tunable bandgap transistors. Sci Rep 6:29184
56. Shi H, Pan H, Zhang Y-W, Yakobson BI (2013) Quasiparticle band structures and optical properties of strained monolayer MoS_2 and WS_2. Phys Rev B 87:155304
57. Zhang C, Johnson A, Hsu C-L et al (2014) Direct imaging of band profile in single layer MoS_2 on graphite: quasiparticle energy gap, metallic edge states, and edge band bending. Nano Lett 14:2443–2447
58. Kaasbjerg K, Thygesen KS, Jacobsen KW (2012) Phonon-limited mobility in n-type single-layer MoS_2 from first principles. Phys Rev B 85:115317
59. Novoselov K, Jiang D, Schedin F et al (2005) Two-dimensional atomic crystals. Proc Natl Acad Sci U S A 102:10451–10453
60. Jena D, Konar A (2007) Enhancement of carrier mobility in semiconductor nanostructures by dielectric engineering. Phys Rev Lett 98:136805
61. Radisavljevic B, Radenovic A, Brivio J et al (2011) Single-layer MoS_2 transistors. Nat Nanotechnol 6:147–150
62. Ghorbani-Asl M, Borini S, Kuc A, Heine T (2013) Strain-dependent modulation of conductivity in single-layer transition-metal dichalcogenides. Phys Rev B 87:235434
63. Conley HJ, Wang B, Ziegler JI et al (2013) Bandgap engineering of strained monolayer and bilayer MoS_2. Nano Lett 13:3626–3630
64. Yang L, Cui X, Zhang J et al (2014) Lattice strain effects on the optical properties of MoS_2 nanosheets. Sci Rep 4:5649
65. Bhimanapati GR, Lin Z, Meunier V et al (2015) Recent advances in two-dimensional materials beyond graphene. ACS Nano 9(12):11509–11539

66. Jaramillo TF, Jørgensen KP, Bonde J et al (2007) Identification of active edge sites for electrochemical H2 evolution from MoS$_2$ nanocatalysts. Science 317:100–102
67. Wang H, Zhang Q, Yao H et al (2014) High electrochemical selectivity of edge versus terrace sites in two-dimensional layered MoS$_2$ materials. Nano Lett 14:7138–7144
68. Schweiger H, Raybaud P, Kresse G, Toulhoat H (2002) Shape and edge sites modifications of MoS$_2$ catalytic nanoparticles induced by working conditions: a theoretical study. J Catal 207:76–87
69. Cao D, Shen T, Liang P et al (2015) Role of chemical potential in flake shape and edge properties of monolayer MoS$_2$. J Phys Chem C 119:4294–4301
70. Lauritsen J, Nyberg M, Nørskov JK et al (2004) Hydrodesulfurization reaction pathways on MoS$_2$ nanoclusters revealed by scanning tunneling microscopy. J Catal 224:94–106
71. Füchtbauer HG, Tuxen AK, Li Z et al (2014) Morphology and atomic-scale structure of MoS$_2$ nanoclusters synthesized with different sulfiding agents. Top Catal 57:207–214
72. Lauritsen JV, Kibsgaard J, Helveg S et al (2007) Size-dependent structure of MoS$_2$ nanocrystals. Nat Nanotechnol 2:53–58
73. Yu S, Zheng W (2016) Fundamental insights into the electronic structure of zigzag MoS$_2$ nanoribbons. Phys Chem Chem Phys 18:4675–4683
74. Heising J, Kanatzidis MG (1999) Structure of restacked MoS$_2$ and WS$_2$ elucidated by electron crystallography. J Am Chem Soc 121:638–643
75. Heising J, Kanatzidis MG (1999) Exfoliated and restacked MoS$_2$ and WS$_2$: ionic or neutral species? Encapsulation and ordering of hard electropositive cations. J Am Chem Soc 121:11720–11732
76. Wang L, Xu Z, Wang W, Bai X (2014) Atomic mechanism of dynamic electrochemical lithiation processes of MoS$_2$ nanosheets. J Am Chem Soc 136:6693–6697
77. Eda G, Fujita T, Yamaguchi H et al (2012) Coherent atomic and electronic heterostructures of single-layer MoS$_2$. ACS Nano 6:7311–7317
78. Chou SS, Sai N, Lu P et al (2015) Understanding catalysis in a multiphasic two-dimensional transition metal dichalcogenide. Nat Commun 6:8311
79. Sandoval SJ, Yang D, Frindt R, Irwin J (1991) Raman study and lattice dynamics of single molecular layers of MoS$_2$. Phys Rev B 44:3955
80. Eda G, Yamaguchi H, Voiry D et al (2011) Photoluminescence from chemically exfoliated MoS$_2$. Nano Lett 11:5111–5116
81. Wypych F, Schöllhorn R (1992) 1T-MoS$_2$, a new metallic modification of molybdenum disulfide. J Chem Soc Chem Commun 19:1386–1388
82. Wypych F, Weber T, Prins R (1998) Scanning tunneling microscopic investigation of 1T-MoS$_2$. Chem Mater 10:723–727
83. Wypych F, Solenthaler C, Prins R, Weber T (1999) Electron diffraction study of intercalation compounds derived from 1T-MoS$_2$. J Solid State Chem 144:430–436
84. Lin Y-C, Dumcenco DO, Huang Y-S, Suenaga K (2014) Atomic mechanism of the semiconducting-to-metallic phase transition in single-layered MoS$_2$. Nat Nanotechnol 9:391–396
85. Calandra M (2013) Chemically exfoliated single-layer MoS$_2$: stability, lattice dynamics, and catalytic adsorption from first principles. Phys Rev B 88:245428
86. Kan M, Wang J, Li X et al (2014) Structures and phase transition of a MoS$_2$ monolayer. J Phys Chem C 118:1515–1522
87. Enyashin AN, Seifert G (2012) Density-functional study of Li$_x$MoS$_2$ intercalates ($0 \leqslant x \leqslant 1$). Comput Theor Chem 999:13–20
88. Yang D, Sandoval SJ, Divigalpitiya W et al (1991) Structure of single-molecular-layer MoS$_2$. Phys Rev B 43:12053
89. Briddon P, Jones R (2000) LDA calculations using a basis of Gaussian orbitals. Phys Status Solidi B 217:131–171
90. Rayson M, Briddon P (2009) Highly efficient method for Kohn-sham density functional calculations of 500-10 000 atom systems. Phys Rev B 80:205104

91. Briddon PR, Rayson MJ (2011) Accurate Kohn-Sham DFT with the speed of tight binding: current techniques and future directions in materials modelling. Phys Status Solidi B 248:1309–1318

92. Hartwigsen C, Gøedecker S, Hutter J (1998) Relativistic separable dual-space Gaussian pseudopotentials from H to Rn. Phys Rev B 58:3641

93. Monkhorst HJ, Pack JD (1976) Special points for Brillouin-zone integrations. Phys Rev B 13:5188

94. Nguyen EP, Carey BJ, Daeneke T et al (2014) Investigation of two-solvent grinding-assisted liquid phase exfoliation of layered MoS_2. Chem Mater 27:53–59

95. Wang K, Wang J, Fan J et al (2013) Ultrafast saturable absorption of two-dimensional MoS_2 nanosheets. ACS Nano 7:9260–9267. https://doi.org/10.1021/nn403886t

96. O'Neill A, Khan U, Coleman JN (2012) Preparation of high concentration dispersions of exfoliated MoS_2 with increased flake size. Chem Mater 24:2414–2421

97. Varrla E, Backes C, Paton KR et al (2015) Large-scale production of size-controlled MoS_2 nanosheets by shear exfoliation. Chem Mater 27:1129–1139

98. Smith RJ, King PJ, Lotya M et al (2011) Large-scale exfoliation of inorganic layered compounds in aqueous surfactant solutions. Adv Mater Weinheim 23:3944–3948. https://doi.org/10.1002/adma.201102584

99. Pagona G, Bittencourt C, Arenal R, Tagmatarchis N (2015) Exfoliated semiconducting pure 2H-MoS_2 and 2H-WS_2 assisted by chlorosulfonic acid. Chem Commun 51:12950–12953

100. Canton-Vitoria R, Sayed-Ahmad-Baraza Y, Pelaez-Fernandez M et al (2017) Functionalization of MoS_2 with 1,2-dithiolanes: toward donor-acceptor nanohybrids for energy conversion. npj 2D Mater Appl 1:13

101. Lee C, Yan H, Brus LE et al (2010) Anomalous lattice vibrations of single- and few-layer MoS_2. ACS Nano 4:2695–2700. https://doi.org/10.1021/nn1003937

102. Laursen AB, Kegnæs S, Dahl S, Chorkendorff I (2012) Molybdenum sulfides—efficient and viable materials for electro-and photoelectrocatalytic hydrogen evolution. Energy Environ Sci 5:5577–5591

103. Chakraborty B, Bera A, Muthu D et al (2012) Symmetry-dependent phonon renormalization in monolayer MoS 2 transistor. Phys Rev B 85:161403

104. Fang H, Tosun M, Seol G et al (2013) Degenerate n-doping of few-layer transition metal dichalcogenides by potassium. Nano Lett 13:1991–1995

105. Kiriya D, Tosun M, Zhao P et al (2014) Air-stable surface charge transfer doping of MoS_2 by benzyl viologen. J Am Chem Soc 136:7853–7856

106. Mao N, Chen Y, Liu D et al (2013) Solvatochromic effect on the photoluminescence of MoS_2 monolayers. Small 9:1312–1315

107. Park H-Y, Lim M-H, Jeon J et al (2015) Wide-range controllable n-doping of molybdenum disulfide (MoS_2) through thermal and optical activation. ACS Nano 9:2368–2376

108. Shokri A, Salami N (2016) Gas sensor based on MoS_2 monolayer. Sensors Actuators B Chem 236:378–385

109. Perkins FK, Friedman AL, Cobas E et al (2013) Chemical vapor sensing with monolayer MoS_2. Nano Lett 13:668–673

110. Li H, Yin Z, He Q et al (2012) Fabrication of single- and multilayer MoS_2 film-based field-effect transistors for sensing NO at room temperature. Small 8:63–67. https://doi.org/10.1002/smll.201101016

111. Late DJ, Huang Y-K, Liu B et al (2013) Sensing behavior of atomically thin-layered MoS_2 transistors. ACS Nano 7:4879–4891. https://doi.org/10.1021/nn400026u

112. Yoon HS, Joe H-E, Kim SJ et al (2015) Layer dependence and gas molecule absorption property in MoS_2 Schottky diode with asymmetric metal contacts. Sci Rep 5:10440

113. Cho B, Yoon J, Lim SK et al (2015) Chemical sensing of 2D graphene/MoS_2 heterostructure device. ACS Appl Mater Interfaces 7:16775–16780

114. Liu B, Chen L, Liu G et al (2014) High-performance chemical sensing using Schottky-contacted chemical vapor deposition grown monolayer MoS_2 transistors. ACS Nano 8:5304–5314

115. Lee K, Gatensby R, McEvoy N et al (2013) High-performance sensors based on molybdenum disulfide thin films. Adv Mater 25:6699–6702
116. Cho B, Hahm MG, Choi M et al (2015) Charge-transfer-based gas sensing using atomic-layer MoS$_2$. Sci Rep 5:8052
117. He Q, Zeng Z, Yin Z et al (2012) Fabrication of flexible MoS$_2$ thin-film transistor arrays for practical gas-sensing applications. Small 8:2994–2999. https://doi.org/10.1002/smll. 201201224
118. Liu G, Rumyantsev S, Jiang C et al (2015) Selective gas sensing with $ h $-BN capped MoS 2 heterostructure thin-film transistors. IEEE Electron Device Lett 36:1202–1204
119. Schotland RM (1969) Some aspects of remote atmospheric sensing by laser radar. Atmos Explor Remote Probes 11:179
120. Inaba H, Kobayasi T (1972) Laser-Raman radar—laser-Raman scattering methods for remote detection and analysis of atmospheric pollution. Opt Quant Electron 4:101–123
121. Kildal H, Byer RL (1971) Comparison of laser methods for the remote detection of atmospheric pollutants. Proc IEEE 59:1644–1663
122. Byer RL (1975) Remote air pollution measurement. Opt Quant Electron 7:147–177
123. Grant W, Hake R Jr, Liston E et al (1974) Calibrated remote measurement of NO2 using the differential-absorption backscatter technique. Appl Phys Lett 24:550–552
124. Seiyama T, Kato A, Fujiishi K, Nagatani M (1962) A new detector for gaseous components using semiconductive thin films. Anal Chem 34:1502–1503
125. Taguchi N (1971) U.S. Patent No. 3,631,436. Washington, DC: U.S. Patent and Trademark Office. (Gas sensor)
126. Morrison SR (1981) Semiconductor gas sensors. Sensors Actuators 2:329–341
127. Jaaniso R, Tan OK (2013) Semiconductor gas sensors. Elsevier, Amsterdam
128. Wopenka B, Pasteris JD (1987) Raman intensities and detection limits of geochemically relevant gas mixtures for a laser Raman microprobe. Anal Chem 59:2165–2170
129. Eckbreth AC (1996) Laser diagnostics for combustion temperature and species. CRC Press, Boca Raton
130. Kiefer J, Seeger T, Steuer S et al (2008) Design and characterization of a Raman-scattering-based sensor system for temporally resolved gas analysis and its application in a gas turbine power plant. Meas Sci Technol 19:085408
131. Buric MP, Chen KP, Falk J, Woodruff SD (2009) Improved sensitivity gas detection by spontaneous Raman scattering. Appl Opt 48:4424–4429
132. Joensen P, Frindt R, Morrison SR (1986) Single-layer MoS$_2$. Mater Res Bull 21:457–461
133. Novoselov KS, Geim AK, Morozov SV et al (2004) Electric field effect in atomically thin carbon films. Science 306:666–669
134. Lu G, Ocola LE, Chen J (2009) Reduced graphene oxide for room-temperature gas sensors. Nanotechnology 20:445502
135. Robinson JT, Perkins FK, Snow ES et al (2008) Reduced graphene oxide molecular sensors. Nano Lett 8:3137–3140
136. Kauffman DR, Star A (2008) Carbon nanotube gas and vapor sensors. Angew Chem Int Ed 47:6550–6570
137. Zhang T, Mubeen S, Myung NV, Deshusses MA (2008) Recent progress in carbon nanotube-based gas sensors. Nanotechnology 19:332001
138. Schedin F, Geim A, Morozov S et al (2007) Detection of individual gas molecules adsorbed on graphene. Nat Mater 6:652–655
139. Malard L, Pimenta M, Dresselhaus G, Dresselhaus M (2009) Raman spectroscopy in graphene. Phys Rep 473:51–87
140. Late DJ, Huang Y-K, Liu B et al (2013) Sensing behavior of atomically thin-layered MoS$_2$ transistors. ACS Nano 7:4879–4891
141. Donarelli M, Prezioso S, Perrozzi F et al (2015) Response to NO$_2$ and other gases of resistive chemically exfoliated MoS$_2$-based gas sensors. Sensors Actuators B Chem 207:602–613
142. Song I, Park C, Choi HC (2015) Synthesis and properties of molybdenum disulphide: from bulk to atomic layers. RSC Adv 5:7495–7514

Chapter 6
Progress of Sensors Based on Hollow Metal Sulfides Nanoparticles

Wenjiang Li, Carla Bittencourt, and Rony Snyders

Abstract Gas sensors are widely used, playing important role in different sectors of daily life ranging from safety and security, environmental monitoring, food safety & control to medical diagnosis. The development of physical, chemical and biological detection systems, triggered the search for more sensitive, reliable, simple and low-cost gas sensors. With this perspective, metal oxides as sensing materials have been widely used in gas sensors due to the high sensitivity, fast response and recovery times. However, their low selectivity, lack of stability, and the high operational temperatures have limited their use as sensing material for applications in gas sensors. Recently, as a promising alternative to metal oxide, metal sulfides have attracted attention as sensing materials because, the activation of intrinsic surface reactions might occur at lower working temperatures, what has the potential to lead to better selectivity and stability during operation. Besides the electronic properties of the gas sensing nanomaterials, their morphology plays an important role for improving their sensing properties. In this context, hollow nanostructures with unique physicochemical property, surface active sites and abundant inner spaces have the potential to improve the gas detection. Here, recent research progress in hollow metal sulfides nanostructures and applications in gas sensors are summarized. The effect of structural and compositional engineering in the sensing properties of hollow metal sulfides is discussed.

Keywords Metal sulfide · Hollow nanostructures · Gas sensor

W. Li (✉)
School of Material Science and Technology, Tianjin University of Technology, Tianjin, China
e-mail: liwj@tjut.edu.cn

C. Bittencourt
Chimie des Interactions Plasma-Surface, University of Mons (UMONS), Mons, Belgium

R. Snyders
Chimie des Interactions Plasma-Surface, University of Mons (UMONS), Mons, Belgium

Materia Nova Research Center, Mons, Belgium

© Springer Nature B.V. 2019
C. Bittencourt et al. (eds.), *Nanoscale Materials for Warfare Agent Detection: Nanoscience for Security*, NATO Science for Peace and Security Series A: Chemistry and Biology, https://doi.org/10.1007/978-94-024-1620-6_6

97

6.1   Introduction

Metal oxides, as a traditional semiconductor sensing materials, have been attracted attention in gas sensors technology because of their low-cost, excellent sensitivity, fast response and recovery times [1–6]. However, with the development of physical, chemical and biological detection systems, the incomplete selectivity and lack of stability of metal oxides limited their applications in gas sensors [7–9]. In order to overcome these drawbacks other materials have been considered as sensing materials. Among them, metal sulfides with low band-gap energies appear as good candidates to replace metal oxides as gas sensor active layer because the activation of intrinsic surface reactions might occur at lower working temperatures, leading to gas sensors with more stable active layers. Moreover, a different catalytic mechanism on the surface reaction is expected due to the absence of oxygen in the crystal lattice. This unique catalytic mechanism may optimally solve the constant drift of the signal observed in metal oxides sensing layers which is associated to the in/out diffusion of oxygen vacancies. The availability of metal sulfides has generated considerable interests in their application for catalysis, photocatalysis, photovoltaic cells, electrochemical sensors, medicine, and tribology [10–14].

An important topic in current research is to control the shape of nanomaterials, as shape can play an important role in the optimization of the performance of functional components. In particular, there is an increasing interest in hollow nanostructures because of their unique physicochemical properties, higher density of surface active sites and abundant inner spaces, which as reported make them potential active materials for delivery vehicles, heterogeneous catalysis, photo-catalysis, surface electron exchanges for DSSC (Dye Sensitized Solar Cells) or semiconductor gas sensors [15–20]. Compared with the non-hollow structure, the response speed of gas sensors fabricated using surface modified metal-oxide or metal-sulfide hollow nanostructures was reported to be increased [21–24]. Here we overview reported progress on gas sensors based on metal sulfides with hollow structures [25–28]. First, we present few important hollow metal sulfides, followed by the characteristics of gas sensors based on these materials.

6.2   Hollow Nanostructures

A wide variety of nanomaterials with hollow structures have been traditionally synthesized by templating, coating and chemical etching methods. More recent reports show that hollow nanostructures can also be be prepared using the Kirkendall

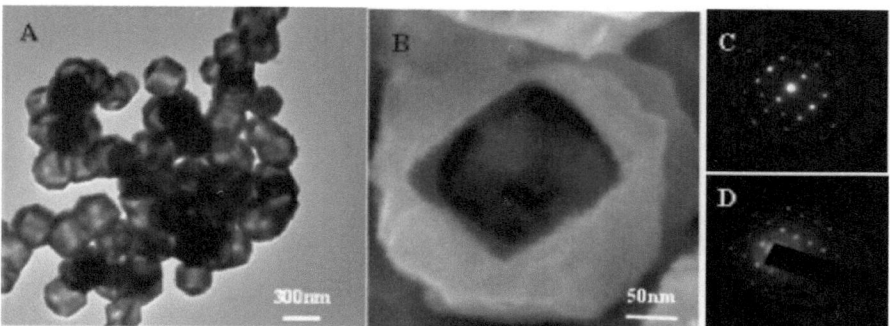

Fig. 6.1 (**a**) TEM image of polyhedral Cu_7S_4 nanocrystals; (**b**) FESEM image of broken polyhedral Cu_7S_4 with regular shape void; (**c**) and (**d**) SAED of two polyhedra with their square and hexagon facets oriented perpendicular to the electron beam. (Reprinted with permission from Ref. [43]. Copyright (2018) American Chemical Society)

effect [29–31] and Ostwald ripening methods [32–34]. In Ostwald ripening large precipitates grow at the expense of smaller precipitates caused by energetic factors; reported to explain the growth of ZnS [35, 36]. Instead, the Kirkendall effect is a classical phenomenon in metallurgy established by Smigelskas and Kirkendall in 1947 [37]. It can be explained as a nonreciprocal atomic diffusion process through an interface of two metals occuring via vacancies. Following the idea of the Kirkendall effect, hollow metal sulfide nanoparticles has been created: CuS [38, 39], MoS_2 [40], and PbS [41]. Yin et al. [42] demonstrated that solid Co nanoparticles can be transformed into hollow nanoparticles through the reaction with sulfur, oxygen and selenium. In another work, Cao et al. [43] report on the preparation of high symmetric 18 facet polyhedron of monoclinic nanocrystalline Cu_7S_4 with regular hollow structure from cubic cuprous oxide nanocrystals. From Fig. 6.1, the obvious difference in terms of contrast between the central and fringe part of each individual particle is clearly observed from the TEM image of Cu_7S_4 nanocrystals. The darker fringe appearance implies that the nanoparticles may have a void inside. The select area electron diffractions (SAED), recorded on the two sets of the Cu_7S_4 polyhedron face, show that both faces are well crystallized.

Wang et al. [44] have demonstrated the synthesis of uniform Pb@PbS core-shell particles by reacting spherical colloids of Pb with S atoms in the vapor phase and in the solution phase. The hollow structure was explained considering the Kirkendall effect that led to the formation of voids inside the PbS system, the Pb cores were completely converted to form hollow spheres made of pure PbS (Fig. 6.2).

The synthesis of micrometer scaled MoS_2 hierarchical hollow cubic cages was reported by Ye et al. [45]. These nanostructures were obtained via a one-step self-assembly method coupled with intermediate crystal templating process without any surfactant, in which an intermediate $K_2NaMoO_3F_3$ crystal was formed in-situ and then served as the self-sacrificed template based on the Kirkendall Effect (Scheme

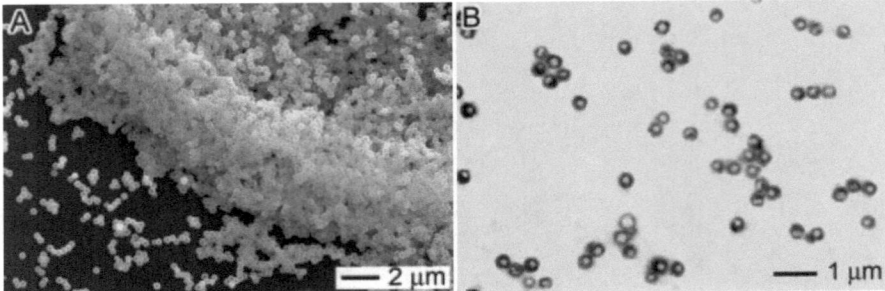

Fig. 6.2 (**a**) low-magnification SEM and (**b**) TEM images of a typical sample of monodisperse PbS hollow spheres that were prepared through the Kirkendall effect

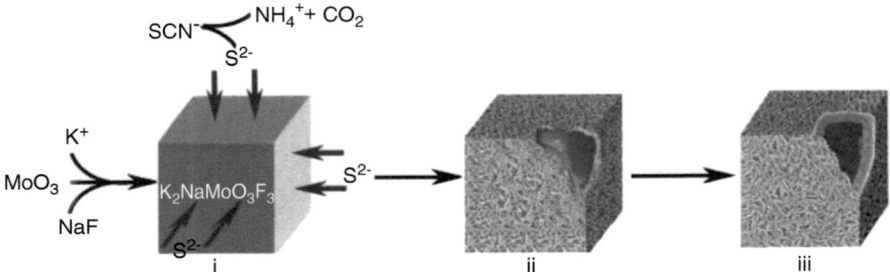

Scheme 6.1 Schematic illustration of the formation of MoS_2 hollow cubic cages: (*i*) intermediate $K_2NaMoO_3F_3$ cubic crystal, (*ii*) the intermediate $K_2NaMoO_3F_3$ core with the layer of MoS_2 shell structure, (*iii*) continuous evacuation of intermediate $K_2NaMoO_3F_3$ core and growth of MoS_2 bilayer shell assembled by many nanoplates. (Reproduced from Ref. [45] with permission from The Royal Society of Chemistry)

6.1). As reported by the authors, FESEM observations showed the formation of uniform MoS_2 cages with cubic morphology (Fig. 6.3a) exhibiting rather rough surfaces (Fig. 6.3b). The hollow MoS_2 cubic cages is clearly revealed on the FESEM image recorded on an incomplete MoS_2 cubic cage (Fig. 6.3c). The MoS_2 hierarchical hollow cubic cages were used for electrochemical hydrogen storage showing high capacity [45].

L. Xu et al. [46] reported on a nanocomposite consisting of silver sulfide (Ag_2S) and hollow Pt nanostructures (Ag_2S-hPtNCs). The nanocomposite was synthesized using core-shell silver (Ag)-platinum (Pt) nanoparticles as starting materials to react with different sulfur sources. The structural conversion from bimetallic core-shell nanostructures to heterogeneous nanocomposites is described in Fig. 6.4.

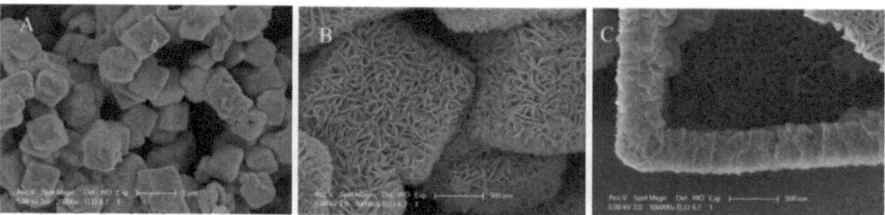

Fig. 6.3 FESEM images of MoS$_2$ hollow cubic cages: (**a**) overall product morphology; (**b**) detailed view of the surface of the cubic cages; (**c**) a detailed view of an incomplete cubic cage, showing its hollow interior and the shell of the cubic cages consisting of irregular-shaped, upright-standing bilayers about 400 nm in thickness. (Reproduced from Ref. [45] with permission from The Royal Society of Chemistry)

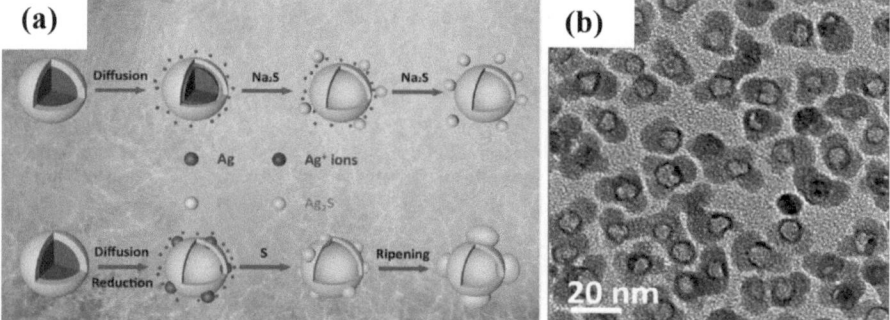

Fig. 6.4 (**a**) Schematic illustration describing the mechanism accounting for the formation of Ag$_2$S-hPtNCs or hollow Pt nanostructures by reacting element sulfur or Na$_2$S with core-shell Ag-Pt nanoparticle. (**b**) TEM image of Ag$_2$S-hPtNCs formed by aging the core-shell Ag-Pt organosol with Ag/Pt molar ratio of 2/1 at 80 °C for 8 h before mixing with aqueous Na$_2$S solution. (Reprinted with permission from Ref. [46])

6.3 Hollow Particles as Sensing Active Layers

The use of hollow metal sulfide nanoparticles as active layer for sensor has been reported [47–53]. For example, ZnS hollow particles that have similar crystal structure with ZnO, show active surface states, sulfur vacancies and interstitial sulfur lattice defects [47], which would facilitate adsorption of oxygen molecules and formation of oxygen negative ion species favoring the enhancement of gas sensing performance. Andrea Gaiardo et al. [48] reported on chemoresistive gas sensors using CdS and SnS$_2$ as sensing functional materials working in thermo- and photo-activation mode on alumina substrates through a screen printing technique. Figure 6.5, shows the sensor operation principle reported in [48] by Gaiardo. Variation of the film resistance is measured using interdigitated Au electrodes at the front-side of alumina substrates, while the back-side is equipped with a heater. As reported by the authors, both SnS$_2$ and CdS showed a high selectivity and a sensitivity comparable

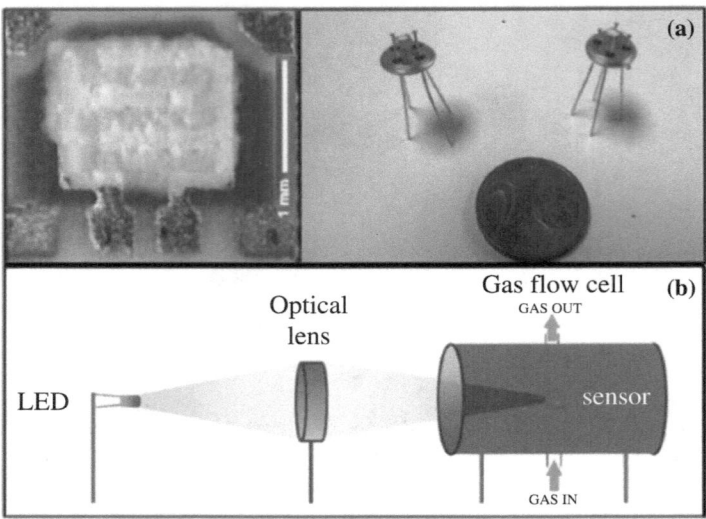

Fig. 6.5 (**a**) Image of the gas sensor device and (**b**) the schematic representation of the gas sensing system used for electrical characterization in photo-activation mode. (Reprinted with permission from Ref. [48])

to metal oxides films [49]. When operated in the photo-activation mode, the CdS active layer exhibited photochemical activity, which allows for possible use at room temperature. Therefore, it can be considered a low-power consumption gas sensing material.

Park et al. [50] reported on a large-area, conformal tactile sensor fabricated with a MoS_2 semiconductor and graphene electrodes synthesized by chemical vapor deposition (CVD). The MoS_2 strain sensor exhibited a variety of advantages, such as good optical transparency, mechanical flexibility, and high gauge factor (GF) compared with conventional strain gauges. In particular, the ultrathin nature of the device ($\approx$75 nm) enabled their integration on unusual substrates, such as a fingertip or leather products. The authors described a prototype using the conformal MoS_2 tactile sensor that can be utilized to achieve a high integration density array, low crosstalk, and high switching speed through active matrix circuitry. Figure 6.6, shows a potential wearable application as E-skin.

Zhang et al. [51] described the synthesis of Au NPs-ZnS Hollow Spheres (HSs) via a simple hydrothermal and deposition-precipitation methods without using any surfactants or toxic organic solvents. Au NPs-ZnS HSs-based gas sensor exhibited excellent gas sensing performances, such as high response, quick response/recovery speed, excellent reproducibility, and simultaneously highly enhanced response to volatile organic pollutants (VOPs). The authors attributed such superior sensing performances to the spillover and catalytic effect of Au NPs, the high accessible surface area, easy gas diffusion and mass transport induced by the unique rough, porous and hollow structure, and the formation of conjugated electron depletion

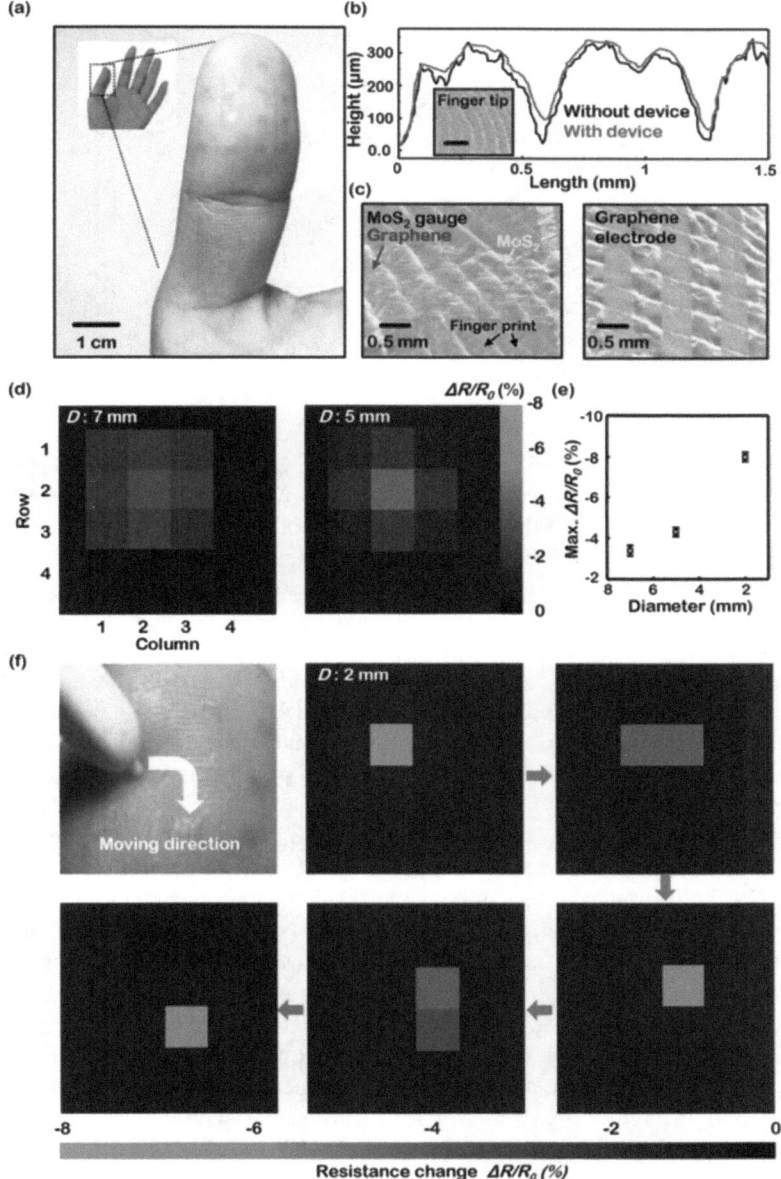

Fig. 6.6 Pressure mapping of MoS$_2$ tactile sensor conformally transferred on the fingertip. (**a**) Photographs of the MoS$_2$ tactile sensor on a fingertip. (**b**) Surface profiles of the fingerprint with the tactile sensor. The inset shows an SEM image of the fingerprint (scale bar: 1 mm). (**c**) Colorized SEM images of the MoS$_2$ strain gauge (left, yellow) and graphene electrode (right, blue). (**d**) Pressure map detected by the MoS$_2$ tactile sensor from contact with pens of different diameters (7 and 5 mm). (**e**) Maximum fractional resistance change of the MoS$_2$ sensor array with different pen-tip sizes (D: 7, 5, and 2 mm). (**f**) Tracking results for the position of a moving, touching pen (D: 2 mm) on the tactile sensor. Color code in the original reference. (Adapted from Ref [50])

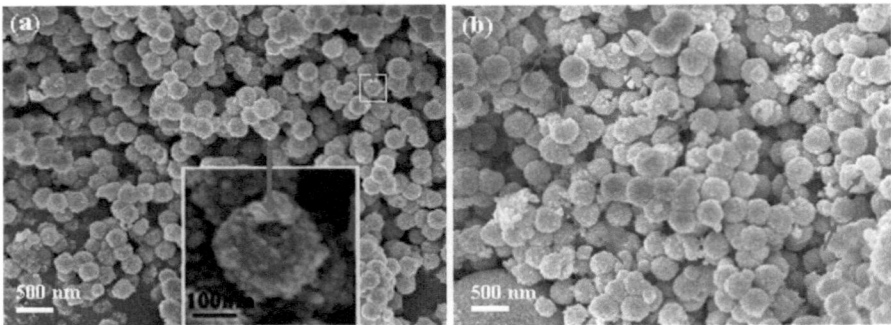

Fig. 6.7 SEM image of (**a**) pristine ZnS HSs (Inset is the enlarged image marked with yellow box) and (**b**) Au NPs-ZnS HSs. (Adapted from Ref. [51])

layers on both the outer and inner surfaces of the HSs. The as-prepared Au NPs-ZnS HSs are a promising candidate material to be used as active layer in gas sensor applications. The authors described that both pristine ZnS (Fig. 6.7a) and Au NPs-ZnS HSs (Fig. 6.7b) consist of numerous nanospheres, openings on some nanospheres (as arrows shown) reveal the hollow structure. It is also clear that the HSs possess relatively rough surface, formed by the accumulation of numerous NPs, which may increase the accessible surface area of the nanomaterials. In Fig. 6.8, recorded on a pristine ZnS HSs sample, the difference in the electron density between the dark edge and pale center indicates a HS morphology, the lattice fringes with an interplanar distance of 0.313 nm correspond to the (002) plane of wurtzite ZnS crystal. In Fig. 6.8, the TEM images of Au NPs-ZnS HSs, show that compared with pristine ZnS HSs, apart from well maintenance of HS structure, there are numerous ultrafine Au NPs attached to the surface of the HSs. In the enlarged image (Fig. 6.8d) Au NPs, marked with circles, can be observed. The sensing mechanism of the HSs is illustrated in Fig. 6.9; Due to the catalytic effect of Au NPs, oxygen molecules from air can easily be adsorbed on Au NPs and be dissociated into $O^{\delta-}$ species (O_2^-, O^- and O^{2-}). Meanwhile, the spillover effect of Au NPs makes the $O^{\delta-}$ species easily be transported and distributed onto the surface of ZnS HSs, which is favorable to accelerate the reactions happening at the interface between Au NPs and ZnS NPs. Additionally, the reported excellent gas sensing performances were related to the unique rough, porous and hollow structure the of Au NPs-ZnS HSs.

Kim et al. [52] developed a high-performance sensor with tunable response and high sensitivity for volatile organic compounds (VOCs) using MoS_2 modified with thiolated ligands (mercaptoundecanoic acid (MUA)). Pristine and MUA-conjugated MoS_2 sensing channels distinctly exhibit different sensor responses toward VOCs. In particular, the pristine MoS_2 sensor presents positive responses for oxygen-functionalized VOCs, while the MUA-conjugated MoS_2 sensor presents negative responses for the same analytes. Such characteristic sensor responses demonstrate that ligand conjugation successfully adds functionality to a MoS_2 matrix. Thus,

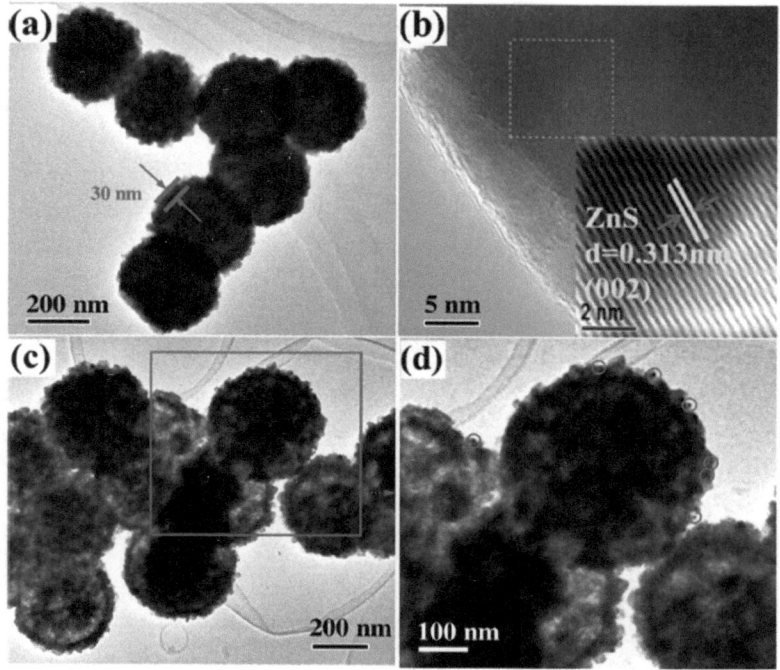

Fig. 6.8 (**a**) TEM and (**b**) HRTEM images of pristine ZnS HSs (Inset in (**b**) is the magnified HRTEM image marked with dashed square); (**c**) and (**d**) TEM images of Au NPs-ZnSHSs. (Adapted from Ref. [51])

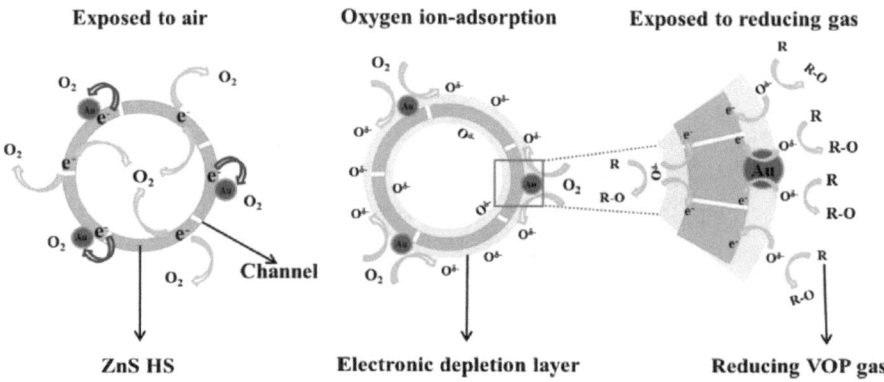

Fig. 6.9 Schematic illustration for the sensing mechanism of the Au NPs-ZnS HSs based sensor. (Reprinted from Ref. [51])

indicating a promising approach to design a versatile sensor array, by conjugating a wide variety of thiolated ligands on the MoS$_2$ surface. Furthermore, as reported by the authors, these MoS$_2$ sensors exhibit high sensitivity to representative VOCs down to a concentration of 1 ppm. This approach for the fabrication of a tunable and sensitive VOC sensor may lead to a valuable real-world application for lung cancer diagnosis by breath analysis. Figure 6.10 illustrates the reported scheme for the fabrication of a MoS$_2$-based gas sensor.

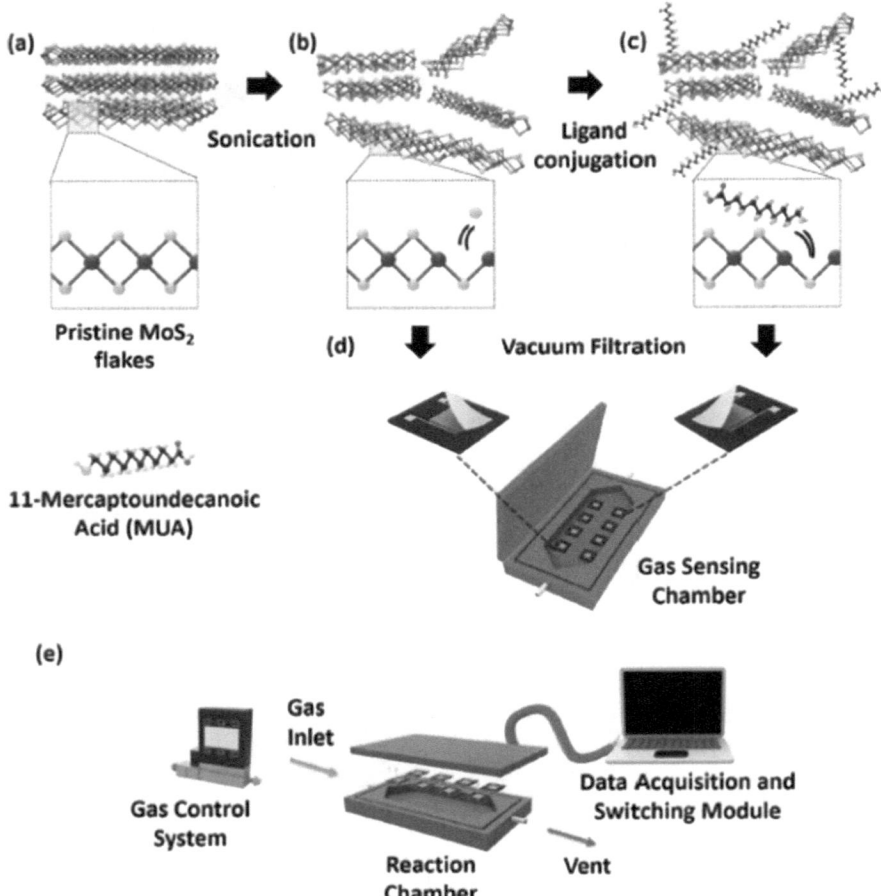

Fig. 6.10 Schematic description for the preparation of MoS$_2$ sensor from powder material. (**a**) MoS$_2$ in pristine powder. (**b**) MoS$_2$ flakes dispersed in ethanol/water solvent through grinding and ultrasonication process. Some of sulfur atoms are naturally detached from the primitive MoS$_2$ flakes. (**c**) Thiolated ligand (MUA) conjugated MoS$_2$ flakes after solution based mixing process. (**d**) Loading the primitive and MUA-conjugated MoS$_2$ films on the microelectrode printed SiO$_2$/Si substrate through vacuum filtration. (**e**) The MoS$_2$ gas sensors are mounted on the customized reaction chamber with a gas delivery system and data acquisition module. (Reprinted with permission from Ref. [52]. Copyright (2018) American Chemical Society)

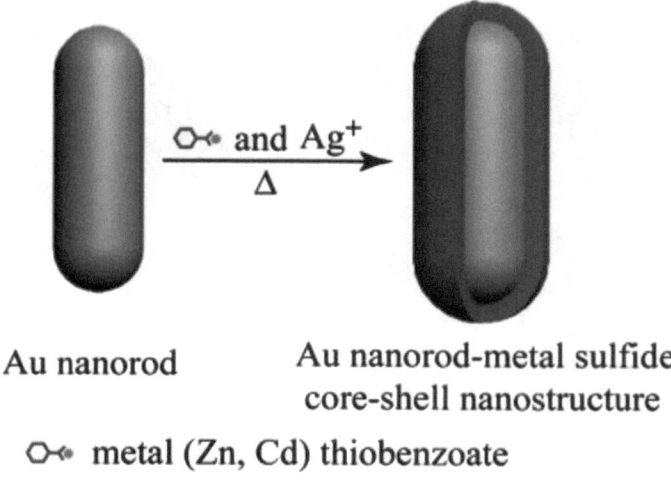

$\overset{\text{O--* and Ag}^+}{\xrightarrow{\hspace{2cm}}}_{\Delta}$

Au nanorod   Au nanorod-metal sulfide
                        core-shell nanostructure

O--* metal (Zn, Cd) thiobenzoate

Scheme 6.2 Schematic illustrating the formation process of AuNR–(metal sulfide) core–shell nanostructures

Wang et al. [53] developed a one-pot hydrothermal method to synthesize Au NR-MS (M = Cd, Zn) core-shell structures, in which first a shell of Ag_2S or AuAgS is formed on the surface of Au nanorods as a wetting layer, and then, it is coated with CdS or ZnS by cation exchange. The authors showed that the extinction spectrum of AuNR-MS (M = Cd, Zn) core–shell structures can be controlled by the Ag+ concentration and the thickness of the shell. As reported, the gas-sensing property of AuNR– CdS core-shell structures was greatly improved, compared with pure CdS nanostructures. Scheme 6.2 shows the formation process of AuNR–(metal sulfide) core–shell nanostructures.

Yue et al. [54] investigated using first-principles calculations the adsorption of various gas molecules (H_2, O_2, H_2O, NH_3, NO, NO_2, and CO) on a monolayer of MoS_2. They reported that all the molecules are weakly adsorbed on the monolayer MoS_2 surface and act as charge acceptors for the monolayer, except NH_3 which is found to be a charge donor. Furthermore, they showed that the charge transfers between the adsorbed molecule and MoS_2 can be significantly modulated by a perpendicular electric field, suggesting that MoS_2 as a potential material for gas sensing application. Tan et al. [55] combined hollow MoS_2 microspheres with nanospheres through a one-step hydrothermal method, presenting photocatalytic activities for photodegradation of methylene blue (MB), and also showing a good humidity sensing properties. Their results revealed high sensitivity at high RH, small humidity hysteresis, fast response and recovery times, and good stability. Figure 6.11 shows the SEM image of the hollow MoS_2 micro@nano-spheres sample, the proposed schematic formation mechanism, the degradation rates of MB with and without the photocatalyst, and the capacitance variations at various RH level in 30 days.

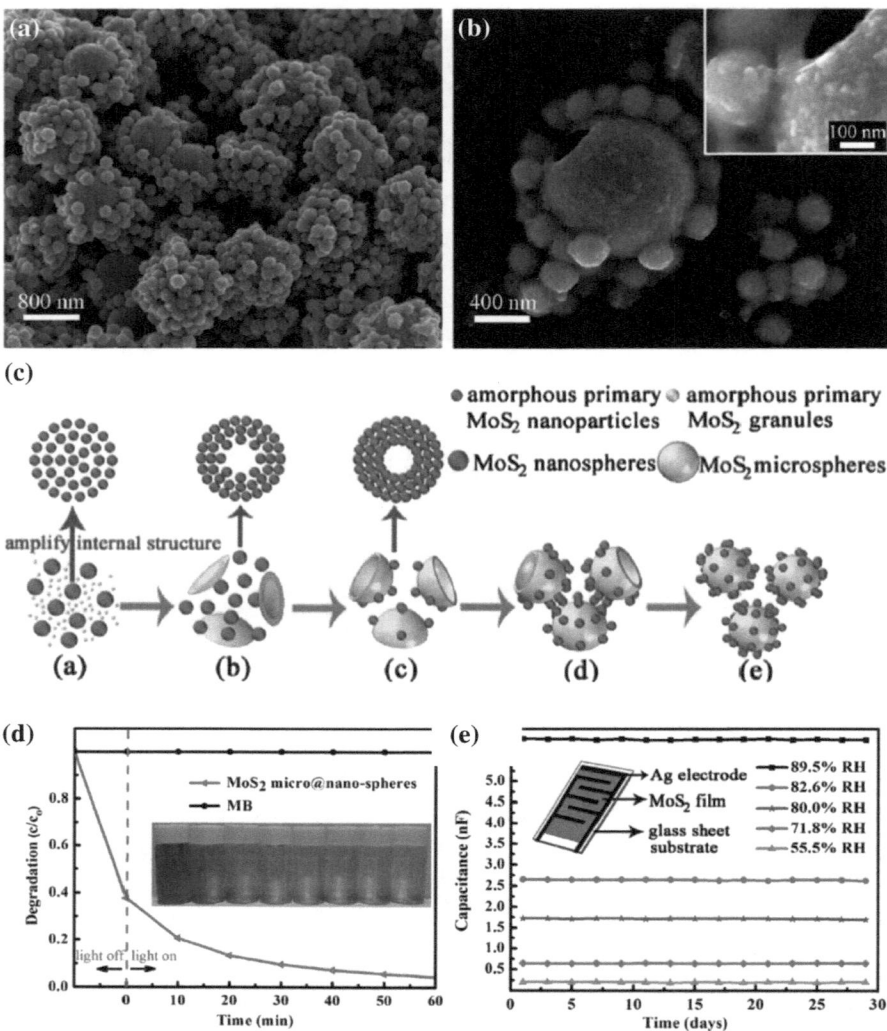

Fig. 6.11 (**a**) Low magnification SEM image. (**b**) The SEM image of a microsphere surface. The inset is the HRSEM of the typical contact interface between a microsphere and a nanosphere. (**c**) The schematic demonstration of the proposed formation mechanism of the hollow MoS₂ micro@nano-spheres. (**d**) Degradation rates of MB with and without the photocatalyst as a function of irradiation time. (**e**) Capacitance variations at various RH levels in 30 days. The inset is a schematic diagram of the fabricated sensor. (Reproduced from Ref. [55] with permission from The Royal Society of Chemistry)

CuS hollow spheres were synthesized via surfactant micelle-template at a large scale by Yu et al. [56]. Due to their special morphology, the hollow spheres showed excellent sensing properties to ethanol, including high sensitivity, rapid response rate and fast recovery. The gain in sensitivity was attributed to both the intrinsic phase-transition property of copper sulfide and their hollow architecture

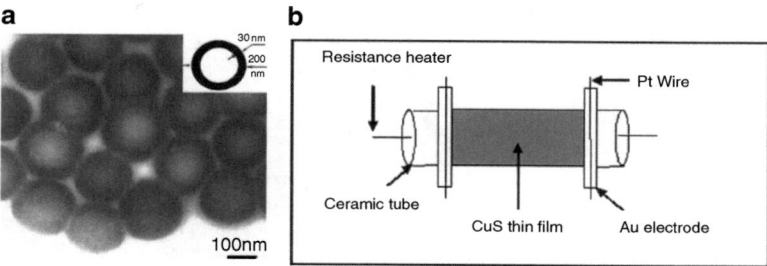

Fig. 6.12 (**a**) TEM image of as-synthesized CuS hollow spheres; (**b**) Schematic structure of gas sensor. (Adapted from Ref. [56])

that promotes analyte diffusion and increases the available active surface area. The authors described the fabrication of CuS-based sensors by dip-coating the as-prepared CuS alcohol colloids on the ceramic tube of the sensor body without an additional annealing process except for aging in the gas sensor system for 240 h. The schematic representation of sensors is shown in Fig. 6.12. A small Ni–Cr alloy coil was placed through the tube as a microheater. Electrical contacts were made with two platinum wires attached to the electrodes.

Guidi et al. [57] investigated the gas sensing properties of cadmium sulfide (CdS) and tin disulfide (SnS_2) semiconductors thick-films in thermo- and photo-activation modes. The performances of both the metal sulfides were compared with the respective oxides. The metal sulfides highlighted a very high selectivity and sensitivity to particular classes of molecules at a working temperature relatively lower than the classic metal oxide. In particular, CdS thick films even allows room temperature operation thanks to its photo-chemical properties. The responses of the sensors were reproducible over time, showing a particular stability of the signal. Therefore, the use of metal sulphide materials, where sulfur atoms replace of the oxygen, would pave the way to a sensitive, selective, and stable sensing material for gas detection.

Large-scale uniform ZnO dumbbells and ZnO/ZnS hollow nanocages were successfully synthesized via a facile hydrothermal route combined with subsequent etching treatment (Fig. 6.13) [58]. The nanocages were formed through preferential dissolution of the twinned (0001) plane of ZnO dumbbells. Due to their special morphology, the hollow nanocages show better sensing properties to ethanol than ZnO dumbbells. The gain in sensitivity is attributed to both the interface between ZnO and ZnS heterostructure and their hollow architecture that promotes analyte diffusion and increases the available active surface area. Figure 6.13 shows the morphologies of the nanostructures before and after sulfurization, the ZnO template with dumbbell shape (Fig. 6.13a) after sulfurization retains its shape, but the inner structure becomes hollow (nanocage Fig. 6.13b). The recorded TEM images suggest that the dumbbell-like ZnO is of high crystallinity and [0001] direction oriented. The strong contrast between the dark edge and the pale center is indicative of the hollow nature of the cage (Fig. 6.13e). The authors showed the existence of ZnS

Fig. 6.13 SEM images of ZnO dumbbells (**a**) and ZnO/ZnS hollow nanocages (**b**); TEM and HRTEM images of ZnO dumbbells (**c**, **d**) and ZnO/ZnS hollow nanocages (**e**, **f**). (Reprinted from Ref. [58])

nanocrystals at the edge of the nanocage (Fig. 6.13f), confirming that the hollow nanocage is a heterostructure composed of ZnO and ZnS.

The growth mechanism of ZnO dumbbells and ZnO/ZnS hollow nanocages are illustrated in Fig. 6.14. For the dumbbell structure, it is mainly related to the hexagonal ZnO crystal structure, which has both polar and nonpolar faces. During the reaction, the growth units, ZnO_2^{2-} and $(CH_2)_6N_4\text{-}4H^+$, adsorb on the different polar faces of the crystal surface, resulting in faster growth along the [0001] and [000–1] directions simultaneously, which is significantly different from the usual

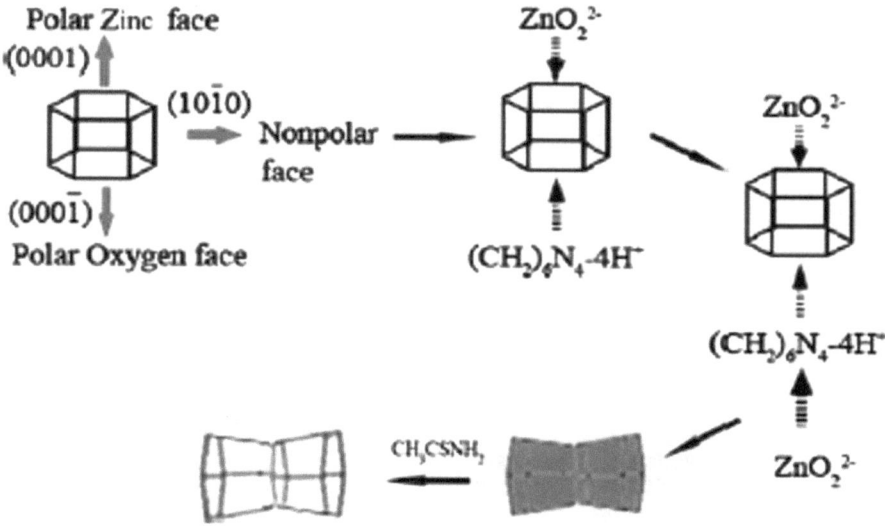

Fig. 6.14 Illustration of the growth mechanism of ZnO dumbbells and ZnO/ZnS nanocages. (Reprinted from Ref. [58])

[0001] growth occurred in the formation of ZnO single crystal. ZnO can react with sulfide ions released from CH_3CSNH_2 and convert into ZnS via the following reactions:

$$CH_3CSNH_2 + 3OH^- = CH_3COO^- + NH_3 \uparrow +S^{2-} + H_2O + ZnO + S^{2-}$$

$$+ H_2O \rightarrow ZnS \downarrow +2OH^-$$

Such reactions can proceed preferentially at the end planes and ridges of the ZnO prism, where many defects were observed in SEM and TEM images, as a result, ZnS nanocrystals can be formed at those sites and connect into a frame; the further growth of ZnS nanocrystals would consume ZnO at the planar surface and the inner of the dumbbells, which results in nanocages with ZnS shell and regular interior space.

6.4   Summary and Outlook

Gas sensors represent the crucial tool in specific analytes detection and olfactory systems for several applications: environmental monitoring, safety and security, quality control of food production, medical diagnosis. In the past few years, hollow micro−/nanostructures as sensing materials have attracted increasing research interest due to their fascinating properties and widespread potential applications. In order to satisfy an expanding demand for novel applications, gas sensors are under continuous evolution. Apart from the specific requirements of the applications, many other parameters have to be taken into account such as selectivity, processing costs, portability, and safety. Recently the use of metal-sulfides as active layer in sensors has come to the scene due to their high stability and design versatility opening vast possibilities.

References

1. Jin C, Kim H, Choi S-W, Kim SS, Lee C (2014) Synthesis, structure, and gas-sensing properties of Pt-functionalized TiO_2 nanowire sensors. J Nanosci Nanotechnol 14:5833
2. Peeters D, Barreca D, Carraro G, Comini E, Gasparotto A, Maccato C, Sada C, Sberveglieri G (2014) Au/ε-Fe_2O_3 nanocomposites as selective NO_2 gas sensors. J Phys Chem C 118:11813
3. Li L, Liu M, He S, Chen W (2014) Free standing 3D mesoporous Co_3O_4@carbon foam nanostructures for ethanol gas sensing. Anal Chem 86:7996
4. Chang SP, Wen CH, Chang SJ (2014) Two-dimensional ZnO nanowalls for gas sensor and photoelectrochemical applications. Electron Mater Lett 10:693
5. Kim HR, Haensch AH, Kim ID, Barsan N, Weimar U, Lee JH (2011) The role of NiO doping in reducing the impact of humidity on the performance of SnO2-based gas sensors: synthesis, strategies, and phenomenological and spectroscopy studies. Adv Funct Mater 21:4456
6. Yamazoe N (2005) Toward innovation of gas sensor technology. Sensors Actuators B Chem 108(2)
7. Cássia-Santos MR, Sousa VC, Oliveira MM, Sensato FR, Bacelar WK, Gomes JW, Longo E, Leite ER, Varela JA (2005) Recent research developments in SnO_2-based varistors. Mat Chem Phys 90:1
8. Barsan N, Koziej D, Weimar U (2007) Metal oxide-based gas sensor research: how to? Sensors Actuators B Chem 121:18
9. Bochenkov VE, Sergeev GB (2010) Sensitivity, selectivity, and stability of gas-sensitive metal-oxide nanostructures. In: Metal oxide nanostructures and their applications. American Scientific Publishers, Valencia, p 31
10. Timoumi A, Bouguila N, Chaari M, Kraini M, Matoussi A, Bouzouita H (2016) Electrical and dielectric properties of In_2S_3 synthesized by solid state reaction. J Alloy Compd 679:59
11. Liu Y, Xu HY, Qian YT (2006) double-source approach to in_2s_3 single crystallites and their electrochemical properties. Cryst Growth Des 6:1304
12. Wu Y, Wadia C, Ma WL, Sadtler B, Alivisatos AP (2008) Synthesis and photovoltaic application of Copper(I) sulfide nanocrystals. Nano Lett 8:2551
13. Li TL, Lee YL, Teng H (2011) $CuInS_2$ quantum dots coated with CdS as high-performance sensitizers for TiO_2 electrodes in photoelectrochemical cells. J Mater Chem 21:5089
14. Bierman MJ, Jin S (2009) Potential applications of hierarchical branching nanowires in solar energy conversion. Energy Environ Sci (2):1050
15. Hornyak GL, Dutta J, Tibbals HF, Rao AK (2008) Introduction to nanoscience. CRC Press, Boca Raton
16. Boudiba A, Zhang C, Umek P, Bittencourt C, Snyders R, Olivier MG, Debliquy M (2013) Sensitive and rapid hydrogen sensors based on Pd–WO_3 thick films with different morphologies. Int J Hydrog Energy 38:2565
17. Wang Y, Wang T, Da P, Wu XM, Zheng HG (2013) Silicon nanowires for biosensing, energy storage, and conversion. Adv Mater 25:5177–5195

18. Desai UV, Xu C, Wu J, Gao D (2012) Solid-state dye-sensitized solar cells based on ordered ZnO nanowire arrays. Nanotechnology 23:205401
19. Bârsan N, Weimar U (2003) Understanding the fundamental principles of metal oxide based gas sensors; the example of CO sensing with SnO_2 sensors in the presence of humidity. J Phys Condens Matter 15:R813–R839
20. Liu Z, Misra M (2010) Dye-sensitized photovoltaic wires using highly ordered TiO_2 nanotube arrays. ACS Nano 4:2196
21. Lee J-H (2009) Gas sensors using hierarchical and hollow oxide nanostructures: overview. Sens Actuators B: Chem 140:319
22. Wang F, Li H, Yuan Z, Sun Y, Chang F, Deng H, Xie L, Li H (2016) High sensitive gas sensor based on CuO nanoparticles synthetized by sol-gel method, vol 6. RSC Advances, Cambridge, p 79343
23. Yu XL, Wang Y, Chan HLW, Cao B (2009) Novel gas sensing materials based on CuS hollow spheres. Microporous Mesoporous Mater 118:423
24. Zhang J, Liu X, Neri G, Pinna N (2016) Nanostructured materials for room-temperature gas sensors. Adv Mater 28:795
25. Zou Z, Qiu Y, Xie C, Xu J, Luo Y, Wang C, Yan H (2015) CdS/TiO_2 nanocomposite film and its enhanced photoelectric responses to dry air and formaldehyde induced by visible light at room temperature. J Alloys Compd 645:17
26. Xu K, Li N, Zeng D, Tian S, Zhang S, Hu D, Xie C (2015) Interface bonds determined gas-sensing of SnO_2-SnS_2 hybrids to ammonia at room temperature. ACS Appl Mater Interfaces 7:11359
27. Shi W, Huo L, Wang H, Zhang H, Yang J, Wei P (2006) Hydrothermal growth and gas sensing property of flower-shaped SnS_2 nanostructures. Nanotechnology 17:2918
28. Fu X, Liu J, Wan Y, Zhang X, Meng F, Liu J (2012) Preparation of a leaf-like CdS micro−/nanostructure and its enhanced gas-sensing properties for detecting volatile organic compounds. J Mater Chem 22:17782
29. Yin YD, Rioux RM, Erdonmez CK, Hughes S, Somorjai GA, Alivisatos AP (2004) Science 304:711
30. Yang YJ, Qi LM, Lu CH, Ma JM, Cheng HM (2005) Angew Chem Int Ed 44:598; (a) Wang YL, Cai L, Xia YN (2005) AdV Mater 17:473
31. Liu B, Zeng HC (2004) J Am Chem Soc 126:16744
32. Ostwald W (1900) On the assumed isomerism of red and yellow mercury oxide and the surfacetension of solid bodies. Z Phys Chem 34:495
33. Pala N, Rumyantsev SL, Sinius J, Talapatra S, Shur MS, Gaska R (2004) Electron Lett 40:273
34. Hara M, Kondo T, Komoda M, Ikeda S, Shinohara K, Tanaka A, Kondo J, Domen K (1998) Chem Common:357
35. Liu B, Zeng HC (2005) Symmetric and asymmetric Ostwald ripening in the fabrication of homogeneous core-shell semiconductors. Small 1:566–571
36. Shao HF, Qian XF, Zhu ZK (2005) The synthesis of ZnS hollow nanospheres with nanoporous shell. J Solid State Chem 178:3522
37. Smigelskas AD, Kirkendall EO (1947) Zinc diffusion in alpha brass. Trans AIME 171:130
38. Wang Q, Li J-X, Li G-D et al (2007) Formation of CuS nanotube arrays from CuCl nanorods through a gas-solid reaction route. J Cryst Growth 299:386
39. Cao H, Qian X, Wang C et al (2005) High symmetric 18-facet polyhedron nanocrystals of Cu7S4 with a hollow nanocage. J Am Chem Soc 127:16024
40. Ye L, Wu C, Guo W et al (2006) MoS_2 hierarchical hollow cubic cages assembled by bilayers: one-step synthesis and their electrochemical hydrogen storage properties. Chem Commun:4738
41. Wang Y, Cai L, Xia Y (2005) Monodisperse spherical colloids of Pb and their use as chemical templates to produce hollow particles. Adv Mater 17:473
42. Yin Y, Rioux RM, Erdonmez CK et al (2004) Formation of hollow nanocrystals through the nanoscale Kirkendall effect. Science 304:711

43. Cao HL, Qian XF, Wang C, Ma XD, Yin J, Zhu ZK (2005) High symmetric 18-Facet polyhedron nanocrystals of Cu_7S_4 with a hollow nanocage. J Am Chem Soc 127:16025

44. Wang YL, Cai L, Xia Y (2005) Monodisperse spherical colloids of Pb and their use as chemical templates to produce hollow particles. Adv Mater 17:473

45. Ye L, Wu Cz, Guo W, Xie Y (2006) MoS_2 hierarchical hollow cubic cages assembled by bilayers: one-step synthesis and their electrochemical hydrogen storage properties. Chem Commun:4738

46. Xu LL, Chen D, Liu H, Yang J (2018) Understanding the formation of nanocomposites consisting of silver sulfide and platinum hollow nanostructures. J Solid State Chem 265:387

47. Wang XF, Xie Z, Huang HT, Liu Z, Chen D, Shen GZ (2012) Gas sensors, thermistor and photodetector based on ZnS nanowires. J Mater Chem 22:6845

48. Gaiardo A, Fabbri B, Guidi V, Bellutti P, Giberti A, Gherardi S, Vanzetti L, Malagù C, Zonta G (2016) Metal sulfides as sensing materials for chemoresistive gas sensors. Sensors 16:296

49. Carotta MC, Benetti M, Ferrari E, Giberti A, Malagù C, Nagliati M, Vendemiati B, Martinelli G (2007) Basic interpretation of thick film gas sensors for atmospheric application. Sensors Actuators B Chem 126:672–677

50. Park M, Park YJ, Chen X (2016) MoS_2-based tactile sensor for electronic skin applications. Adv Mater 28(13):2556

51. Zhang LP, Dong R, Zhu ZhY, Wang ShR (2017) Au nanoparticles decorated ZnS hollow spheres for highly improved gas sensor performances. Sens Actuators B 245:112

52. Kim JS, Yoo HW, Choi HO (2014) Tunable volatile organic compounds sensor by using thiolated ligand conjugation on MoS_2. Nano Lett 14:5941

53. Wang H, Sun Z, Lu Q et al (2012) One-pot synthesis of (Au Nanorod)–(Metal Sulfide) core–shell nanostructures with enhanced gas-sensing property. Small 8(8):1167–1172

54. Yue Q, Shao ZZ, Chang SL, Li JB (2013) Adsorption of gas molecules on monolayer MoS_2 and effect of applied electric field. Nanoscale Res Lett 8:425

55. Tan YH, Yu K, Yang T, Zhang QF, Cong WT, Yin HH, Zhang ZL, Chen YW, Zhu ZQ (2014) The combinations of hollow MoS_2 micro@nano-spheres: one-step synthesis, excellent photocatalytic and humidity sensing properties. J Mater Chem C 2(27):5422

56. Yu XL, Wang Y, Chan HLW, Cao CB (2009) Novel gas sensing materials based on CuS hollow spheres. Microporous Mesoporous Mater 118:423

57. Guidi V, Fabbri B, Gaiardo A, Gherardi S, Giberti A, Malagù C, Zonta G, Bellutti P (2015) Metal sulfides as a new class of sensing materials. Procedia Eng 120:138

58. Yu XL, Ji HM, Wang HL, Sun J, Du XW (2010) Synthesis and sensing properties of ZnO/ZnS nanocages. Nanoscale Res Lett 5:644

Chapter 7
Synthesis of Tridimensional Ensembles of Carbon Nanotubes

Miro Haluska

Abstract The topic of this chapter is the synthesis of both aligned and unaligned carbon nanotube bulk ensembles by different methods such as electric arc discharge, laser ablation and chemical vapor deposition methods. First, general requirements for the CNTs synthesis are introduced. Utilization of different types of nucleation centers for nanotube synthesis as well as the role of CNT growth promoters and inhibitors is reviewed. Particular attention is paid to CVD methods which are most easily scalable, they offer a relatively good control over synthesis conditions and a high quality of as produced CNTs. Two general approaches for formation of catalyst for the CVD nanotube synthesis are discussed, namely methods utilizing pre-deposited catalysts or their precursors and methods exploiting an injection of catalyst precursors during the nanotube synthesis. Examples of breakthrough synthesis approaches, fundamental studies and those with best known results are given. The different nanotube fabrication methodologies are reviewed and discussed in details. This may assist readers to select the proper method and synthesis conditions with regards to nanotube targeted application.

Keywords SWCNT and MWCNT synthesis · CNT synthesis methods · CVD · Catalysts · Alignment

7.1 Introduction

Nowadays, carbon nanotubes (CNTs) can be synthesized by a broad variety of techniques. Initially, the synthesis of CNTs was hardly controlled and barely reproducible; serendipity played an important role. In the last few decades, a considerable amount of time and other resources have been invested into gaining scientific understanding of CNT growth and in enabling the reproducible synthesis

M. Haluska (✉)
Micro and Nanosystems, D-MAVT, ETH Zürich, Zürich, Switzerland
e-mail: haluska@micro.mavt.ethz.ch

© Springer Nature B.V. 2019   115
C. Bittencourt et al. (eds.), *Nanoscale Materials for Warfare Agent Detection: Nanoscience for Security*, NATO Science for Peace and Security Series A: Chemistry and Biology, https://doi.org/10.1007/978-94-024-1620-6_7

of CNTs with predictable and even pre-defined properties. Although significant progress has been made in this field, there are still many open questions regarding synthesis mechanisms and process engineering.

In this chapter, we provide an overview of CNT synthesis methods commonly used to fabricate bulk ensembles of CNTs for application as well as examples of those used for basic research on nanotube formation mechanisms.

7.2   General Consideration for Synthesis of Carbon Nanotubes

The synthesis of a CNT requires a source of carbon, nucleation center, energy, and controlled reaction conditions. For the synthesis of a SWCNT, a nucleation center with nanometer-scale curvature is necessary. To support the synthesis of a MWCNT, a nucleation center with smaller curvature (larger diameter) can be used.

7.2.1   Carbon & Energy Sources

The *carbon feedstock* necessary for the synthesis of CNTs can be provided by sublimation of solid carbon, for example from a graphitic electrode in the electric arc discharge (EAD) method [1–3], or from a graphitic target in experiments utilizing concentrated light, i.e. laser ablation method (LA) [4] or focused sunlight method [5]. Other carbon materials like carbon black or coal [6–8], or even carbon based waste [9, 10] have been used as carbon sources too. Another class of carbon source materials are carbon containing gasses/vapors in chemical vapor deposition (CVD) [11–13] or in thermal plasma (TP) [14] methods.

Energy is needed to supply the carbon feedstock, activate catalysts, and build up the nanotube crystal structure. Because solid-state carbon material has the highest sublimation temperature of all known elements (> 3200 °C), utilizing it as a carbon source requires high energy consumption. In the case of the CVD method, catalytically driven dissociation of carbon-containing gas/vapor reduces the energy consumption compared to their thermal dissociation and, of course, much lower energy is required then for sublimation of carbon. Plasma used in plasma enhanced CVD (PE CVD) can contribute to the dissociation of carbon containing gasses [15–18] and can even further reduce the thermal energy requirement.

7.2.2 Nucleation Centers and Promoters for Carbon Nanotube Formation

SWCNTs can only be produced when nucleation centers with curvatures approaching a few nm or less serve as catalysts in the nanotube nuclei formation process (here we do not count fullerenes as short nanotubes, and we exclude the rolling of graphene as a nanotube synthesis method). These nucleation centers usually have convex curvatures, such as the commonly used nanoparticles [19]. There are, however, also approaches utilizing nucleation centers with concave curvatures, such as hollow templates [20].

A process to elongate existing nanotubes (not to form the nanotube nucleus itself) is the so-called "cloning" [21, 22]. In this case, an open nanotube (a nanotube with one or both closed-ends removed by physical or chemical "cutting") can continue to grow without the presence of heterogeneous concave nanoparticles if carbon feedstock is provided and the experimental conditions prevent the nanotube ends from closing. Nano-rings or nano-belts (for example cycloparaphenylenes) that represent the shortest sidewall segment of CNTs can be "elongated" to form longer nanotubes too [23].

7.2.2.1 Nanoparticles

A suitable nanoparticle at suitable conditions can support the initial formation of a semi-fullerene cup which is gradually separated from the curved particle surface. The carbon cup belt-like circumference stays at the catalyst particle [24] [25] [26]. Additional carbon atoms (or dimers) can join this circumference from the surface or bulk of the particle, they can be incorporated into the honeycomb carbon structure and cause the prolongation of the nanotube. A nanotube nucleated on a particle can either have nearly the same outer diameter as the particle (so-called tangential mode of growth), or its outer diameter can be significantly smaller than the diameter of the catalyst particle (perpendicular growth mode) [27]. The tangential growth mode is obviously more effective for controlling the nanotube diameter by controlling the particle diameter. With the support of tight-binding Monte Carlo (TBMC) simulations, Fiawoo et al. [27] concluded that the tangential growth mode is favored under "slow" growth conditions close to local thermodynamic equilibrium, whereas the perpendicular mode requires a larger kinetic activation energy. Amara et al. [28] and He et al. [29] demonstrated by atomistic simulations that low carbon concentration in the catalyst particle causes tangential growth mode and the high ratio of carbon to metal (Ni) leads to the perpendicular mode. Both formation modes have been experimentally observed.

Nanoparticles for nanotube synthesis are most often formed by one type of metal or by metal compounds (for a review see [30]). Other materials (Al_2O_3, SiO_2, SiC, diamond, . . .) have also been shown to act as nucleation centers for SWCNT synthesis by CVD (for a review, see [31]). It appears that most of the reasonable elements from the periodic table have been tested already in the context of CNT synthesis.

A large variety of metals and their mixtures can serve as nucleation centers for the formation of SWCNTs and simultaneously catalyze the dissociation of carbon containing gasses (in CVD), or dissolve carbon clusters (in EAD, LA, TP) [32]. Most common, however, are Fe, Ni, Co and their mutual mixtures or their alloys or compounds with other elements (Y, Mo, Pt...), especially because of the resultant superior nanotube yield and quality.

A very special case of synthesis of CNTs is MWCNT formation by the EAD, where the presence of heterogeneous nucleation centers (from an elemental point of view) is not required. According to [33], carbon nanoparticles formed at the cathode serve as seeds for further MWCNT growth. It was suggested that the strong electric field in EAD determines the tube alignment, prevents closing of the ends of the tubes, and allows addition of new carbon atoms or other carbon species (dimers ...) to the tube honeycomb lattice.

For other MWCNT synthesis methods (by CVD), the presence of heterogeneous nucleation centers is required. These nucleation centers are typically larger in size than those used for the nucleation of SWCNTs.

7.2.2.2 Hollow Templates

Specific hollow templates can also be used for the synthesis of SWCNTs by CVD without additional presence of convex-shaped nanoparticles. The inner diameters of the templates determine the outer diameters of the CNTs. Various materials have been tested as templates; for example porous aluminophosphate crystal, $AlPO_4$ (AFI) [20], or porous silicon [34] using tripropylamine or methane gases as sources of carbon.

The growth of a SWCNT inside another larger SWCNT can be seen as a special case of the hollow template method. After opening the carbon nanotubes, filling them with carbon source material, and heating them to high temperature ($>1100\,°C$), the formation of an inner SWCNT can take place. Fullerenes [35], benzene, or toluene [36] were used as carbon source materials in these experiments. Until now, the template method has not been used for the synthesis of "larger" amounts of CNTs. Therefore, it will not be discussed in more detail here.

7.2.3 Carbon Nanotube Synthesis Conditions

The synthesis of CNTs is affected by many conditions specific to each method. The observation that CNT growth recipes cannot be directly transferred to different synthesis equipment, even of the similar type, and sometime observed lack of long-term reproducibility on a given type of equipment indicates insufficient control of the growth conditions. These include control of catalysts, catalyst support layer, temperature, surrounding atmosphere composition and pressure. Important properties of catalyst include both chemical and physical characteristics such as composition, type of elements involved, oxidation level, shape, structure, size, melting and evaporation temperatures, and particle (atom) mobility on/into the support substrate [19, 30]. Similarly, relevant properties of catalyst support materials includes their composition, structure and type of surface plane in crystalline materials, presence of steps or grooves, homogeneity and (thermal) stability [37, 38]. The composition of the surrounding atmosphere, its pressure, flow rate, their local fluctuations as well as presence of so-called growth inhibitors and promoters in the environment of catalysts are important parameters involved in growth conditions too.

Inhibitors can deactivate the catalyst and terminate nanotube growth. Carbon material itself can inhibit CNT synthesis if it covers the surface of the catalyst. Inorganic salts, in particular sodium chloride, can hinder CNT synthesis as well [39]. So-called promoters added into the reaction, can on other hand, promote nanotube synthesis or cause the growth of nanotubes with different characteristics, or prolong the real growth time of the nanotubes. The presence of promoters can support the formation of SWCNTs from particles, which otherwise barely catalyze the formation of SWCNTs. Examples of promoters are sulfur (added to Co [40], FeY [41], PtRh [42]), bismuth (added to Fe [40]), selenium (added to NiY) [43] in EAD experiments. Sulfur (often in a form of thiophene precursor) can be used as a promoter in CVD processes [44–46].

Promoters applied under improper conditions or in wrong amounts become inhibitors for nanotube formation. For example, in [47] it was observed that if water vapor concentration in the system using ethylene and $Fe/Al_2O_3/SiO_2/Si$ for SWCNT synthesis increased to 140 ppm, the rate of nanotube forest growth increased (and simultaneously increased the lifetime of catalysts). When the concentration of water increased over the optimum value, the synthesis of nanotubes was slowed down (and simultaneously the lifetime of the catalysts was shortened). Similar results were observed in [48] for the synthesis of CNTs using cobalt-based mixed-oxide catalysts (containing Co, Mn, Mg, and Al) at 650 °C. H_2O vapor was shown to be a promoter for the CNT growth at low concentrations ($\leq$250 ppm), but an inhibitor at high concentrations ($\geq$300 ppm). In the same system, water vapor added to C_2H_4 at 550 °C behaves as an inhibitor.

Promoters can affect the solidification/melting temperature of particles, their surface tension, the solubility of carbon, and can remove inhibitors from catalysts . . . .

Possible reasons for the termination of nanotube growth should be considered to set the proper synthesis conditions for growing CNTs possessing a preferred length for actuator applications. Catalytic CVD (CCVD) methods can virtually provide, if "optimal" conditions for nanotube growth are ensured, "infinitely" long CNTs as indicated in [49]. As already mentioned above, the presence of inhibitors could be one of the growth termination reasons, thus their presence should be avoided. Ostwald ripening (growth of larger catalyst particles at the cost of smaller particles by atomic inter-particle diffusion) is another possible nanotube growth termination reason. This phenomenon can be blocked for example by adding water vapor to the system to form an OH-functionalized support surface, and thereby hindering catalyst atoms (or particles) movement on the substrate surface [50]. Catalyst particles could also diffuse into the supporting layer causing interruption of CNT growth [51].

7.3 Methods of CNT Synthesis with Respect to Alignment Criterion

CNTs could be synthetized by methods producing their 3D "bulk" assemblies with arbitrarily oriented individual or bundled nanotubes or by methods producing aligned and in defined direction oriented CNTs. Oriented or not oriented individual CNTs (CNTs with very low density = no or minimal nanotube-nanotube inter-actions) could be synthetized as well but these methods are not presented here. "Bulky" materials of unaligned nanotubes can be fabricated for example by electric arc discharge (EAD), laser ablation (LA), induction thermal plasma (TP), fluidized bed chemical vapor deposition (FB CVD), direct injection pyrolytic synthesis (DIPS) CVD. Plasma enhanced CVD (PE CVD) can be used for preparation of vertically aligned forests (VAFs) of CNTs and laser assisted CVD for microscopic configurations at predefined surface locations. Individual CNTs could be prepare by CVD by controlling the density of the nucleation centers.

7.3.1 Synthesis of Aligned 3D CNT Ensembles

7.3.1.1 Synthesis of Aligned 3D CNT Ensembles: Electric Arc Discharge Method (MWCNTs)

The electric arc discharge (EAD) method for the production of MWCNTs is an adaptation of the carbon arc method used by W. Krätschmer and Huffman in 1990 [52] for the synthesis of fullerenes. In 1991 Iijima observed MWCNTs in a deposit collected from the negative graphitic electrode (cathode) in an electric discharge experiment [1]. The nanotube yield was relatively poor. Thus the EAD conditions were further optimized for an increased yield of MWCNTs [53] finally enabling the production of gram quantities.

The principle of MWCNT synthesis by the EAD method is as follows: an arc discharge is ignited between anode and cathode graphitic electrodes in a buffer gas atmosphere. Electrons are accelerated by the applied electric field towards the anode where they lose their kinetic energy. This causes heating of the anode. Plasma ignited between electrodes can reach very high temperature of a few thousand Kelvin – with a maximum temperature of ~6500 K at a distance of ~1 mm from the anode surface [54]. If a dc electric field is applied, material from only one electrode will evaporate. A portion of the carbon vapor condenses on the cathode. Another

Fig. 7.1 (**a**) EAD's stub deposit on the cathode. (**b**) Cross section of the portion of the stub with hard shell and fibrous core material containing MWCNTs. (**c**) TEM image of one MWCNT separated from the core of the stub. (**d**) Raman spectrum obtained from the core material measured with 633 nm laser light, RBM peaks originated from inner tubes are recognizable

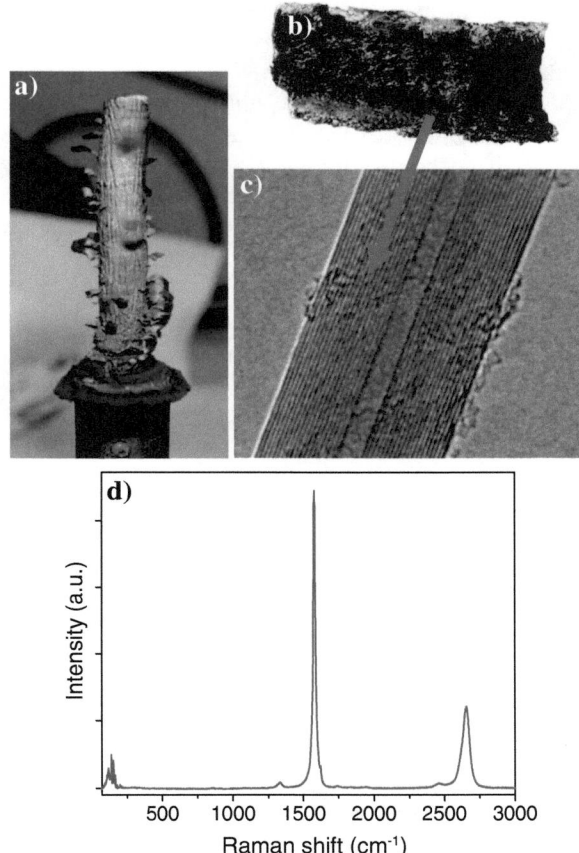

portion creates a deposit on the reactor walls. In [53] it was reported that MWCNTs were only observed in a deposit formed at the cathode inside a cylindrical stub with hard shell containing macroscopically fibrous material with fibers oriented in parallel. An image of a similar EAD cathode deposit fabricated by an author of this review is shown in Fig. 7.1.

The cathode deposit had usually a rod-like shape with a hard cylindrical grey shell (Fig. 7.1a, b). The inner core of the stub can be soft and contain MWCNTs (Fig. 7.1c) as well as undesired graphitic or onion like carbon particles. The yield of MWCNTs is influenced by arc conditions such as current density, applied voltage between electrodes, buffer gas pressure, and type. Electrode properties such as size, size ratio, geometry, and distance of electrodes, crystallinity and density of anode are important also. In [53] the maximum yield of synthesized MWCNTs was obtained at ∼500 torr of inert gas (He or Ar), ∼18 V of potential difference between electrodes and a current ∼100 A.

Reduced production costs can be achieved by synthesizing MWCNTs (in the deposited stub) in an air atmosphere at a pressure of 300 torr as reported in [55]. An additional advantage of this "air" approach is the reduced deposition of amorphous carbon and other carbon nanoparticles on the chamber walls.

MWCNTs produced by EAD method are usually of very high structural quality. They are straight and rigid with small structural defect densities. Under TEM imaging, (Fig. 7.1c) parallel nanotube walls are visible. The Raman spectrum (Fig. 7.1d) exhibits a high G (at about 1580 cm^{-1}) to D ($\sim$1335 cm^{-1}) intensity ratio ($\sim$ 55). The presence of Raman RBM modes (here below 215 cm^{-1}) indicates the occurrence of (inner) tubes with diameters below 3.1 nm. The length of the tubes and the abundance of impurities (graphitic and onion like particles, etc.) in the fibrous core are a related to plasma stability.

Because of the high temperature used in the process, the electrode holders and reactor walls are usually water-cooled. The chamber can be equipped with a window to observe the electrodes and the arc between them (Fig. 7.3a), or to study optical emissions in order to characterize the process [56].

Since the reduction of the anode length is different from the cathode length gain, their distance will change during arcing. This change will affect the conditions of nanotubes formation and the arc process will eventually stop. To obtain high quality nanotubes, the plasma stability is crucial and thus the electrode gap must be kept constant by moving one or both electrodes compensating the shortening of the anode by evaporation of material.

7.3.1.2 Synthesis of Aligned CNT Ensembles: CVD Using Pre-Deposited Catalyst Precursors

An attempt to grow vertically aligned nanotubes utilizing a mesoporous silica substrate containing iron catalyst nanoparticles embedded in the pores was presented by [57] in 1997. In their approach, orientation of tubes was caused by constraints from the orientation of pores (with pore walls aligned perpendicularly to the surface). Tubes nucleated in the pores and continued growing along the pore axis direction even above the substrate surface. Tubes nucleated from iron particles on the substrate surface, grew in non-preferable (different) directions. Through this method, a large area of highly ordered, isolated, $\sim$100 μm long CNTs were synthetized at 700 °C utilizing acetylene as carbon source. In 1998, [16] presented an approach to produce vertically aligned CNTs by plasma enhanced CVD (PE CVD) without using substrate with "aligning" pores. CNTs were synthetized from plasma pretreated Ni catalysts deposited on glass, at a temperature below 666 °C using acetylene as carbon source. The tube length was shorter than 50 μm. Advantage of the PE CVD method (for review on PECVD see [58]) is that carbon nanotubes forests could be synthetized at relatively low temperatures what allows to use flexible plastic substrates. In [59] VAF of MWCNTs was fabricated at Kapton polyimide foil at 200 °C, using Cr/Ni film as catalyst precursor in 1.5 mbar C_2H_2/NH_3 (30:200 sccm) atmosphere.

The above mentioned approaches produced low density VAFs. Such nanotube ensembles are for example not directly suitable for actuator fabrication by the spinning method, but could be used as electron emitters [59]. VAFs with higher densities and heights have been obtained on porous silicon substrates with a thin nanoporous layer (with pore diameters of $\sim$3 nm) on top of a macroporous layer (with sub-micrometer pores) [60]. It was assumed that nanotubes have grown from the top of the substrate (not from the pores). The porous structure of the surface influenced the iron catalysts size, density, and activity. Clean, and up to 240 μm high VAFs were grown at 700 °C from ethylene. Even higher CNT VAF density was synthetized by [61] utilizing high density of catalysts (Fe on Al_2O_3) and 15 mbar C_2H_2/H_2 at 700 °C. Their VAF grown with 1.48×10^{13} CNT/cm^{-2}.

Later, a so-called super-aligned carbon nanotube array of vertically aligned MWCNTs kept together by van der Waals forces to form bundles allowed for the first time preparation of yarns by continuous dry-drawing from the VAF as presented by [62]. Their VAFs have been fabricated on polished flat surfaces of a silicon wafer, a thermal-oxidized silicon wafer, or a quartz plate utilizing iron catalysts and acetylene at T < 720 °C [63].

Similarly, in [64], the growth of clean VAFs of $\sim$10 nm diameter MWCNTs suitable for the fabrication of yarns have been achieved at 680 °C on silicon wafer substrates using a 5 nm iron film as a catalyst precursor and acetylene as a carbon source. A study on the influences of temperature, acetylene and hydrogen flow rates, and reaction time on the height of the MWCNT forest has been presented in [65]. Their spinnable VACNT arrays grew up to a 630 μm height under optimum reaction conditions (60 sccm C_2H_2, 0.5 SLPM H_2, 750 °C and 10 min growth run) with an initial growth rate of $\sim$465 μm/min.

A breakthrough in the synthesis of high SWCNT VAFs were reported by Hata et al. [66] in 2004. Forests with heights of up to 2.5 mm were synthetized by water assisted CVD with ethylene (on substrate covered by sputtered 1 nm Fe/10 nm Al_2O_3 onto SiO_2/Si). The method was further optimized by Futaba et al. [47] by investigation of the dependence of VAF height on ethylene flow rate and water vapor concentration. The forest with high $\sim$970 μm (in 20 min of growth) has been synthesized for a $\sim$140 ppm of water vapor concentration in a gas flow of He/ethylene at water to ethylene ratio of 1/1000. The synthesis was performed at the same substrate type as [66] at 750 °C growth temperature.

In [67], Hata's group presented an improvement of VAF growth by controlling the gas flow direction utilizing a gas shower system. The gases were delivered into the reaction chamber in a direction perpendicular to the substrate surface. With this improved growth method, they demonstrated the synthesis of a 1 cm tall forest on a 1 × 1 cm area and the fabrication of SWCNT VAFs on an A4 size substrate.

The results published in [13] demonstrate that SWCNT VAFs can be fabricated by water-assisted CVD on Fe/Al_2O_3 at 750 °C from a very broad range of carbon sources. They proved that the resulting SWCNTs possessed similar average diameters indicating that the diameter was determined predominantly by the catalyst size (same catalyst preparation procedure applied) rather than by the carbon feedstock within the same synthetic system. The tallest nanotube forest was approx. 1.5 mm height and was synthesized from butane.

In [68] was showed that the SWCNT diameter distribution and nanotube density in VAFs can be tuned by controlling the catalyst formation temperature (from 1.5 nm Fe film deposited on a 30 nm AlO_x layer covering a Si/SiO$_2$ substrate) and exposure to H$_2$ without sacrificing the yield and quality of the nanotubes. The nanotube diameter tuning ranged from 1.9 to 3.2 nm at densities of 0.03–0.11 g/cm^3. Diameters and mass densities were inversely correlated. An approach for independent control of nanotube diameter distributions and their densities on the substrate was presented in [69, 70]. By embedding size-controlled catalysts into a carbon matrix their agglomeration was inhibited. Another approach allowing independent control of catalyst diameters (predefined by ultracentrifuged ferritin particles) and their density on the support surface was presented in [71].

The fabrication of CNT VAFs at micrometer scale is possible by localized heating using laser light in laser assisted CVD (for the method review see [72]). Vertically aligned CNT hills, as shown in Fig. 7.2a, were fabricated on laser heated areas using an iron film as catalyst precursor deposited on Al$_2$O$_3$/SiO$_2$/Si substrate and ethylene as carbon source similarly to [73, 74].

Recent reports showed that VAFs of CNTs can be fabricated by growing nearly perfectly oriented nanotubes (Herman's orientation factor 0.85) [75]; VAFs heights can reach centimeters (MWCNTs ~1.2 cm [76], SWCNTs ~1 cm [67]); SWCNTs area densities can reach ~$1.5*10^{13}$ cm^{-2} (corresponding to 0.36 g/cm^3) [61]; The area density of MWCNT VAF reaching 1.2 g/cm^3 was reported in [77] (for 300 nm high VAF). VAFs can cover areas up to ~700 cm^2 (Fig. 7.2d) and CNT diameter distributions can be tuned [68, 78].

These results show that CNT VAFs seem to be a very promising source material for the fabrication of actuators in form of either "yarns", based on spinnable VAFs [65, 79] or thin paper-like nanotube assemblies based on transferable nanotubes [80], even if those great achievements have not been reached simultaneously until now.

7.3.1.3 Synthesis of Aligned CNT Ensembles: CVD Using Injected Catalyst Precursors

The injection of catalyst precursors into the reactor removes the necessity of their predeposition on the surface. Pyrolysis of metallocenes such as ferrocene, cobaltocene, and nickelocene in the presence of hydrocarbons can produce high yields of nanotubes [81]. CNT VAFs can be fabricated utilizing this approach as well. Substrate-site selective growth of CNT VAFs from ferrocene and xylene precursor mixtures at 800 °C was reported in [82]. Using this method, only predefined SiO$_2$ areas were covered by VAFs, as shown by synthesis experiments with patterning of silica on the silicon surface [82]. The effect was explained by the preferential formation of active catalyst particles for nanotube synthesis on specific substrates because of their specific adhesion to and/or diffusivity on different

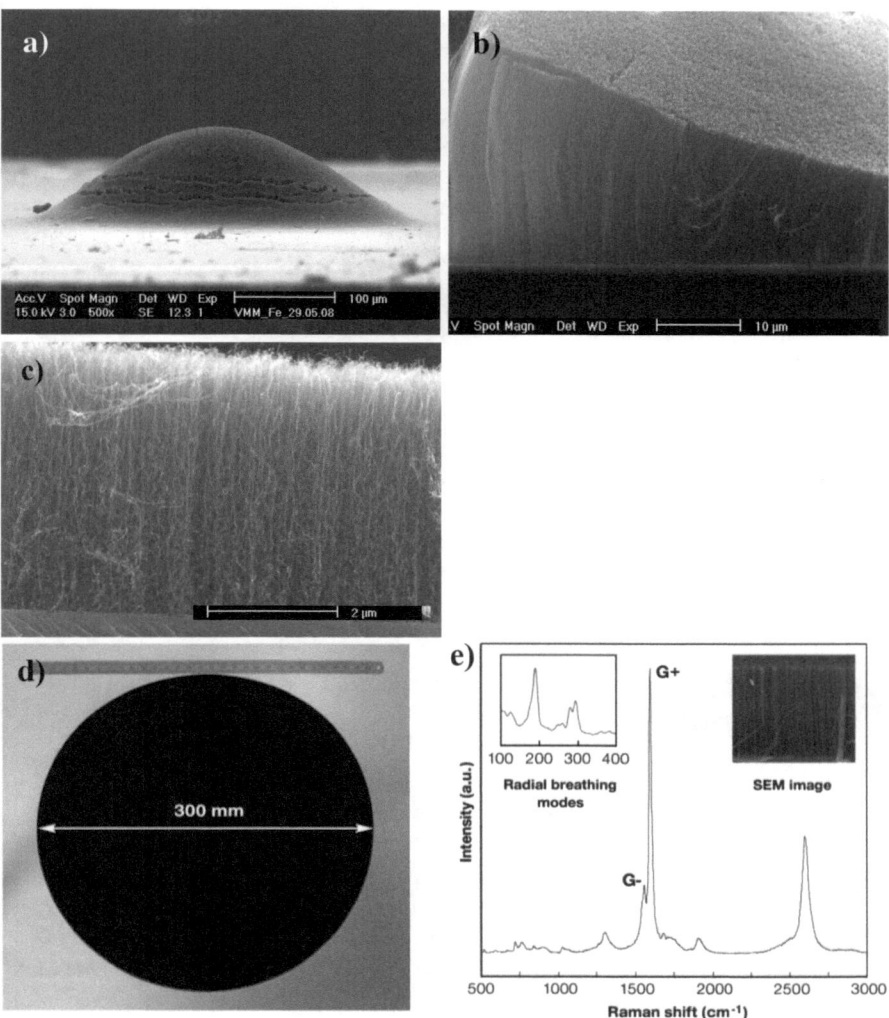

Fig. 7.2 SEM images of CNT VAFs prepared by (**a**)–(**c**) laser assisted CVD and (**e**) PE CVD. (**a**) As grown "hill" of aligned MWCNTs. (**b**)–(**c**) cross section of "foot of the hill". (**d**) Photo of 30 cm wafer covered by VAF. (**e**) Raman spectra from the VAF on the wafer. Inset. SEM of VAF cross section from (**d**). (**d**)–(**e**) Courtesy of K.B.K. Teo, AIXTRON

substrates. The systematic study of synthesis temperature effect on VAF height, CNT diameter, nanotube structural quality and presence of amorphous carbon have been investigated in [83] for SiO_2 substrates, injecting ferrocene precursor to $C_2H_4/Ar/H_2$.

The continuous process of producing MWCNT VAFs on metallic (stainless steel) substrates coated by Al_2O_3 by thermal plasma at atmospheric pressure was demonstrated in [84]. Ferrocene and ethylene were used as catalyst precursor and carbon source. There was not use other heat source as the microwave plasma which heated substrate and decomposed carbon source.

The method of injection of catalyst precursors could be used for direct spinning of CNT fibers out of a CVD reactor as shown in [44]. In the work, ferrocene (as catalyst precursor), thiophene (as precursor of sulfur promoter), and ethanol (carbon source) were injected into the hot zone of a CVD reactor. Nanotubes formed on unsupported catalysts were captured and wound out of the hot zone continuously as a fiber with a rotating spindle. Li et al. [44] showed that such a continuous spinning process is possible using different carbon sources containing oxygen (for example diethylether, polyethylene glycol, acetone, etc). Aromatic-hydrocarbons, however, did not enable a sustained continuous spinning process unless they were mixed with another oxygen containing precursor such as methanol. As prepared spun fibers are typically composed of MWCNTs with diameters of about 30 nm and lengths of a few tens of micrometers and concentration of metal particles of $\sim$10 wt%. Recently, fibers with a dominating abundance of metallic SWCNTs were fabricated by such a continuous direct spinning approach [85].

7.3.2   Synthesis of 3D Ensembles with Unaligned CNTs

7.3.2.1   Synthesis of 3D Ensembles with Unaligned CNTs: Electric Arc Discharge (SWCNTs)

In 1993, two groups detected the presence of SWCNTs in a soot produced by the electric arc discharge (EAD) method evaporating a metal enriched graphitic anode. A group at NEC [2] used Fe enrichment. Researchers at IBM [3] utilized Co. Soon after this discovery, an increase of the SWCNT yield with mixtures of two metals, e.g. Fe/Ni or Co/Ni [86] and Co/Pt [87], for anode enrichment was reported. This approach enabled relatively simple synthesis of SWCNTs in sufficient amounts for the quickly developing research field of SWCNTs. Due to the strong light emitted by the electric arc, the high fluctuations of the anode ablation, the relatively small gap between electrodes, and the possible presence of carbon microcrystallites in the gas phase (which causes scattering of the radiation) [56] this approach did not provide a straight forward opportunity for understanding the mechanisms of nanotube formation. Nevertheless, in order to better understand and optimize nanotube synthesis, various in-situ characterizations of the plasma process have been performed. For example, C_2 Swan bands of arc emission were detected by optical emission spectroscopy to determine the species distribution of excited C_2 species and their rotational and vibrational temperatures in the active plasma zone [88, 89].

In contrast to the preparation of MWCNTs in the cathode deposit of EAD as mentioned above, SWCNTs form a web-like material (shown in Fig. 7.3c). This web can be found in various places of the reactor, typically between the cathode and reactor walls. Additionally, a collar-shaped deposit around the cathode was

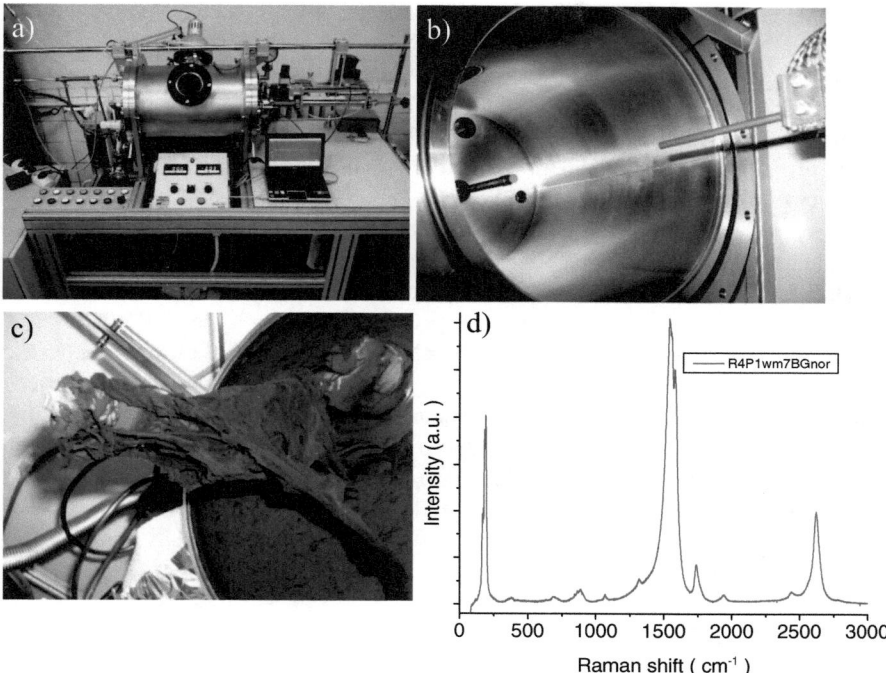

Fig. 7.3 (**a**) EAD reactor. (**b**) Anode and cathode. (**c**) Web like deposit containing SWCNTs, hanging on the cathode. (**d**) Raman spectrum from the web like assembly synthetized by using Pt/Rh/S enriched anode (Raman line 632 nm). Photos in (**a**) and (**b**) courtesy of V. Vretenar, Danubia Nanotech

found at the same growth conditions. The different localization of as-synthesized SWCNTs in the reactor points on differences in nanotube formation mechanisms in experiments evaporating either metal-enriched or pure graphitic anode. SWCNTs are formed on small metal particles condensing from metal vapors out of the hot part of plasma of the arc discharge. P. Bernier's group [90] achieved a breakthrough in increasing the amount of produced SWCNTs by EAD using Ni and Y in a 4.2: 1 atomic ratio (and using $p(He) = 660$ mbar, $V = 40$ V, $I_{DC} = 100$ A, electrode distance $d_{AC} = 1$ mm) as carbon-anode additives. According to Saito et al. [91], the efficiency of an Rh-Pt bimetal mixture to catalyze SWCNT synthesis could reach one for Ni-Y.

For specific applications, Fe metal "impurity" in CNTs can be preferred rather than using other metals because of its nontoxicity, low cost, and the existing experience with its broad application in industrial production of vapor grown carbon nanofibers. Anodes solely containing Fe and graphite produce SWCNTs with very low yields [2] and, therefore, as an anode additive in its pure form iron plays a limited role in SWCNT production by EAD. The addition of promoters to Fe, for example 3% of Bi [40] increases the nanotube yield.

Typically, as-grown EAD products, except of SWCNTs, also contain metal particles, graphitic particles, amorphous carbon, and possible traces of fullerenes. Haddon's group [92] optimized the bimetal composition in the anode with respect to the purity of as-synthetized SWCNTs, utilizing their self-developed optical solution-phase NIR spectroscopy method. They found that a relatively wide range of catalyst concentrations (3–6 atom % Ni and 1–2 atom % Y), allow efficient production of high-purity as-prepared SWCNT soot [93]. The relatively broad window of acceptable synthetic parameters found in the study indicated the robustness of the method and suggested an opportunity to scale-up the EAD synthesis of SWCNTs.

The yield and characteristics of the CNTs fabricated by EAD are influenced by the anode composition and its pretreatment, the buffer gas type and its pressure, the arc current density and its stability, electrical potential between electrodes, electrodes size, size ratio and shape, their relative movement (distance between electrodes, rotation), size of the reactor chamber, presence of magnetic field, time of arcing, and some other parameters. The optimization of process parameters with respect to the tuning of nanotube characteristics is as a result often related only to the specific setup and/or the synthesis conditions.

Nanotube diameter distributions can be, for example, influenced by changing the Ar/He ratio in the arc chamber [94, 95]; the mean diameter of SWCNTs decreased from ~1.4 nm for an Ar/He ratio of 0.2 to 1.23 nm for pure Ar. A change in the nanotube diameter distribution was observed also for a change of He pressure from 0.95 nm at 50 torr to 1.4 nm at 1520 torr for the cathode soot synthetized by utilizing a Pt/Rh/C anode [96]. A change in pressure in another type of EAD set-up with vertically oriented electrodes, utilizing NiY bimetals showed insignificant pressure influence on the diameters of the fabricated nanotubes in a pressure range of 500 to 1000 torr of helium. In this range, the average diameter of nanotubes was approx. 2 nm. In an experiment with 100 torr of He, slightly smaller diameters of about 1.7 nm were observed [97].

Using a Bi promoter enriched Fe/C anode [40] caused a nearly 2x increase of the mean diameter of SWCNTs from ~1 nm for those produced using pure Fe catalysts.

The lengths of as-prepared EAD SWCNTs, could be influenced by the presence of a magnetic field [98, 99]. In their experiments, the length of SWNTs increased to above ~5 μm, when a magnetic field of 0.4 T was applied during an arc discharge, in comparison with ~2 μm long tubes synthesized without a magnetic field. The synthesis of longer tubes was explained by the fact that the magnetic field ensures a significant enlargement of the high-density plasma area. Thus, the catalyst particles and/or nanotubes reside in the area in which growth is possible for a longer time resulting in the growth of longer SWCNTs.

The length of as-produced SWCNTs can be influenced by the diameter of the cathode (for a constant size of anode) as well. In [100], the dependence of the SWCNT length on the cathode diameter was analyzed. While increasing the cathode diameter by a factor of almost three, the proportion of long SWNTs increases. The mean length changed from ~1.2 to 1.8 μm and reached a maximum length of ~ 5 μm.

Modifications of the EAD method mimicking the laser ablation (LA) method (introduced below), have been developed [101, 102], as the quality of SWCNTs produced in LA is excellent, but the setup is expensive. Examples include electrodes enclosed in a quartz tube that have been placed into an oven. Similarly to LA, gas flow, and Ni/Co as catalysts were used in [101]. In [102] pulse arcs have been used. The yield and quality of as-grown nanotubes approached those fabricated by laser ablation under this method. The latter modification was claimed to be easily up-scalable.

An important modification of the EAD method was achieved by immersing the electrodes in liquids, for example, immersions into liquid nitrogen [103], water [104], or hydrocarbons [105] were reported. The main advantages of this approach are an easier collection of the products and the possibility of continuous nanotube synthesis. Another advantage is related to the simplicity of the setup with no requirements for a vacuum system and inert gasses. In general, such a system is significantly less expensive to build-up and operate. High-quality nanotubes were produced at a rate of 5 g/min [105].

The popularity of producing CNTs by EAD has rapidly decreased in the past few years because of the difficulties in precisely controlling the synthesis parameters, including the temporal changes of the metal concentration distribution in the anode [106] during the run. As-grown nanotubes are contaminated with impurities such as metal and graphitic particles as well as amorphous carbon and for most applications their purification is necessary. Additionally, collecting dry products containing nanoparticles in powder form could bear higher health risks than, for example, working with CNT VAFs "sitting" on substrates as obtained by CVD. Nevertheless, the EAD method can "offer" possibility for cheap & relatively simple fabrication of source material for many types of applications and research in form of fibers (as synthesized) and buckypapers as shown in the examples in Fig. 7.4 or for preparation of chiral selected SWCNTs.

7.3.2.2 Synthesis of 3D Ensembles with Unaligned CNTs: Laser Ablation

Trying the laser ablation method to fabricate carbon nanotubes was a very natural choice of Smalley's group at Rice University as the method had already been explored for the production of fullerenes [107]. Similar to the results of arc discharge, mixtures of bimetals with graphite have proven to be efficient source materials for the synthesis of SWCNTs; a mixture of Ni/Co/graphite has been used as a promising target material in [4].

The setup for SWCNT synthesis by laser ablation can consist of a vacuum sealed quartz tube mounted inside a tube furnace. The laser beam enters the quartz tube through a window selected specifically for a used laser wavelength. Inert gas (often argon) is flown in the same direction as the laser beam. Before the gas exits the system, it passes a water-cooled collector (Fig. 7.5b) and a filter. A target consisting of a carbon source and metal(s) mixture is attached to a rotating rod (Fig. 7.5a). This is to ensure a more homogeneous ablation of the target. The carrier gas flow sweeps up most of the carbon species produced by laser evaporation out of the furnace zone, depositing it on a water-cooled rod. In [4] and many other works, the ablation laser

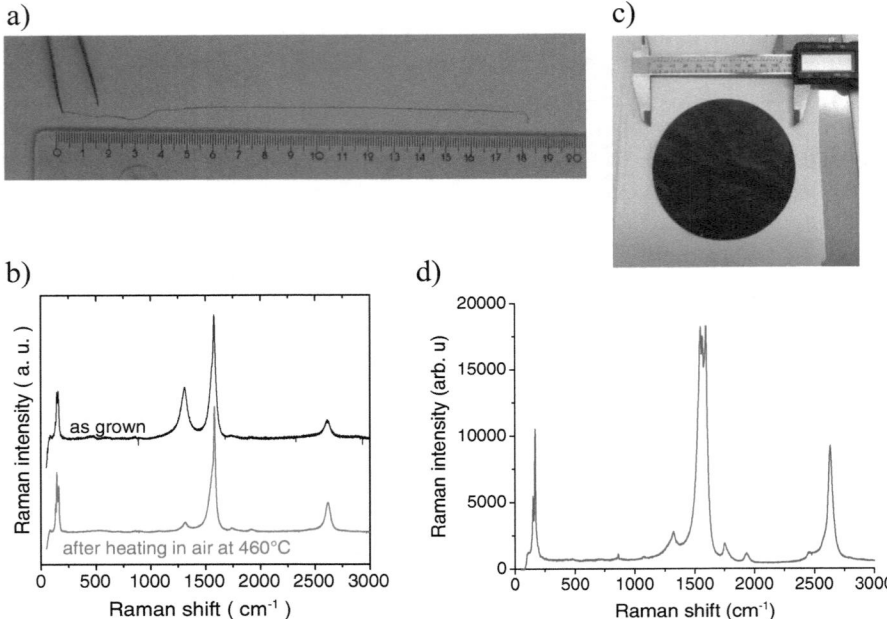

Fig. 7.4 EAD products. (**a**) As grown fiber synthesized from Pt/Rh/S/C anode containing SWCNTs, metal particles and amorphous carbon. (**b**) Raman spectrum from as grown fiber. Raman spectrum after fiber heating at 460 °C in air. Increased G/D ratio demonstrates purification of the fiber by oxidizing amorphous carbon. (**c**) Buckypaper from web material (courtesy of V. Vretenar.) (**d**) Raman spectra, from a buckypaper prepared from EAD SWCNTs

was a Nd:YAG type laser operating at 532 nm or 1064 nm providing 300 mJ to1.5 J of energy per pulse at <10 ns pulse width. The beam can be focused to a spot of a few mm in diameter. In [108] are two lasers used, Nd:YAG and cw CO_2 laser. The role of the CO_2 laser is to increase the temperature of the condensing plume. In [4], the SWCNT yield increased with increasing oven temperature up to the oven limit of 1200 °C. SWCNT formation has been confirmed for oven temperatures above ∼ 780 °C [109].

The SWCNT production was scaled up by modification of the original method [4] by a dual pulsed laser vaporization process [110, 111]. The target was first exposed to laser pulses with a 532 nm wavelength and energies of 250–490 mJ /pulse, then to beams with a 1064 nm wavelength and 300–550 mJ/pulse with a 42–50 ns delay between the individual pulses. Laser spots of diameter ∼6–7 mm were localized at the same place. The dual pulsed laser vaporization process enabled the production of gram-per-day quantities of 60–70 vol% pure SWCNTs.

Kingston et al. [112] enhanced the SWCNT yield considerably (exceeding production rates of 400 mg/h) by using two Nd:YAG 1064 nm lasers, one (1.5 J/pulse) operating in the pulsed-mode to ablate the target and the other one in continuous-

a)

b)

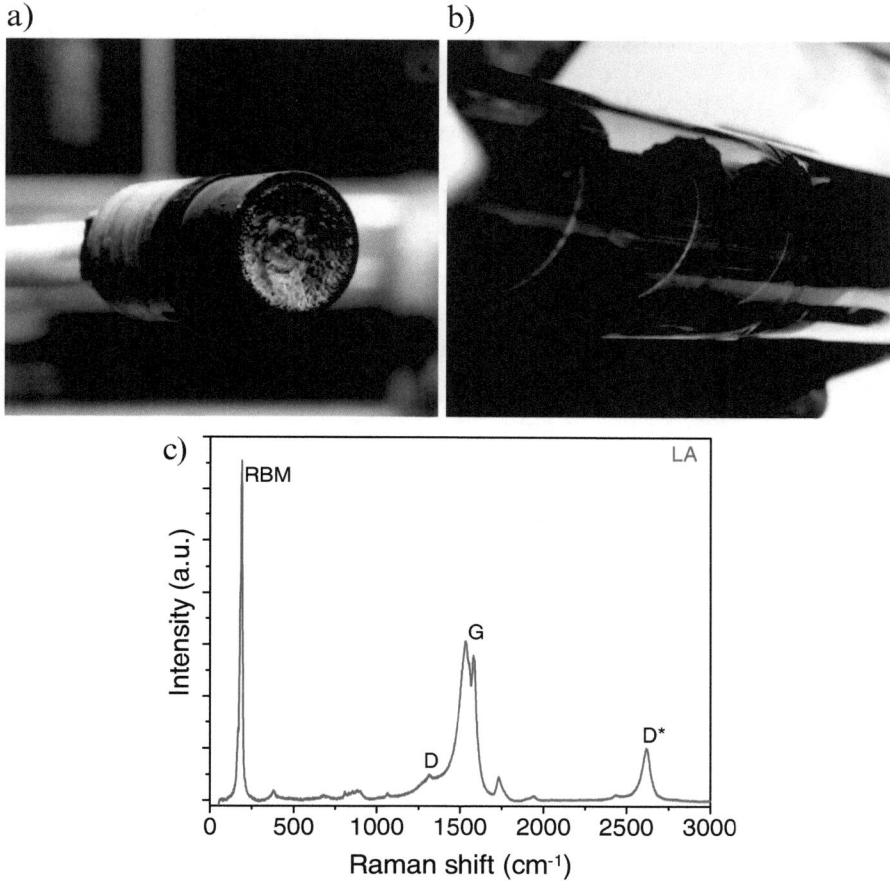

c)

Fig. 7.5 (**a**) LA target composed of Ni/Co/graphite. (**b**) Collector with deposited SWCNTs. (**c**) Raman spectra of the LA deposit (Raman line 633 nm). (Photos were taken from a LA system built by B. Hornbostel from S. Roth group, MPI Stuttgart)

mode at 50 W to alter the rate of cooling of the condensing plume. The synthesis was performed in a hot oven.

The diameter distributions of the synthesized SWCNTs can be tuned by the oven temperature [109, 113, 114]. In general, for different catalysts and buffer gases the mean diameter increases with increasing oven temperature. In the case of NiCo in He, the mean SWCNT diameter increases linearly from about 1.1 nm at 950 °C to about 1.2 nm at about 1150 °C of oven temperature [113]. The inert gas type influences the diameter distribution as well. With increasing thermal conductivity of the gas, the mean diameter of nanotubes decreases under otherwise identical synthesis conditions, for example at oven temperature 1200° the mean diameter of SWCNTs synthesized using Pt/Rh/Re catalyst precursors in He (∼0.15 W/m K) was below 1 nm and in Xe (0.005 W/m K) reached about 1.4 nm [114]. The effect of various synthesis conditions including catalyst type and composition was investigated in [113].

A different approach for the synthesis of SWCNT by the laser ablation is the use of a continuous wave CO_2-laser with a wavelength of 10.6 μm [115]. In this case, a heated furnace is not necessarily needed. The energy required for the evaporation of the target can be delivered completely by the laser beam itself. Additionally, a part of the incoming laser energy continuously dissipates as heat into the target. The target itself thus, acts as a local furnace. The highest SWCNT yields were achieved with Ni/Y (4.2/1) and Ni/Co (2/2) (ratio in at %) additives to a graphitic target using Ar or N_2. Using He, no SWCNTs were produced by this method [116]. The gas effect was related to the change of the local temperature gradients. Helium is more efficient for rapid cooling than Ar and N_2. Therefore, the temperature gradient becomes too large to favor the growth for SWCNTs. On the other hand, argon and nitrogen keep the temperature gradient small enough for the production of SWCNT.

Maser et al. [116] found no big difference between Nd:YAG and CO_2-laser systems concerning yields and characteristics of as-produced SWCNTs. However, they consider the scaling-up possibility of CO_2-laser systems to be easier by far.

The above mentioned synthesis of SWCNTs based on laser vaporization of source material has been carried out using lasers emitting either in the visible or in the infrared range. The latter predominately caused photo-thermal ablation. In [117, 118] it was shown that SWCNTs can be synthesized by using a laser emitting in the UV range known to be effective for photo-chemical ablation. Additionally, heated environments were necessary. In [117], a pulsed KrF laser with a 248 nm wavelength, pulse width of 15 ns, and a laser intensity 350 MW/cm^2 was used for the ablation of CoNi-doped graphite pellets in an Ar flow at 500 torr. The formation of SWCNTs has been detected for oven temperatures higher than 550 °C.

A XeCl excimer laser with a 308 nm wavelength, 16 ns pulse width, and 58 mJ/pulse ablating a NiCo enriched graphitic target with a laser power density $\sim$140 MW/cm^2 has been shown to produce SWCNTs in an Ar environment if heated in an oven to temperatures of 800–1400 °C [118, 119]. For both experiments with excimer lasers mentioned above [117, 119] the shift in the SWCNT diameter distribution shows that the increase of the furnace temperature favors the growth of nanotubes with larger diameters, similar to experiments with a Nd:YAG laser.

LA utilizing pulsed and continuous lasers have been explored to investigate mechanisms of SWCNT synthesis and to try to determine how, where, when, why, at which conditions, from which species and how quickly SWCNTs are formed.

A few groups utilized pulsed lasers for the in-situ investigation of SWCNT synthesis. The NASA Johnson Space Center [120] presented optical diagnostic measurements of a laser ablation plume during the production of SWCNTs. The method was extended by [121] and further elaborated by the group of D.B. Geohegan [122–125], which determined the temperature profile of the target surface and the time evolution of vapors and carbon clusters. They observed that during the first 0.1 ms after the end of the laser vaporization pulse, the plasma is very hot, and emission from excited atoms and molecules dominates. Later, at $\sim$0.2 ms, carbon in the presence of Ni and Co vapors condenses and forms clusters. The metal atoms condense much later, at $\sim$2 ms at a temperature of around 1400 °C. Between 1400 > T > 1320 °C the onset of nanotube growth was detected. If growth was stopped after 25 ms, only short nanotubes were found with a length of $\sim$240 nm. If the growth was continuing for a longer time, then after $\sim$100 ms tubes up to a few μm long were found.

A dedicated synthesis reactor based on the continuous CO_2 laser vaporization of a metal-enriched graphitic target under fixed conditions was used for the excellent investigation of SWCNT synthesis processes by the ONERA group [126–128]. *In-situ* diagnostics of the gas phase and of the carbon soot in this reactor was combined with *ex-situ* TEM analysis of the synthesis products. The group showed complementary roles of Ni and Co on condensation of carbon vapor in separated experiments with mono-metal doped graphite and explained why a NiCo mixture is favorable for SWCNT synthesis. At T > 2500 K, Ni vapor has no effect on the presence of carbon dimers (C_2), in contrast to Co vapor, which interacts in the gas phase with C_2, forming CoC_n clusters. For 2200 K > T > 1600 K, the dissolution of primary carbon soot into Ni particles was detected, but its density was not affected in presence of Co. An ex-situ TEM study confirms this different behavior observed in-situ, finding a broad diameter distribution of Ni particles, in contrast to the case of Co. The Co-C_2 reaction inhibits the formation of larger Co particles, while Ni particles can grow, as there is no relevant Ni-C_2 reaction. These observations support a vapor-liquid-solid model of nanotube formation by LA.

The fundamental limitations, which are inherent to laser ablation systems, are their restrictions regarding mass production, price, length of individual tubes and safety issues related to work with deposits containing nanopowders. Typically, no more than a gram quantity of SWCNTs contaminated with metal and other carbon impurities can be produced per day.

7.3.2.3 Synthesis of 3D Ensembles with Unaligned CNTs: CVD Using Unsupported Catalysts

The advantage of CVD method with injected unsupported catalysts is its scalability to a continuous fabrication method and previous experience with synthesis of carbon fibers (for review see [129]).

To synthesize CNTs, Sen et al. [81] used two stage furnace system. In the first stage ferrocene was evaporated at 200 °C, its vapors were carried by Ar/H_2/benzene into the second furnace stage where benzene and ferrocene were pyrolysed at 900 °C. In [130] was shown that both gas-phase pyrolysis of acetylene along with a metallocene(s), or pyrolysis of Fe(CO)$_5$-acetylene mixtures at 1100 °C yields SWCNTs. SWCNTs have also been synthesized utilizing Fe(CO)$_5$ and CO in an oven heated up to 1100 °C [131]. In [132], the effects of sulfur promoter, and growth conditions on the nanotube diameters were investigated. MWCNTs with 10–100 nm diameters were obtained.

Probably the most famous method of producing SWCNTs by injection of catalyst precursors into a CVD reactor and formation of nanotubes on floating, unsupported catalysts is the high pressure carbon monoxide method known as the HiPco process, established in 1999 by R. Smalley's group at Rice University Houston [133–135]. The method involves the thermolysis of $Fe(CO)_5$ in the presence of CO. SWCNTs grown at 800–1200 °C in a CO flow at pressures <50 atm on iron catalytic particles. The HiPco process yielded ~450 mg/h of SWCNTs [135]. As collected nanotube products typically contained no more than 7 mol% of Fe impurities.

The ability of tuning the SWCNT's diameter distribution in HiPco processes was observed with a dependence on the pressure in the synthesis reactor zone. A narrower distribution with a smaller mean diameter for higher CO pressures was obtained. For example: SWCNT with diameters from 0.6 to 1.3 nm and the mean diameter ~0.8 nm at 10 atm were synthesized [133].

The possibility to change the nanotube diameter distributions in the so-called DIPS method (direct injection pyrolytic synthesis) have been shown by S. Iijima's group [136] using two kinds of carbon sources with different decomposition properties. Toluene with dissolved ferrocene was injected directly to a vertical ceramic tube reactor kept at 1200 °C through a spray-nozzle under a hydrogen carrier gas flow. In order to control the nanotube diameter distribution, ethylene with a flow rate ranging from 0 to 200 sccm was added to the reactor. An increase of the ethylene flow rate caused a decrease in nanotube mean diameter.

Another approach to inject catalyst particles have been used by Nasibulin et al. [137]. They used a hot wire method to evaporate iron and inject its vapor simultaneously with CO for nanotube synthesis into the CVD reactor heated to 1500 °C. Progress in the aerosol method has been reviewed recently in [138].

Important progress in the synthesis of unaligned SWCNT assemblies by CVD using unsupported catalysts was reported in [139]. Bimetallic nanoparticles with controlled size and composition were prepared by precise control of the precursor ratio and synthesis conditions in a micro-plasma reactor. They were injected into a tubular flow furnace heated to 600 °C together with acetylene as a carbon source. The semiconducting content in as-grown mixtures of SWCNTs was estimated to reach 90% for nanotubes catalyzed by $Ni_{0.27}Fe_{0.73}$ nanoparticles of 2.0 nm mean diameter [140].

Continuous and scalable fabrication of carbon nanotube aerogels with very small volumetric density (minimal value was 0.55 mg/cm^3) and surface area up to 170 m^2/g was reported by Mikhalchan et al. [141] . They continuously injected ferrocene and thiophene into the reactor together with CH_4/H_2 flow. The reactor was kept at 1200 °C.

7.3.2.4 Synthesis of 3D Ensembles with Unaligned CNTs by CVD Using Supported Catalysts / Fluidized Bed Method

The fluidized bed CVD (FB CVD) method is an industrially well-established method [142]. Typically, pressurized gas enters the FB CVD reactor through a distributor plate containing many holes, located at the bottom of the vessel. The fluid flows upwards causing the solid particles (often porous, made from oxides based on Al, Si, Mg, Ca, Ti, etc.) to float in the reaction zone. To reach a high yield of CNTs, these particles are covered by metal catalyst nano-particles. In contrast to the above mentioned method of unsupported floating catalysts, in the fluidized bed method, the residence time in the hot reaction zone of the catalysts deposited at supporting particles is much longer and well-controllable. This fact, together with uniform particle mixing, uniform temperature field, excellent contact between the catalyst surface and flowing carbon source gas, promises to fabricate large amounts of CNTs. The FB CVD method can be used in a continuous process regime if the setup provides continuous feeding as well as continuous removal of products. These properties of the FB CVD approach naturally attracted the attention of researches in the CNT field.

In [143] different carbon sources and support materials for iron catalysts have been used and the advantages of a fluidized bed over a fixed bed have been demonstrated for MWCNT fabrication. The effects of selected parameters (reaction temperature and time, presence of hydrogen and iron content) on MWCNT production utilizing C_2H_4 and iron/silica particles have been presented in [144]. Li et al. [145] reported high yields in synthesis of SWCNTs by FB CVD over nickel/porous silica particles from CH_4 at 760 °C.

Progress in increasing the yield of SWCNTs has been reported recently using a coupled downer and turbulent fluidized bed (TFB) CVD reactor [146]. Fe/MgO particles spend short residence times of about 1.2–2.4 s in the first part of setup where they move downward, to reduce iron oxide (to prepare active catalysts), then they are continuously introduced into the TFB where they spend ~3–6 min to yield SWCNTs in CH_4 gas. Both reactor parts were kept at 900 °C.

The topic of FB CVD was reviewed in [147, 148]. The FB CVD method has already been scaled up and used for industrial production of carbon nanotubes by a few companies because it offers the possibility of mass production at relatively low cost. However, the as-grown product does not contain pure CNTs. It always contains catalyst supports, metal catalysts, and possibly some reaction byproducts, such as amorphous and/or graphitic carbon particles. Therefore, FB CVD product requires a purification process before using it in many possible applications.

7.3.2.5 Synthesis of 3D Ensembles with Unaligned CNTs: Thermal Plasma

Unlike low-pressure plasmas used in CVD (PE CVD) to activate catalysts or to support the dissociation of carbon containing gases, the hot plasma can cause evaporation of solid carbon and metal materials (catalyst precursors). For CNT fabrication, thermal plasma is typically ignited either by an inductively-coupled torch or by an arc (DC or RF).

In the method using hot arc plasma, electrodes are not the primary source of carbon and metal catalyst precursors. Different carbon containing materials, for example CO (with $Fe(CO)_5$ catalyst precursor) [14], coal [8] or carbon black [6] can be used as the carbon source material.

In [14], Ar thermal plasma was generated by applying a high DC voltage of 3 kV between a zirconium-containing tungsten cathode and a copper anode. CO and $Fe(CO)_5$ source materials were atomized in the thermal plasma jet. The product passed through a helical extension of the reactor and was harvested in a collector. The authors of the research claim the production of high purity ($\sim$80%) CNTs with this continuous process.

SWCNTs with comparable quality to those produced by laser ablation method have been synthesized at large scales (100 g/h) by the induction thermal plasma method from carbon black, and Ni, Co and Y_2O_3 catalyst precursors [6]. An RF plasma torch can generate a plasma jet of extremely high temperature ($\sim$15,000 K). In comparison to other high density thermal plasma generators, RF plasma torches exhibit a larger hot thermal plasma volume, longer reaction time, and a greater flexibility in controlling the operating parameters [149], and are thus promising for large scale SWCNT fabrication. As-produced carbon soot contains approximately 40 wt% of SWCNTs. Its purity can be further enhanced by oxidation in air [150].

7.4 Conclusion

The reproducible synthesis of CNTs which simultaneously fulfill multiple specifications on their properties is still a challenging task. Further progress in: understanding the CNT formation process, optimization and control of CNT synthesis conditions, standardization of the CNT product quality determination, development of suitable process flows for nanotube integration into targeted applications, as well as minimization of the health and environmental risks are prerequisites for full usability and exploitation of nanotube's outstanding properties in practical life.

Nevertheless, companies and research labs producing SWCNTs and MWCNTs for different applications already demonstrated possibilities for large scale nanotube production especially achievable by the fluidized bed and the thermal plasma methods. The products should, in most cases, be purified from support particles (in the case of FB CVD), catalysts and other impurities. Injected catalyst CVD methods are promising as well, working nowadays at a lower fabrication scale compared to FB CVD. Laser ablation and electric arc discharge methods could be used as well despite their lower popularity. Their disadvantages include the short length of individual nanotubes (typically below 5 μm), presence of impurities in as-grown products, difficulty in scaling to an industrial scale, health risks and the high price of setups in the case of LA. The high structural quality of the produced nanotubes, the availability of well-established equipment with functional recipes in many research labs and small companies, and the relatively low operational price of EAD are obvious advantages of these two methods.

Vertically aligned ensembles of clean, acceptable long and dense CNTs ("on request" SWCNTs or MWCNTs) can nowadays be synthesized by the CVD method presented in this chapter on a more or less "semi-industrial" scale. These VAFs can be used directly without any purification in further applications.

Acknowledgement I would like to thank for help, support, fruitful discussions and cooperation to all my colleagues and students from S. Roth group at FKF MPI Stuttgart, D. Carroll group at WFU-Winston-Salem, A. Dietzel group at TU Eindhoven, and Ch. Hierold group at ETH Zurich, who joined my excitement from CNT synthesis and characterization. The valuable suggestions to manuscript from Viera Skakalova, Valentin Döring, Kiran Chikkadi, Matthias Muoth, Christina Wouters, and Stuart Truax are acknowledged. The support from ETH-FIRST and BRNC (Binnig and Rohrer Nanotechnology Center, Ruschlikon/Zurich) operation teams is highly appreciated.

References

1. Iijima S (1991) Helical microtubules of graphitic carbon. Nature 354:56–58
2. Iijima S, Ichihashi T (1993) Single-shell carbon nanotubes of 1-nm diameter. Nature 363:603–605
3. Bethune DS, Kiang CH, de Vries MS, Gorman G, Savoy R, Vazquez J et al (1993) Cobalt-catalysed growth of carbon nanotubes with single-atomic-layer walls. Nature 363:605–607
4. Guo T, Nikolaev P, Thess A, Colbert DT, Smalley RE (1995) Catalytic growth of single-walled nanotubes by laser vaporization. Chem Phys Lett 243:49–54
5. Laplaze D, Bernier P, Maser WK, Flamant G, Guillard T, Loiseau A (1998) Carbon nanotubes: the solar approach. Carbon 36:685–688
6. Kim KS, Cota-Sanchez G, Kingston CT, Imris M, Simard B, Soucy G (2007) Large-scale production of single-walled carbon nanotubes by induction thermal plasma. J Phys D Appl Phys 40:2375–2387
7. Tian Y, Zhang Y, Wang B, Ji W, Zhang Y, Xie K (2004) Coal-derived carbon nanotubes by thermal plasma jet. Carbon 42:2597–2601
8. Moothi K, Iyuke SE, Meyyappan M, Falcon R (2012) Coal as a carbon source for carbon nanotube synthesis. Carbon 50:2679–2690
9. Mishra N, Das G, Ansaldo A, Genovese A, Malerba M, Povia M et al (2012) Pyrolysis of waste polypropylene for the synthesis of carbon nanotubes. J Anal Appl Pyrolysis 94:91–98
10. Bazargan A, McKay G (2012) A review – synthesis of carbon nanotubes from plastic wastes. Chem Eng J 195-196:377–391
11. Oberlin A, Endo M, Koyama T (1976) Filamentous growth of carbon through benzene decomposition. J Cryst Growth 32:335–349
12. Dai HJ, Rinzler AG, Nikolaev P, Thess A, Colbert DT, Smalley RE (Sep 27 1996) Single-wall nanotubes produced by metal-catalyzed disproportionation of carbon monoxide. Chem Phys Lett 260:471–475
13. Kimura H, Goto J, Yasuda S, Sakurai S, Yumura M, Futaba DN et al (2013) The infinite possible growth Ambients that support Single-Wall carbon nanotube Forest growth. Sci Rep 3:3334
14. Hahn J, Han JH, Yoo J-E, Jung HY, Suh JS (2004) New continuous gas-phase synthesis of high purity carbon nanotubes by a thermal plasma jet. Carbon 42:877–883
15. Qin LC, Zhou D, Krauss AR, Gruen DM (1998) Growing carbon nanotubes by microwave plasma-enhanced chemical vapor deposition. Appl Phys Lett 72:3437
16. Ren ZF, Huang ZP, Xu JW, Wang JH, Bush P, Siegal MP et al (1998) Synthesis of large arrays of well-aligned carbon nanotubes on glass. Science 282:1105–1107
17. Kato T, Hatakeyama R (2010) Growth of single-walled carbon nanotubes by plasma CVD. J Nanotechnol 2010:1–11

18. Lim SH, Luo Z, Shen Z, Lin J (2010) Plasma-assisted synthesis of carbon nanotubes. Nanoscale Res Lett 5:1377–1386
19. Harutyunyan AR (2009) The catalyst for growing single-walled carbon nanotubes by catalytic chemical vapor deposition method. J Nanosci Nanotechnol 9:2480–2495
20. Sun HD, Tang ZK, Li G (1999) Synthesis and Raman characterization of mono-sized single-wall carbon nanotubes in one-dimensional channels of AlPO4-5 crystals. Appl Phys A Mater Sci Process 69:381–384
21. Yao Y, Feng C, Zhang J, Liu Z (2009) Cloning of single-walled carbon nanotubes via open-end growth mechanism. Nano Lett 9:1673–1677
22. Liu J, Wang C, Tu X, Liu B, Chen L, Zheng M et al (2012) Chirality-controlled synthesis of single-wall carbon nanotubes using vapour-phase epitaxy. Mol Ther 3:1199
23. Omachi H, Nakayama T, Takahashi E, Segawa Y, Itami K (2013) Initiation of carbon nanotube growth by well-defined carbon nanorings. Nat Chem 5(7):572–576
24. Ding F, Rosén A, Bolton K (2004) The role of the catalytic particle temperature gradient for SWNT growth from small particles. Chem Phys Lett 393:309–313
25. Hofmann S, Sharma R, Ducati C, Du G, Mattevi C, Cepek C et al (2007) In situ observations of catalyst dynamics during surface-bound carbon nanotube nucleation. Nano Lett 7:602–608
26. Yoshida H, Takeda S, Uchiyama T, Kohno H, Homma Y (2008) Atomic-scale in-situ observation of carbon nanotube growth from solid state Iron carbide nanoparticles. Nano Lett 8:2082–2086
27. Fiawoo MFC, Bonnot AM, Amara H, Bichara C, Thibault-Pénisson J, Loiseau A (2012) Evidence of correlation between catalyst particles and the single-wall carbon nanotube diameter: a first step towards chirality control. Phys Rev Lett 108
28. Amara H, Bichara C (Jun 2017) Modeling the growth of single-wall carbon nanotubes. Top Curr Chem (J) 375:55
29. He M, Magnin Y, Amara H, Jiang H, Cui H, Fossard F et al (2017) Linking growth mode to lengths of single-walled carbon nanotubes. Carbon 113:231–236
30. Jourdain V, Bichara C (2013) Current understanding of the growth of carbon nanotubes in catalytic chemical vapour deposition. Carbon 58:2–39
31. Tan L-L, Ong W-J, Chai S-P, Mohamed AR (2013) Growth of carbon nanotubes over non-metallic based catalysts: a review on the recent developments. Catal Today 217:1–12
32. Robertson J, Hofmann S, Cantoro M, Parvez A, Ducati C, Zhong G et al (2008) Controlling the catalyst during carbon nanotube growth. J Nanosci Nanotechnol 8:6105–6111
33. Gamaly EG, Ebbesen TW (1995) Mechanism of carbon nanotube formation in the arc discharge. Phys Rev B 52:2083–2089
34. Ismagilov RR, Shvets PV, Zolotukhin AA, Obraztsov AN (2013) Growth of a carbon nanotube Forest on silicon using remote plasma CVD. Chem Vap Depos 19:332–337
35. Bandow S, Takizawa M, Hirahara K, Yudasaka M, Iijima S (2001) Raman scattering study of double-wall carbon nanotubes derived from the chains of fullerenes in single-wall carbon nanotubes. Chem Phys Lett 337:48–54
36. Simon F, Kuzmany H (2006) Growth of single wall carbon nanotubes from 13C isotope labelled organic solvents inside single wall carbon nanotube hosts. Chem Phys Lett 425:85–88
37. de los Arcos T, Gunnar Garnier M, Oelhafen P, Mathys D, Won Seo J, Domingo C et al (2004) Strong influence of buffer layer type on carbon nanotube characteristics. Carbon 42:187–190
38. Amama PB, Putnam SA, Barron AR, Maruyama B (2013) Wetting behavior and activity of catalyst supports in carbon nanotube carpet growth. Nanoscale 5:2642–2646
39. Forrest GA, Alexander AJ (2008) Quantitative inhibiting effect of Group I–III cations on the growth of carbon nanotubes. Carbon 46:818–821
40. Kiang C-H (2000) Growth of large-diameter single-walled carbon nanotubes. J Phys Chem A 104:2454–2456
41. Haluška M, Skakalova V, Carroll D, Roth S (2005) The influence of sulfur promoter on the production of SWCNTs by the arc-discharge process. AIP Conf Proc 786:87–91

42. Haluška M, Hulman M, Hornbostel B, Čech J, Skákalová V, Roth S (2006) Synthesis of SWCNTs for C82 peapods by arc-discharge process using nonmagnetic catalysts. Phys Status Solidi (b) 243:3042–3045
43. Huang L, Wu B, Chen J, Xue Y, Liu Y, Kajiura H et al (2011) Synthesis of single-walled carbon nanotubes by an arc-discharge method using selenium as a promoter. Carbon 49:4792–4800
44. Li YL, Kinloch IA, Windle AH (2004) Direct spinning of carbon nanotube fibers from chemical vapor deposition synthesis. Science 304:276–278
45. Motta MS, Moisala A, Kinloch IA, Windle AH (2008) The role of Sulphur in the synthesis of carbon nanotubes by chemical vapour deposition at high temperatures. J Nanosci Nanotechnol 8:2442–2449
46. Jung Y, Song J, Huh W, Cho D, Jeong Y (2013) Controlling the crystalline quality of carbon nanotubes with processing parameters from chemical vapor deposition synthesis. Chem Eng J 228:1050–1056
47. Futaba D, Hata K, Yamada T, Mizuno K, Yumura M, Iijima S (2005) Kinetics of water-assisted single-walled carbon nanotube synthesis revealed by a time-evolution analysis. Phys Rev Lett 95
48. Xie K, Muhler M, Xia W (2013) Influence of water on the initial growth rate of carbon nanotubes from ethylene over a cobalt-based catalyst. Ind Eng Chem Res 52:14081–14088
49. Zhang R, Zhang Y, Zhang Q, Xie H, Qian W, Wei F (2013) Growth of half-meter long carbon nanotubes based on Schulz-Flory distribution. ACSnano 7:6156–6161
50. Amama PB, Pint CL, McJilton L, Kim SM, Stach EA, Murray PT et al (2009) Role of water in super growth of single-walled carbon nanotube carpets. Nano Lett 9:44–49
51. Kim SM, Pint CL, Amama PB, Zakharov DN, Hauge RH, Maruyama B et al (2010) Evolution in catalyst morphology leads to carbon nanotube growth termination. J Phys Chem Lett 1:918–922
52. Krätschmer W, Lamb LD, Fostiropoulos K, Huffman DR (1990) Solid C60: a new form of carbon. Nature 347:354–358
53. Ebbesen TW, Ajayan PM (1992) Large-scale synthesis of carbon nanotubes. Nature 358:220–222
54. Farhat S, Hinkov I, Scott CD (2004) Arc process parameters for single-walled carbon nanotube growth and production: experiments and modeling. J Nanosci Nanotechnol 4:377–389
55. Kim HH, Kim HJ (2006) Preparation of carbon nanotubes by DC arc discharge process under reduced pressure in an air atmosphere. Mater Sci Eng B 133:241–244
56. Lange H, Huczko A, Sioda M, Louchev O (2003) Carbon arc plasma as a source of nanotubes: emission spectroscopy and formation mechanism. J Nanosci Nanotechnol 3:51–62
57. Li WZ, Xie SS, Qian LX, Chang BH, Zou BS, Zhou WY et al (1996) Large-scale synthesis of aligned carbon nanotubes. Science 274:1701–1703
58. Gangele A, Sharma CS, Pandey AK (2017) Synthesis of patterned vertically aligned carbon nanotubes by PECVD using different growth techniques: a review. J Nanosci Nanotechnol 17:2256–2273
59. Hofmann S, Ducati C, Kleinsorge B, Robertson J (2003) Direct growth of aligned carbon nanotube field emitter arrays onto plastic substrates. Appl Phys Lett 83:4661–4663
60. Fan S (1999) Self-oriented regular arrays of carbon nanotubes and their field emission properties. Science 283:512–514
61. Zhong G, Warner JH, Fouquet M, Robertson AW, Chen B, Robertson J (2012) Growth of ultrahigh density single-walled carbon nanotube forests by improved catalyst design. ACSnano 6:2893–2903
62. Jiang K, Li Q, Fan S (2002) Spinning continuous carbon nanotube yarns. Nature 419:801
63. Zhang X, Jiang K, Feng C, Liu P, Zhang L, Kong J et al (2006) Spinning and processing continuous yarns from 4-inch wafer scale super-aligned carbon nanotube arrays. Adv Mater 18:1505–1510
64. Zhang M, Atkinson KR, Baughman RH (2004) Multifunctional carbon nanotube yarns by downsizing an ancient technology. Science 306:1358–1361

65. Cui Y, Wang B, Zhang M (2013) Optimizing reaction condition for synthesizing spinnable carbon nanotube arrays by chemical vapor deposition. J Mater Sci 48:7749–7756

66. Hata K, Futaba DN, Mizuno K, Namai T, Yumura M, Iijima S (2004) Water-assisted highly efficient synthesis of impurity-free single-walled carbon nanotubes. Science 306:1362–1364

67. Yasuda S, Arakawa K, Futaba DN, Yumura M, Yamada T, Satou J et al (2009) Improved and large area single-walled carbon nanotube Forest growth by controlling the gas flow direction. ACSnano 3:4164–4170

68. Sakurai S, Inaguma M, Futaba DN, Yumura M, Hata K (2013) Diameter and density control of single-walled carbon nanotube forests by modulating Ostwald ripening through decoupling the catalyst formation and growth processes. Small

69. Krause M, Haluška M, Abrasonis G, Gemming S (2012) SWCNT growth from C:Ni nanocomposites. Phys Status Solidi (b) 249:2357–2360

70. Melkhanova S, Haluska M, Hubner R, Kunze T, Keller A, Abrasonis G et al (2016) Carbon : nickel nanocomposite templates - predefined stable catalysts for diameter-controlled growth of single-walled carbon nanotubes. Nanoscale 8:14888–14897

71. Chikkadi K, Mattmann M, Muoth M, Durrer L, Hierold C (2011) The role of pH in the density control of ferritin-based catalyst nanoparticles towards scalable single-walled carbon nanotube growth. Microelectron Eng 88:2478–2480

72. van de Burgt Y (2014) Laser-assisted growth of carbon nanotubes—a review. J Laser Appl 26:032001–032001

73. Haluška M, Bellouard Y, Dietzel A (2008) Time dependent growth of vertically aligned carbon nanotube forest using a laser activated catalytical CVD method. Phys Status Solidi (b) 245:1927–1930

74. Haluska M, Bellouard Y, van de Burgt Y, Dietzel A (2010) In situ monitoring of single-wall carbon nanotube laser assisted growth. Nanotechnology 21:75602

75. Xu M, Futaba DN, Yumura M, Hata K (2012) Alignment control of carbon nanotube Forest from random to nearly perfectly aligned by utilizing the crowding Effec. ACSnano 6:5837–5844

76. Shanov V, Cho W, Malik R, Alvarez N, Haase M, Ruff B et al (2013) CVD growth, characterization and applications of carbon nanostructured materials. Surf Coat Technol 230:77–86

77. Sugime H, Esconjauregui S, D'Arsie L, Yang J, Robertson AW, Oliver RA et al (Aug 05 2015) Low-temperature growth of carbon nanotube forests consisting of tubes with narrow inner spacing using co/Al/Mo catalyst on conductive supports. ACS Appl Mater Interfaces 7:16819–16827

78. Youn SK, Frouzakis CE, Gopi BP, Robertson J, Teo KBK, Park HG (2013) Temperature gradient chemical vapor deposition of vertically aligned carbon nanotubes. Carbon 54:343–352

79. Foroughi J, Spinks GM, Wallace GG, Oh J, Kozlov ME, Fang S et al (2011) Torsional carbon nanotube artificial muscles. Science 334:494–497

80. Kobashi K, Hirabayashi T, Ata S, Yamada T, Futaba DN, Hata K (2013) Green, scalable, binderless fabrication of a single-walled carbon nanotube nonwoven fabric based on an ancient Japanese paper process. ACS Appl Mater Interfaces 5:12602–12608

81. Sen R, Govindaraj A, Rao CNR (1997) Carbon nanotubes by the metallocene route. Chem Phys Lett 267:276–280

82. Zhang ZJ, Wei BQ, Ramanath G, Ajayan PM (2000) Substrate-site selective growth of aligned carbon nanotubes. Appl Phys Lett 77:3764

83. Pham QN, Larkin LS, Lisboa CC, Saltonstall CB, Qiu L, Schuler JD et al (2017) Effect of growth temperature on the synthesis of carbon nanotube arrays and amorphous carbon for thermal applications. Phys Status Solidi (a) 214:1600852

84. Szymanski L, Kolacinski Z, Wiak S, Raniszewski G, Pietrzak L (2017) Synthesis of carbon nanotubes in thermal plasma reactor at atmospheric pressure. Nanomaterials (Basel) 7(2):45

85. Sundaram RM, Koziol KK, Windle AH (2011) Continuous direct spinning of fibers of single-walled carbon nanotubes with metallic chirality. Adv Mater 23:5064–5068

86. Seraphin S, Zhou D (1994) Single-walled carbon nanotubes produced at high yield by mixed catalysts. Appl Phys Lett 64:2087
87. Lambert JM, Ajayan PM, Bernier P, Planeix JM, Brotons V, Coq B et al (1994) Improving conditions towards isolating single-shell carbon nanotubes. Chem Phys Lett 226:364–371
88. Huczko A, Lange H, Bystrzejewski M, Baranowski P, Ando Y, Zhao X et al (2006) Effect of graphitization of Fe-doped anode and optical Emision studies. J Nanosci Nanotechnol 6:1319–1324
89. Li J, Kundrapu M, Shashurin A, Keidar M (2012) Emission spectra analysis of arc plasma for synthesis of carbon nanostructures in various magnetic conditions. J Appl Phys 112:024329
90. Journet W, Maser K, Bernier P, Loiseau A, de la Chapelle ML, Lefrant S et al (1997) Large-scale production of single-walled carbon nanotubes by the electric-arc technique. Nature 388:756–758
91. Saito Y, Tani Y, Miyagawa N, Mitsushima K, Kasuya A, Nishina Y (1998) High yield of single-wall carbon nanotubes by arc discharge using Rh–Pt mixed catalysts. Chem Phys Lett 294:593–598
92. Itkis ME, Perea DE, Niyogi S, Rickard SM, Hamon MA, Hu H et al (2003) Purity evaluation of as-prepared single-walled carbon nanotube soot by use of solution-phase near-IR spectroscopy. Nano Lett 3:309–314
93. Itkis ME, Perea DE, Niyogi S, Love J, Tang J, Yu A et al (2004) Optimization of the Ni-Y catalyst composition in bulk electric arc synthesis of single-walled carbon nanotubes by use of near-infrared spectroscopy. J Phys Chem B 108:12770–12775
94. Hinkov I, Grand J, Lamy De La Chapelle M, Farhat S, Scott CD, Nikolaev P et al (2004) Effect of temperature on carbon nanotube diameter and bundle arrangement: microscopic and macroscopic analysis. J Appl Phys 95:2029–2037
95. Farhat S, Lamy De La Chapelle M, Loiseau A, Scott CD, Lefrant S, Journet C et al (2001) Diameter control of single-walled carbon nanotubes using argon–helium mixture gases. J Chem Phys 115:6752
96. Saito Y, Tani Y, Kasuya A (2000) Diameters of single-wall carbon nanotubes depending on helium gas pressure in an arc discharge. J Phys Chem B 104:2495–2499
97. Waldorff EI, Waas AM, Friedmann PP, Keidar M (2004) Characterization of carbon nanotubes produced by arc discharge: effect of the background pressure. J Appl Phys 95:2749–2754
98. Keidar M, Levchenko I, Arbel T, Alexander M, Waas AM, Ostrikov K (2008) Increasing the length of single-wall carbon nanotubes in a magnetically enhanced arc discharge. Appl Phys Lett 92:043129
99. Keidar M, Levchenko I, Arbel T, Alexander M, Waas AM, Ostrikov KK (2008) Magnetic-field-enhanced synthesis of single-wall carbon nanotubes in arc discharge. J Appl Phys 103:094318
100. Su Y, Zhang Y, Wei H, Qian B, Yang Z, Zhang Y (2012) Length-controlled synthesis of single-walled carbon nanotubes by arc discharge with variable cathode diameters. Physica E 44:1548–1551
101. Takizawa M, Bandow S, Torii T, Iijima S (1999) Effect of environment temperature for synthesizing single-wall carbon nanotubes by arc vaporization method. Chem Phys Lett 302:146–150
102. Roch A, Jost O, Schultrich B, Beyer E (2007) High-yield synthesis of single-walled carbon nanotubes with a pulsed arc-discharge technique. Phys Status Solidi (b) 244:3907–3910
103. Ishigami M, Cumings J, Zettl A, Chen S (2000) A simple method for the continuous production of carbon nanotubes. Chem Phys Lett 319:457–459
104. Antisari MV, Marazzi R, Krsmanovic R (2003) Synthesis of multiwall carbon nanotubes by electric arc discharge in liquid environments. Carbon 41:2393–2401
105. Ryzhkov VA (2002) Carbon nanotube production by a cracking of liquid hydrocarbons. Phys B Condens Matter 323:324–326
106. Zhao X, Kadoya T, Ikeda T, Suzuki T, Inoue S, Ohkohchi M et al (2007) Development of Fe-doped carbon electrode for mass-producing high-yield single-wall carbon nanotubes. Diam Relat Mater 16:1101–1105

107. Kroto HW, Heath JR, O'Brien SC, Curl RF, Smalley RE (1985) C60: Buckminsterfullerene. Nature 318:162–163
108. Hornbostel B, Haluska M, Cech J, Dettlaff U, Roth S (2006) Arc discharge and laser ablation synthesis of single-walled carbon nanotubes. NATO Science Series II 2:1–19
109. Bandow S, Asaka S, Saito Y, Rao AM, Grigorian L, Richter E et al (1998) Effect of the growth temperature on the diameter distribution and chirality of single-wall carbon nanotubes. 80:3779–3782
110. Thess A, Lee R, Nikolaev P, Dai H, Petit P, Robert J et al (1996) Crystalline ropes of metallic carbon nanotubes. Science 273:483–487
111. Rinzler AG, Liu J, Dai H, Nikolaev P, Huffman CB, Rodriguez-Macias FJ et al (1998) Large-scale purification of single-wall carbon nanotubes: process, product, and characterization. Appl Phys A Mater Sci Process 67:29–37
112. Kingston CT, Jakubek ZJ, Dénommée S, Simard B (2004) Efficient laser synthesis of single-walled carbon nanotubes through laser heating of the condensing vaporization plume. Carbon 42:1657–1664
113. Jost O, Gorbunov A, Liu X, Pompe W, Fink J (2004) Single-walled carbon nanotube diameter. J Nanosci Nanotechnol 4:433–440
114. Ruemmeli MH, Kramberger C, Loeffler M, Jost O, Bystrzejewski M, Grueneis A et al (2007) Catalyst volume to surface area constraints for nucleating carbon nanotubes. J Phys Chem B 111:8234–8241
115. Maser WK, Munoz E, Benito AM, Martinez MT, de la Fuente GF, Maniette Y et al (1998) Production of high-density single-walled nanotube material by a simple laser-ablation method. Chem Phys Lett 292:587–593
116. Maser WK, Benito AM, Munoz E, de Val GM, Martinez MT, Larrea A et al (2001) Production of carbon nanotubes by CO2-laser evaporation of various carbonaceous feedstock materials. Nanotechnology 12:147–151
117. Braidy N, El Khakani MA, Botton GA (2002) Single-wall carbon nanotubes synthesis by means of UV laser vaporization. Chem Phys Lett 354:88–92
118. Kusaba M, Tsunawaki Y (2006) Production of single-wall carbon nanotubes by a XeCl excimer laser ablation. Thin Solid Films 506-507:255–258
119. Kusaba M, Tsunawaki Y (2007) Raman spectroscopy of SWNTs produced by a XeCl excimer laser ablation at high temperatures. Appl Surf Sci 253:6330–6333
120. Arepalli S, Scott CD (1999) Spectral measurements in production of single-wall carbon nanotubes by laser ablation. Chem Phys Lett 302:139–145
121. Kokai F, Takahashi K, Yudasaka M, Iijima S (2000) Laser ablation of graphite-co/Ni and growth of Single-Wall carbon nanotubes in vortexes formed in an Ar atmosphere. J Phys Chem B 104:6777–6784
122. Puretzky AA, Geohegan DB, Fan X, Pennycook SJ (2000) In situ imaging and spectroscopy of single-wall carbon nanotube synthesis by laser vaporization. Appl Phys Lett 76:182
123. Puretzky A, Schittenhelm H, Fan X, Lance M, Allard L, Geohegan D (2002) Investigations of single-wall carbon nanotube growth by time-restricted laser vaporization. Phys Rev B 65
124. Geohegan DB, Puretzky AA, Styers-Barnett D, Hu H, Zhao B, Cui H et al (2007) In situ time-resolved measurements of carbon nanotube and nanohorn growth. Phys Status Solidi (b) 244:3944–3949
125. Puretzky AA, Styers-Barnett DJ, Rouleau CM, Hu H, Zhao B, Ivanov IN et al (2008) Cumulative and continuous laser vaporization synthesis of single wall carbon nanotubes and nanohorns. Applied Physics A 93:849–855
126. Cochon JL, Gavillet J, de la Chapelle ML, Loiseau A, Ory M, Pigache D (1999) A continuous wave CO2 laser reactor for nanotubes synthesis. In: Kuzmany H, Fink J, Mehring M, Roth S (eds.), AIP conference proceedings, electronic properties of novel materials– Science and technology of molecular nanostructures, vol. 486, pp. 237–240,

127. Dorval N, Foutel-Richard A, Cau M, Loiseau A, Attal-Trétout B, Cochon JL et al (2004) In-Situ optical analysis of the gas phase during the formation of carbon nanotubes. J Nanosci Nanotechnol 4:450–462
128. Cau M, Dorval N, Attal-Trétout B, Cochon JL, Foutel-Richard A, Loiseau A et al (2010) Formation of carbon nanotubes: in situ optical analysis using laser-induced incandescence and laser-induced fluorescence. Phys Rev B 81:165416
129. Dresselhaus MS, Dresselhaus G, Sugihara K, Spain IL, Goldberg HA (1988) Synthesis of graphite fibers and filaments. In: Graphite fibers and filaments, springer series in materials science, vol 5. Springer, Berlin/Heidelberg, pp 12–34
130. Satishkumar BC, Govindaraj A, Sen R, Rao CNR (1998) Single-walled nanotubes by the pyrolysis of acetylene-organometallic mixtures. Chem Phys Lett 293:47–52
131. Bladh K, Falk LKL, Rohmund F (2000) Onthe iron-catalysed growth of single-walled carbon nanotubes and encapsulated metal particles in the gas phase. Appl Phys A Mater Sci Process 70:317–322
132. Ci L, Wei B, Liang J, Xu C, Wu D (1999) Preparation of carbon nanotubules by the foating catalyst method. J Mater Sci Lett 18:797–799
133. Nikolaev P, Bronikowski MJ, Bradley RK, Rohmund F, Colbert DT, Smith KA et al (1999) Gas-phase catalytic growth of single-walled carbon nanotubes from carbon monoxide. Chem Phys Lett 313:91–97
134. Bronikowski MJ, Willis PA, Colbert DT, Smith KA, Smalley RE (2001) Gas-phase production of carbon single-walled nanotubes from carbon monoxide via the HiPco process: a parametric study. J Vac Sci Technol A: Vac Surf Films 19:–1800
135. Nikolaev P (2004) Gas-phase production of single-walled carbon nanotubes from carbon monoxide: a review of the HiPco process. J Nanosci Nanotechnol 4:307–316
136. Saito T, Ohshima S, Okazaki T, Ohmori S, Yumura M, Iijima S (2008) Selective diameter control of single-walled carbon nanotubes in the gas-phase synthesis. J Nanosci Nanotechnol 8:6153–6157
137. Nasibulin AG, Moisala A, Brown DP, Jiang H, Kauppinen EI (2005) A novel aerosol enlargethispage*-12ptmethod for single walled carbon nanotube synthesis. Chem Phys Lett 402:227–232
138. Nasibulin AG, Shandakov SD, Timmermans MY, Kauppinen EI (2011) Aerosol synthesis and applications of single-walled carbon nanotubes. Russ Chem Rev 80:771–786
139. Chiang W-H, Sankaran RM (2009) Linking catalyst composition to chirality distributions of as-grown single-walled carbon nanotubes by tuning Ni_xFe_{1-x} nanoparticles. Nat Mater 8:882–886
140. Chiang W-H, Sakr M, Gao XPA, Sankaran RM (2012) Nanoengineering Ni_xFe_{1-x} catalysts for gas-phase, selective synthesis of semiconducting single-walled carbon nanotubes. ACSnano 3:4023–4032
141. Mikhalchan A, Fan Z, Tran TQ, Liu P, Tan VBC, Tay T-E et al (2016) Continuous and scalable fabrication and multifunctional properties of carbon nanotube aerogels from the floating catalyst method. Carbon 102:409–418
142. Tavoulareas ES (1991) Fluidized-bed combustion technology. Annu Rev Energy Environ 16:25–57
143. Hernadi K, Fonseca A, Nagy JB, Bernaerts D, Lucay AA (1996) Fe-catalyzed carbon nanotube formation. Carbon 34:1249–1257
144. Venegoni D, Serp P, Feurer R, Kihn Y, Vahlas C, Kalck P (2002) Parametric study for the growth of carbon nanotubes by catalytic chemical vapor deposition in a fluidized bed reactor. Carbon 40:1799–1807
145. Li Y-L, Kinloch IA, Shaffer MSP, Geng J, Johnson B, Windle AH (2004) Synthesis of single-walled carbon nanotubes by a fluidized-bed method. Chem Phys Lett 384:98–102
146. Yun S, Qian W, Cui C, Yu Y, Zheng C, Liu Y et al (2013) Highly selective synthesis of single-walled carbon nanotubes from methane in a coupled downer-turbulent fluidized-bed reactor. J Energy Chem 22:567–572

147. MacKenzie KJ, Dunens OM, Harris AT (2010) An updated review of synthesis parameters and growth mechanisms for carbon nanotubes in fluidized beds. Ind Eng Chem Res 49:5323–5338

148. Dasgupta K, Joshi JB, Banerjee S (2011) Fluidized bed synthesis of carbon nanotubes – a review. Chem Eng J 171:841–869

149. Kim KS, Seo JH, Nam JS, Ju WT, Hong SH (2005) Production of hydrogen and carbon black by methane decomposition using DC-RF hybrid thermal plasmas. IEEE Trans Plasma Sci 33:813–823

150. Shahverdi A, Kim KS, Alinejad Y, Soucy G (2012) In situ purity enhancement/surface modification of single-walled carbon nanotubes synthesized by induction thermal plasma. J Nanopart Res 14

Chapter 8
Challenges on the Production and Characterization of B-Doped Single Walled Carbon Nanotubes

Paola Ayala

Abstract This chapter is mainly devoted to give a fundamental insight on the concepts behind the wall modification, doping, and general formation of single-walled nanotubes that involve the presence of boron as heteroatoms within the nanotube structure. Research on carbon nanotubes has matured in various fields reaching real possibilities for applications. However, in structures like substitutionally doped nanotubes, the full application potential can only be reached if bonding environments, doping levels and overall morphology can somewhat be controlled. This is not the case for boron doped single-walled carbon nanotubes and it will be taken as example of discussion throughout the following sections. The bulk and local characterization tools employed with these materials are here discussed regarding their suitability and limitations. Furthermore, focusing on applications, the theoretical approaches confirming the physical and chemical properties are objectively analyzed versus the materials available at this moment.

Keywords Single-walled carbon nanotubes · Functionalization · Doping · Spectroscopy

8.1 Introduction

After almost three decades of intensive research on carbon nanotubes, it is clear that their physical properties have placed them at a leading position in the nanotechnology related research. These structures are well–known by their outstanding structural and electronic properties and it is not surprising that significant research has been pointed towards gaining their control via various functionalization methods [1]. In this context, using the word functionalization can involve different conceptualizations when used to describe carbon nanotubes [2]. Here we only

P. Ayala (✉)
Faculty of Physics, University of Vienna, Vienna, Austria
e-mail: paola.ayala@univie.ac.at

© Springer Nature B.V. 2019   145
C. Bittencourt et al. (eds.), *Nanoscale Materials for Warfare Agent Detection: Nanoscience for Security*, NATO Science for Peace and Security Series A: Chemistry and Biology, https://doi.org/10.1007/978-94-024-1620-6_8

focus on wall doping (substitutional doping and vacancies due to the presence of heteroatoms) [3], although it will be useful to have in mind other methods like topological wall changes, exohedral doping (for instance intercalation on bundles), filling (with molecules, molecular structures, nanoclusters, etc.) and the formation of novel heterostructures. Covalent functionalization and molecular wrapping are also extremely important methods, especially for biological applications [4], but this is out of the scope of this chapter.

Turning back to carbon nanotubes, they have been the focus of very intensive research for years but there are some subfields where partial work has been left aside or it has been less intensively investigated for different reasons. For some applications, bulk quantities of the material are required. Especially in those where the tube's mechanical properties play a crucial role. In this context, the production and optimization are extremely important and this has been nicely achieved over the last decades for pristine carbon nanotubes in multi and single-walled form [5]. Fairly assessing substitutionally doped nanotubes, also the research with nitrogen doped tubes has reached some milestones but the control of N content and morphology is still elusive [6–8]. Starting to think about applications with other substitutionally doped tubes is still not realistic. Other elements that are potential candidates as substitutional dopants are Si and P besides B, which follows N from afar [9, 10]. The work with P and Si has been rather limited. Also the reasons why B falls so far behind are not trivial. The research in the field of carbon nanotubes grew exponentially since there discovery until greater focus turned the community's eyes to graphene.

As seen on Fig. 8.1, the very specific and small field devoted to heteronanotubes and doped tubes where B and C play a role has had a very symmetric evolution that went down in activity due to specific reasons related to the production feasibility. If

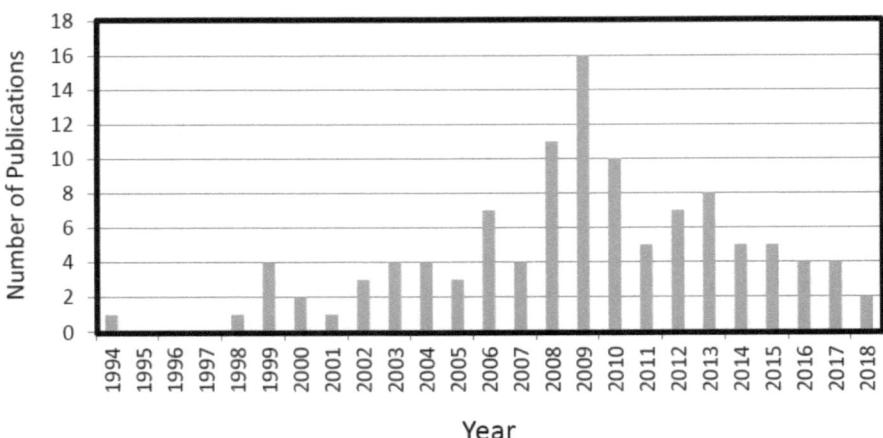

Fig. 8.1 Field activity on synthesis, properties and applications on boron doped nanotubes from the last three decades

factually overcome, these materials could have enormous potential for applications. Boron has a smaller atomic size compared to carbon and intuitively it should be more stable than the other mentioned substitutional dopants. The graphitic wall of the C-SWNTs incorporates the B atoms directly at the C sites without additional atomic reorganization or creation of vacancies. Also nanodomains of BC_3 can be expected. This could be seen as a practical advantage for the development of purification methods (and implicitly for applications) of doped nanotubes in general, which have been very scarcely explored but the first publications towards the development of such methods have started to appear [10, 11]. The research status has started to reach a level at which perspective applications can start to be envisaged.

The next sections of this chapter briefly cover the theoretical concepts and experimental achievements from the research on carbon nanotubes with low B through high substitutional doping (Fig. 8.2a) to the new physics behind boron and carbon heteronanotubes (like the ones with BC_3 stoichiometry-Fig. 8.2b). Furthermore, special attention is given to review the bulk and local characterization tools commonly employed with these materials, their suitability and limitations.

8.2 From Low Doping to Stable Heteronanotubes

The properties of a system like a nanotube depend not only on its dimensionality, but the position of each atom influences the structural behavior as a whole. It is well understood and documented in the literature that the electronic properties of C-SWNTs vary upon their chirality and diameter, rendering metallic or semi-conducting 1D objects. However, when atoms from a different element occupy the C sites a different scenario must be envisaged [1]. In the case of study of this chapter, the stability of boron-doped single-walled carbon nanotubes (CBx-SWNTs) is among their most promising characteristics and accounts for a tube band gap that may be modified independently from its radius and chirality according to the theory [12]. Also the potential enhancement of their mechanical [13, 14], electrical [15], optical [16, 17], transport [18, 19] and chemical properties has been suggested. There are different types of heteronanotubes with a hexagonal wall structure that contain C and B as described in the next section, and these exhibit from slightly modified physical behavior to radically different properties.

Carbon nanotubes and their heteronanotubes also have similarities. The electronic structure of any heteronanotube can be constructed starting from the electronic structure of a graphene single sheet (or the corresponding planar material from the stoichiometric point of view), via the well known zone-folding procedure. Also, the band structure of the tubular material can be approximated by imposing periodic boundary conditions on the band structure of the planar counterpart.

B has one electron less than C. Unlike the N incorporation in C-SWNTs, where two or more bonding environments can be expected (i.e. substitutional, pyridinic, pyrrolic), in the case of boron doped SWNTs, mainly direct substitution

Fig. 8.2 Molecular models of a boron doped single-walled carbon nanotube with a random substitution of B atoms at C sites (**a**), and a BC₃ nanotube (**b**)

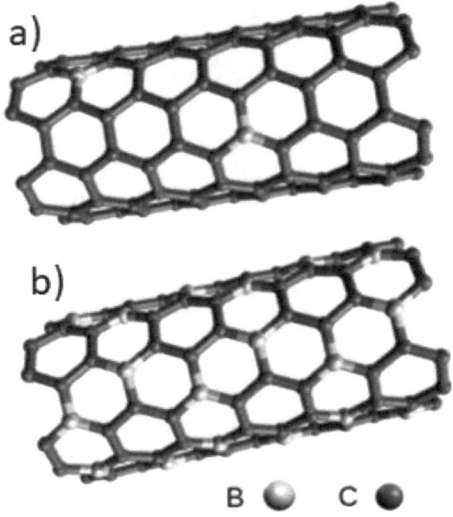

of C atoms with B is expected with a triple coordinated bond. Furthermore, the potential defects on tubes with BC₃ stoichiometry can generate different topological arrangements [20]. It is expected that boron substitutional atoms shift the Fermi level leading in turn to an augmented electronic alteration due the special quantum confinement of the quasi 1D structure. However, this has neither been proved nor quantified because any shift or reorganization of the band structure of the material also depends on the doping levels of B on the carbon nanotube structures.

8.2.1 Low Doping

The transformations occurred on single-walled carbon nanotubes upon the concentration of dopants are scarcely analyzed in the literature . However, this is a decisive parameter for some type of applications. For instance, substitutional doping can introduce strongly localized electronic features in the valence or conduction bands, respectively, and can increase the number of electronic states at the Fermi level depending on the location and concentration of dopants. For instance, substitutional doping by heteroatoms is a well established key technique in Si technology since doping with electron donors or acceptors leads to a very sensitive shift of the Fermi level. In all semiconductor applications the properties of a device depend on the control of the electronic states in the valence and conduction bands for design and optimization purposes. However, the doping levels taken into account for such applications are in the orders of parts per million [21], which allows working under the regime of a rigid band model. Its applicability for the case of nanotubes was proven by field effect doping (p- and n- type) and intercalation with alkali metals even to very high doping levels (n- type) i.e. up to 10%, as well as (p-type) for Br_2, I_2, $FeCl_3$ doping [22–24]. However, the doping levels at which substitutional heteroatoms in CNTs allow the rigid band model to be valid is still unknown. It can most likely be considered in a similar way to doped Si but it is necessary to quantify how much corresponds to a small number of dopant related defect centers. When analyzing the properties of the nanotubes, doping amounts in order of parts per million will not necessary be the range to consider a rigid band model, which assumes that the electronic density of states of a structure can be inferred from that of the host. The goal is that the band structure of the nanotubes is only very weakly modified upon doping so that the state of the electronic system is obtained only by the slight shifts of the Fermi level in the band structure. In other words, the effect of a band structure modification by doping should be almost negligible and the band structure of the doped system should be obtained by simply shifting the Fermi level in the band structure of the pristine system towards the valence or conduction bands respectively.

Nevertheless, carbon nanotubes, which have been very well studied in the last years, are strong candidates for applications in nano-based semiconductor technology. However, an optimized control of the electronic properties is still far from being reached. It is not surprising that the first studies analyzing the doping effects on nanotubes appeared in 1993 [25], not much latter than Iijima's famous first reports on TEM observations were published [26, 27]. Heteroatoms like boron and nitrogen may lead to electronic properties that are more controlled by the chemistry than the specific geometry of the tubes. Keeping this concept in mind, the heteroatoms inserted in the single-walled carbon nanotube (C-SWNT) structure must not only be present up to certain doping levels but their bonding environments must be controlled. This will be discussed later in this chapter.

8.2.2 High Doping and Heteronanotubes

Setting an offset for a nanotube to be considered highly doped is still done by a rule of thumb. However, in general terms higher doping, implies modifications in the density of states where the localization, local concentration and overall amounts of substitutional atoms have to be taken into account. This means that a simple rigid band shift model is no longer applicable because the changes in of the electronic properties are not anymore inferrable from the pristine system, or because novel heteronanotubes are born. The B content in carbon nanotube samples reported within the years of research on single-walled boron doped carbon nanotubes (CBx-SWNTs) range from 0.1% to nearly 4%. However, two ambiguities have not been ruled out to date because the literature on this topic does not disregard synthesis byproducts that can also incorporate part of the boron atoms. Whatever the case may be, it is important to recall that for pristine C-SWNTs the density of states is symetric around the band gap for the first and second van Hove singularities, whereas the introduction of B generates asymmetries that appear around the band gap due to the mixing of the p and s orbitals (Fig. 8.3). When high doping occurs

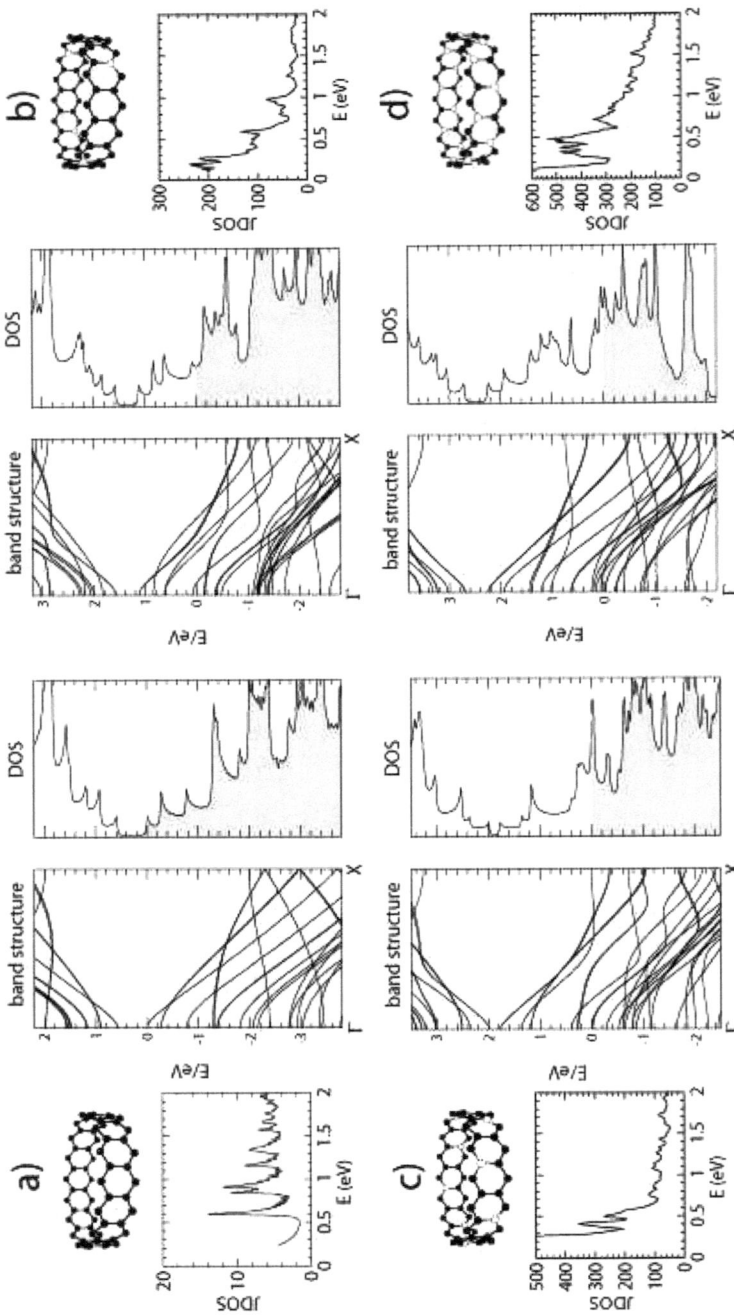

Fig. 8.3 The electronic configuration of B-doped nanotubes depends clearly on the amount of heteroatomic incorporation in the nanotube structure. The figure depicts the band structure and DOS of a C(16,0) tube with (**a**) 0%, (**b**) 6.25%, (**c**) 12.5%, and (**d**) 25% boron doping. Zero energy denotes the Fermi edge. Filled states are indicated by grey shadowing. (Reproduced from [33])

this is critical because doping induces a lowering of the Fermi level into the valence band of the pristine tube and this can take very different shapes.

One can think about increasing numbers of boron atoms in the structure until a homogeneous stoichiometry along the tube like BC_3 is reached (Fig 8.2b). Such a nanotube is not a highly B-doped SWNTs strictly speaking, it is a heteronanotube. BC_3 can take the form of a planar graphite-like stable sheet. In this sense, the tubular counterparts can be regarded within the framework of heteronanotubes rather than doped nanotubes. On these grounds, it is reasonable to expect incorporation of B beyond 15% as the experimental reports show, disregarding the solubility limits for boron in carbons as some authors suggest [28, 29]. The synthesis of BC_3-SWNT has been reported using substitution reactions [30–32]. Higher local boron concentration observed on selected transmission electron microscopy (TEM) spots have been described, which is consequent with the total amounts of B identified on such samples where pristine C nanotubes are the starting material. Furthermore, different concentrations of BC_3 tubes at different substitution levels have been studied. Results on BC_3 nanotubes reported in [33], have shown the formation of a uniform energy gap of 0.4 eV for B doped nanotubes at very high doping. They confirmed the existence of BC_3 nanotubes owing an acceptor-like band of ~0.1 eV above the Fermi level. The minimum energy gap predicted was of about 0.2 eV [34].

8.2.3 Heteronanotubes

There are other types of single-walled nanotubes where C and its closest neighbors B and N, and their combination generate interesting properties. Among the B, C and N combinations that crystalize in graphite-like planar structure is BC_2N (see Fig. 8.4a). This is one of the most stable geometries. Miyamoto [35] performed first-principles and tight-binding band-structure calculations to analyze the stability of families of nanotubes originating from hexagonal stable BC_2N sheets [36]. In these studies a preferential chiral nanotube formation was reported. It is expected that these systems show intermediate stabilities between those of graphite and hexagonal boron nitride. The BC_2N arrangement is one among other possibilities of heteronanotubes of B, C and N that have been investigated (see Fig. 8.4b). Furthermore, first principles calculations have suggested that a structure with $B_3C_2N_3$ could be more stable than BC_2N. Nevertheless, there is still lacking experimental evidence for this tubular structure.

8.2.4 Nanodomains

While applying different synthesis methods aiming at the incorporation of dopants, the possibility that only regions incorporate the heteroatoms is certainly high. Carroll et al. [37] proposed that B substitution in CNTs leads to the formation of

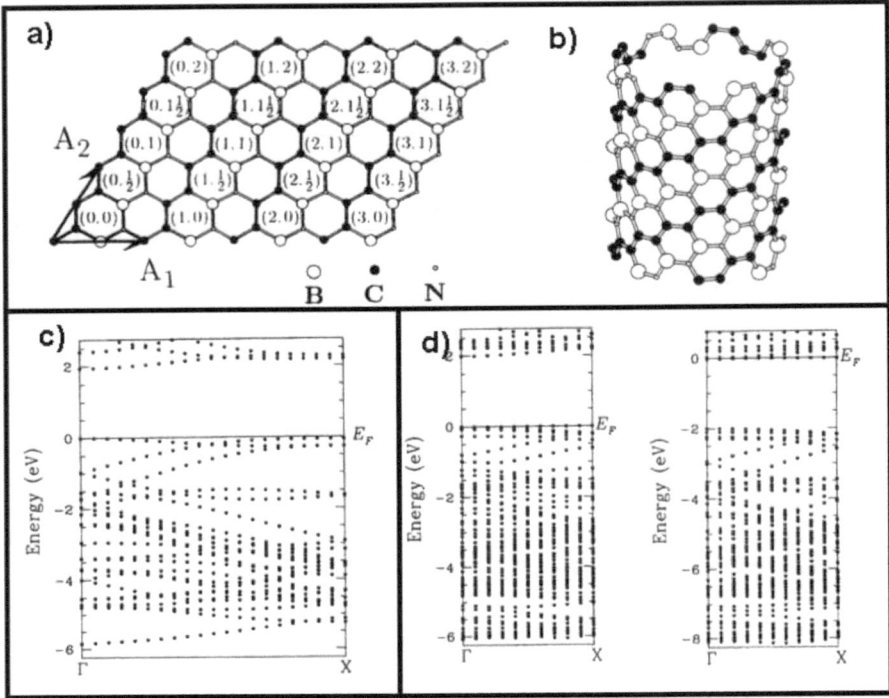

Fig. 8.4 One possible configuration of a BC_2N sheet (**a**) and the corresponding $BC_2N(4,4)$ semiconducting nanotube obtained rolling the hexagonal stable structure (**b**). The lower panels correspond to the band structure of $(2,2)BC_2N$-SWNTs (**c**) and the cases where a C atom is substituted as impurity on the N site(left-(**d**)) and the B site (right-(**d**)). (From [1, 35])

BC_3 nanodomains (see Fig. 8.5). The formation of these island-type arrangements was reported to give rise to an acceptor state near the Fermi level due to the realignment of the superstructure with the pristine graphitic network (see Fig. 8.3). Nanodomains of B along any CBx-SWNT cannot be discarded. Regarding the changes induced in the electronic structure, only the low concentrations of B atoms should lead to the formation of the acceptor state. Further studies by Wirtz et al. [38] discussed the evolution of this acceptor-like level with the degree of B-doping. The authors suggested recalled the importance related to low doping and high doping, which in this case also entails critical differences in the physical properties (Fig. 8.3).

8.3 Synthesis

Hundreds of papers related to the synthesis, purification and separation according to metallicity of carbon single-walled nanotubes have appeared to date. Although some

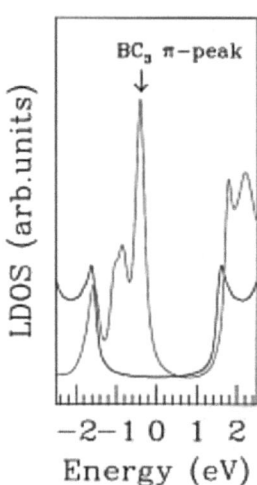
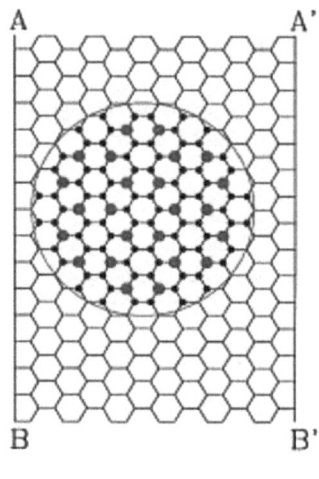

Fig. 8.5 The theoretical LDOS of BC$_3$ nanodomain against LDOS associated to a pure C region. (From Carrol et al. [37].)

practical issues still exist, a great number of methods have been optimized [39]. The state of the research, if we only take into account the carbon tubular structures, demands now high quality material rather than mass production because some applications indeed require large amounts of relatively clean material. At the beginning of this decade the CoMoCAT method using bimetallic catalyst was able to produce samples rich in (6,5) and (7,5) chiralities [40, 41], followed by the introduction of ultracentrifugation and electrophoresis, which made possible the separation of single walled carbon nanotubes according to metalicity [42–44].

The tremendous improvements on the synthesis of doped carbon nanotubes and heteronanotubes (not only of C and B, but even considering N) have not gone side by side with the quality improvements for pristine material. Most experimental studies started late in the 1990s, when several from the now well known techniques applied for the synthesis of pristine C-NTs had been at least tested straight after the first reports from Iijima in 1991 and 1993 correspondingly. What made things complicated was mainly the difference in the reaction environment where these different kinds of nanotubes grow. The production of multi-walled N and B doped nanotubes was already achieved in 1994 [45] but developing methods to produce single-walled species has required significantly more time [1]. Of course there are drawbacks encountered in each technique. In general terms the synthesis and use of multi-walled nanotubes vs single-walled equivalents must be treated separately. Review features and contributions to collected editions, which cover thoroughly doped multi-walled tubes topic have appeared in the literature [29, 46, 47]. This whole chapter is mainly oriented to the single-walled nanotube case with special attention on B doping and the scope of heteronanotubes associated to these atoms.

As above mentioned, the rich literature found for pristine tubes is by far more extensive than everything found for the case here discussed. A representative sample of the synthesis with the different techniques is seen in Table 8.1. It is the best feasible attempt to summarize each B-doped nanotube material alike that was first synthesized per technique or inspired later work published in the literature. Note

Table 8.1 Sample of the work published to date related to the production of B-doped carbon nanotubes sorted by the different methods. Note that those reports on tubes containing B, C and N have intentionally been excluded since it is not straightforward to discern the pure formation of CB_x structures. The maximum boron content that is incorporated in the samples is shown in the table

Arc discharge			
Material	B source	B content	Reference
CB_x-MWNTs	B in electrodes + N2 atmosphere	Possible BCN	Stephan et al. [45]
BC_3-NTs	B in electrodes	Possible BC_3	Weng-Sieh et al. [48]
CB_x-MWNTs	BC4N	1–5	Redlich et al. [37, 49–51]
CB_x-MWNTs	B in electrodes	Undefined	Babanejad et al. [52]
CB_x-SWNTs	B in electrodes	–	Wang et al. [53]
Substitution reactions			
Material	B source	Max.B content	Reference
CB_x-MWNTs	CNT+BxC	0.02–0.1	Han et al. [54, 55]
CB_x-MWNTs	CNT+B2O3	0.4–3.92	Chiang et al. [56]
CB_x-SWNTs	BNNT + B2O3	0.1–10	Golberg et al. [57]
CB_x-SWNTs	SWCNT + B2O3	BC_3 tubes	Borowiak-Palen et al. [30]
Chemical vapor deposition			
Material	B source	Max.B content	Reference
CB_x-MWNTs	C3H9B	0.5–2	Chen et al. [58]
CB_x-MWNTs	Methane+Diborane	1	Panchakarla et al.
CB_x-MWNTs	Toluene+C3H9B	1	Koos et al. [59]
CB_x-MWNTs	Acetylene+Diborane	1	Panchakarla et al. [60]
CB_x-MWNTs	Tetrahydrofuran+Triisopropyl Borate	0.5–3.1	Lyu et al. [61]
CB_x-MWNTs	Triethyl Borate	0.46	Cao et al [62].
CB_x-MWNTs	B(C6H5)3	0.4–2.5	Keru et al. [63]
CB_x-SWNTs	Triisopropyl Borate	1	Ayala et al. [64–66]
CB_x-SWNTs	Triisopropyl Borate	0.5	Ruiz-Soria et al. [11]
CB_x-SWNTs	Triethyl Borate	1.2	Monteiro et al. [67, 68]
CB_x-SWNTs	Triethyl Borate	–	Ruiz-Soria et al. [69]
Laser Ablation			
Material	B Source	Max.B content	Reference
CB_x-SWNTs	C-Paste + elemental B	1.5–3	Gai et al. [70]
CB_x-SWNTs	C-Paste + elemental B	1.5–10	Mc Guire et al. [71]
CB_x-SWNTs	C-SWNTs + NiB	1.8	Blackburn et al. [72]
CB_x-SWNTs	C-Paste + elemental B	0.3	Ayala et al. [73]
CB_x-SWNTs	C-SWNTs + B2O3	3	Panchakarla et al. [60]

that this is a very short list although it actually includes reports related to CBx-MWNTs. It is clearly inferable that this low amount of reports must be associated to difficulties during production. From the solubility of B in C, to the competing byproduct on each synthesis method, many variables seem to have played a role on the different techniques to hinder further studies. The production of CBx-SWNTs has drawbacks associated to each specific technique but the difficulties have been even more evident in chemical vapor deposition related methods due to costs and possibilities for up-scalability that are in fact highly desirable.

8.3.1 Arc-Discharge

Arc discharge is a high temperature method based on the ignition of an arc between two graphite electrodes in a gaseous background. The arcing evaporates the carbon and it condenses during the cooling process. The first multiwalled nanotubes containing carbon, boron, and nitrogen were grown with arc discharge. Graphite cathodes and amorphous boron-filled inside graphite anodes in a nitrogen atmosphere were used in those first experiments [45]. It is well known that the derived product and the yields are mainly dependent on the atmosphere and catalysts utilized. In order to obtain doped nanotubes two modifications have been applied: the use of solid mixtures in the electrodes with compounds containing the desired doping agent [74], and experiments using pure graphite electrodes in an gaseous atmosphere [75]. As far as one can tell from the literature, even though arc discharge has been used in attempts to produce CBx-MWNTs, no experimental report of B doped CBx-SWNTs is available, as also summarized in Table 8.1. Nevertheless, there is a very important fact to consider related to this method. The experiments have attempted a higher incorporation of the dopant upon higher relative concentrations of B (and N) in the electrodes, but it still remains an open question if the incorporation of heteroatoms in the nanotube walls corresponds linearly to the electrodes composition.

8.3.2 Substitution Reactions

This technique, also known as carbothermal process, was envisioned keeping in mind the structural similarities of the h-BN and graphite networks. The substitution reactions employ carbon nanotubes as template materials in which presence a chemical reaction should occur. In fact, the carbothermal process is a modification of one of the most widely used synthesis methods to obtain h-BN, which comes from the reaction boron oxide (or boric acid) with ammonia, at temperatures close to 900 °C. In the case of nanotubes, this "substitution reaction" assumes a continuous atomic substitution between C and the heteroatoms (B and N) present in the reaction. Han et al. [76] were the first to successfully achieve the production of BCN-MWNTs. These experiments were performed with mixtures of B_2O_3 and C-MWNTs at elevated temperatures (1200–1700 °C) in N_2 and NH_3 atmospheres. Shortly after this, Golberg et al. [77] reported the first results of BN-SWNTs produced with this method. A later study from the same group [32] reported a large-scale synthesis and HRTEM analysis of B- and N-doped C-SWNT bundles where up to 10at% of B and 2at% of N were detected.

Later, Borowiak-Palen et al. [31] employed a substitution reaction similar to the one employed by [32]. They used bundles of C-SWNTs as templates and heated them in presence of boron oxide at 1150 °C whilst a flux of NH_3 was continuously used as etching agent. They observed a high content of boron (between 10% and 15%) which was significantly higher than the content reported by [32]. At that time it was believed that such a high B content could not be possible given the low solubility of B in C. However, this was further studied by the same group [33] in a combined theoretical and experimental study, where the authors were able to demonstrate that the stability of those allegedly CBx-SWNTs was justified by the actual formation of other type of stable tubes. The high content of B in the nanotube network was a result of the formation of the BC_3-nanotubes. Indeed, it was not adequate to qualify those nanotubes as boron-doped tubes but as BC_3 heteronanotubes as already explained in Sect. 8.2.

8.3.3 Chemical Vapor Deposition

One of the most popular methods explored over the last three decades is chemical vapor deposition (CVD), but for B-doped nanotubes it has faced very important difficulties compared to processes aimed for the synthesis of pure C nanotubes. Whereas CVD is seen as one of the most cost efficient and upscalable techniques from the technological point of view, it has had major problems producing doped material when B is involved in the process. The chemistry associated to the ceramic materials used for the furnaces in presence and diffusion of B at the required high synthesis temperatures changes radically, making their melting temperature lower. In Table 8.1, the precursors used for CVD require enough lab safety conditions and this is likely to be one of the major reasons why synthesis using this method has not been inspected as a large scale procedure. Nevertheless, reports on high vacuum chemical vapor deposition (HV-CVD) have shown so far that using adapted methods with low vapor pressure precursors make this synthesis feasible [64, 65, 68]. To date two liquid C/B feedstocks have been successfully used to make CBx-SWNTs, with yield limitations. It has been found that although high quality C-SWNTs can be formed in a wide temperature range, B does not always get incorporated in the carbon nanotube network. According to the synthesis temperatures, boron carbide and boron oxide can have more than 80at% of the B content in a sample. Other competing byproduct is boric acid. These two are consequently observed also in samples produced form substitution reactions [30, 65]. It is also worth mentioning that making CBx-MWNTs is not much easier. Adapted mechanisms have had to be

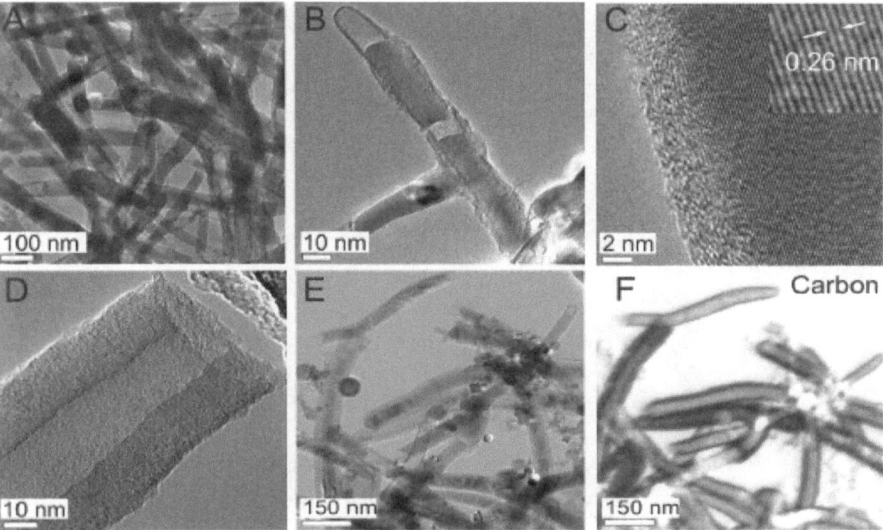

Fig. 8.6 Multiwalled B-doped nanotubes grown via CVD using a templated ceramic instead of traditional metallic catalysts in the process [78]

used for these purpose as the material synthesized by a CVD modification where a templated ceramic had to be designed to assist the growth instead of using traditional metallic catalysts (see Fig. 8.6) [78]

8.3.4   Laser Ablation/Vaporization

The laser ablation technique uses either pulsed or continuous lasers to vaporize a target inside a furnace at high temperatures [79]. The conventional graphite targets are usually mixed with a small amount of a transition metal catalyst in order to condense the SWNTs in presence of different inert gas environments that keep the pressure of the reaction chamber at certain pressure levels. In this process, all the parameters can be varied in order to obtain either doped C-SWNTs or heteronanotubes.

In the work reported with this method, the ablation targets have been prepared with different kinds of compounds containing the dopant agents. Given the technological difficulties to detect low dopant concentrations in nanotube samples, the first estimations of the B dopant concentration when using this method were directly related to the target mixtures [70, 71]. A sample of the morphology of laser ablation reported in [71] is shown in Fig. 8.7. At the beginning, laser ablation was the only technique to produce CBx-SWCNTs. The introduction of B into graphite-Co-Ni targets made possible the formation of single walled nanotubes [70], which overall

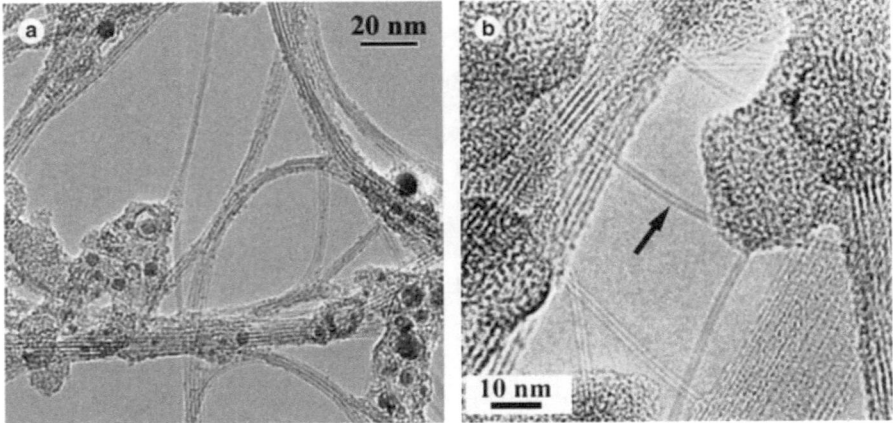

Fig. 8.7 Laser ablation B-doped nanotubes. The TEM images correspond to material produced from targets with nominal boron concentrations of (**a**) 0at.%, (**b**) 2.5at.%. (Reprinted from [71], Copyright (2005), with permission from Elsevier)

morphology can be seen in Fig. 8.7. Although the single-walled material is very nice, as well as for the previously described methods, the difficulty of incorporating B in the graphitic network was decisive. Using analytical transmission electron microscopy they found minimal amounts of B although when the B content in the target material was up to 3at%. For higher B concentrations in the graphite target (3.5at%), graphite and metal-encapsulated particles appeared and the actual growth of SWNTs was hindered.

8.4   Structural Characterization

With the availability of various microscopy and spectroscopy tools, the structural characterization of CBx-SWNTs has evolved with very interesting steps, mostly linked to the development of the techniques. Transmission electron microscopy (TEM) offers the possibility to study the morphology and atomic structure of nano-objects and the latest years advances have brought along significant improvements on the spatially resolved associated techniques converting this instrument into a very powerful analytic tool. Recalling the material published on B-doped nanotubes shown in Table 8.1, ideally, besides using TEM as a tool to see the morphology of the tubes, identifying at local level the position of the dopants would be ideal but none of those works have succeeded on such task. One of the main problems when working with such structures is the sample damage due to its sensitivity to the focused electron beam and dedicated microscopes are necessary. An alternative to this has already shown to be successful to study doped graphene and corresponding hybrids, which corresponds to contrast imaging. However, for substitutionally doped nanotubes, the image contrasts could not be proved so far to be sensitive enough to the different nature of substitutional heteroatoms such as boron, since they also behave as weak phase objects and display a phase contrast very close to that of carbon. This gets additionally more complicated in small diameter tubes taking into account their structural curvature.

Scanning probe techniques such as atomic force microscopy as well as scanning tunneling microscopy (spectroscopy) allow to get atomic resolution and to map the electronic structure of the nanotube geometry. Scanning tunneling spectroscopy has been used to determine the morphology and the local electronic structure of carbon nanotubes [80–85]. Additionally, the presence of local defects or atoms can be identified tracking changes in the local density of states. Typical examples are the correlation between the energy gap in semiconducting CNT and the tube diameter [86]. In fact, the great advances in simulation of STM imaging for carbon nanotubes [87, 88] represent also a great advantage in the inspection of wall imperfections and their effects on transport properties [81]. However, this has also been unsuccessful for substitutionally doped tubes mainly given the technical challenges of localizing the doped tubes because of the low atomic incorporation.

From the spectroscopy point if view, optical absorption, photoluminescence, and Raman spectroscopy can provide a quick and reliable characterization in terms structure, morphology and physical properties in general. For instance, an overall idea of the diameter distribution and population can be extracted from these techniques, even though they are able to provide a lot more information on the electronic and optical properties of the materials as explained in next section.

8.5 Optical Spectroscopy

For this section it is understood that the reader has prior knowledge of the use of optical spectroscopy techniques and their most important characteristics associated to the studies on carbon nanotubes.

Optical absorption (OA) spectroscopy allows extracting different pieces of information from the quality of a nanotube sample. For instance, the area under the first interband transition peak of the C-SWNT in the OA spectrum provides a direct measurement of the nanotube abundance. The OA spectrum provides the optical transitions between the van Hove singularities of the valence and conduction bands in the density of states, which are in turn correlated with the tube's diameter and metallicity. Although very few has been reported on OA of CBx-SWNTs, interesting features have been seen such as an acceptor-like band [30] in samples with high boron content. As mentioned in previous sections, those samples most likely contain a great portion of BC_3 nanotubes. For such heteronanotubes it has been reported a uniform energy gap of about 1.8 eV and an optically forbidden gap around 0.4–0.5 eV (see Fig. 8.8). The first optically allowed transition is at about 1.8 eV, similar to hexagonal BC_3 sheets [34]. The fact that an optically forbidden transition is observed was explained by symmetry breaking due to defects which

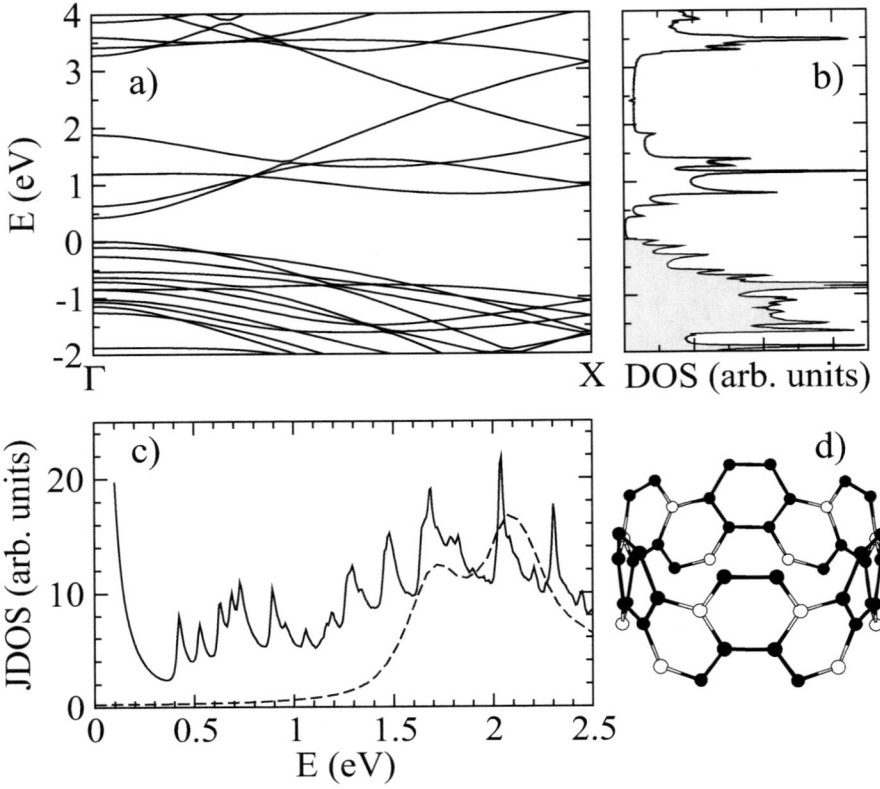

Fig. 8.8 (**a**) Band structure of BC$_3$(3,3) nanotube (Fermi energy at 0 eV). (**b**) Density of states (filled states shadowed with grey). (**c**) Joint density of states divided by E^2 (solid line) and calculated optical absorption (dashed line) of a perfect infinite BC$_3$(3,3) tube with light polarization along the tube axis. (**d**) Sketch of the unit-cell. The black dots represent C atoms. (Reproduced from [33])

are created during the substitution process. Therefore, due to the strong dependence of the band structure on the degree of substitution and the particular arrangement of the boron atoms, Fuentes et al. [33] concluded that those OA spectra could be best explained assuming that only BC$_3$ nanotubes were preferentially formed for such high boron substitution. This means that OA spectroscopy is a clear proof that at such high B contents, BC$_3$ nanotubes are most likely to form rather than B-doped C-SWNT with random B dopant distribution as suggested. In the same contribution from Fuentes et al., information about the dielectric function was retrieved from the normalized optical absorption of both B doped C-SWNT and pristine C-SWNT between 0.05 and 2.5 eV. It has been observed that metallic tubes are less affected by the substitution reaction than the semiconducting ones.

Additionally to OA spectroscopy, photoselective resonance Raman spectroscopy and photoluminescence spectroscopy are among the most powerful tools to analyze C-SWNTs [41, 89] as fast and reliable tools to asses the structural, vibrational and electronic properties. The strong characteristic Raman response of C-SWNTs as well as the strong near infrared luminescence of individual semiconducting C-SWNTs should imply that defects are easily detected in Raman spectroscopy because they locally break the symmetry of the crystal and relax momentum conservation rules that govern Raman scattering processes. Additionally, the intensity of the D-band is often used to estimate defect densities. Caudal et al. [90] proposed a model to determine the doping level from the value of Kohn anomaly (KA) induced frequency shift and from the line width of the G- peak and suggested that the KA should be observed in the D, 2D and 2G peaks. Maciel et al. [91] suggested that the 2D peak in p- and n- doped nanotubes could be a method to identify doping via the splitting of the 2D peak at higher or lower energy values in the Raman shift according to the nanotubes charge state. Additional results by McGuire et al. [71] have shown analysis of the shift of the G-line. However, the few results on B-doped SWNT at low doping by [71] suggest that the G-line is hardly affected by the B doping. In the case of these samples, no local proof of doping has been reported so far, so the Raman studies on these tubes still cannot be considered as a pattern for other doped tubes. Regarding the G' the publication of Maciel et al. [91] also reported near-field measurements on B doped SWNT unpurified bundles where they showed that a split of the 2D peak is upshifted with p-type doping. However, this has not been reproduced with any other type of sample alike. On the other hand, in BC_3 nanotubes the features in Raman are somewhat different since p-type doping usually leads to a blue shift of the first order tangential modes. However, no significant changes have been seen in Raman. One explanation is the loss of the resonance of these tubes and in Raman the preferential observation of the remaining thin diameter SWCNT, in full agreement with the optical observations. Only a subset of tubes are B rich, while the other tubes remain almost pure carbon tubes. Nevertheless, further studies sensitive to the electronic properties of either metallic or semiconducting tubes should be carried out in order to understand the observed differences.

8.6 Boron Content and Bonding Environments

The identification of the B atoms that get incorporated in a sample can be done at local and bulk level. However, this has critical conditions imposed quality of the sample and the byproducts. The technique to be used for such purposes must carefully be understood instead of using it as a tool to generate a number that should allegedly represent the content of foreign atoms. From the spectroscopy point of view, the sensible methods to carry out elemental analysis are electron energy-loss spectroscopy (EELS) and X-ray photoelectron spectroscopy (XPS). X-ray absorption spectroscopy (XAS) can also be used but due to the requirement of a variable energy X-ray source it imposes limitations for lab-based research so studies with this technique are scarcely reported. In all these techniques it is important to take into account instrumental precautions as it is to understand the underlying physical processes behind the method's choice. In bulk systems finding dopants is not anymore technologically limited. However, in molecular nanostructures the determination of doping amounts and dopant distribution is not a straightforward task and this is why most of the initial studies listed in Table 8.1 used either Raman as indirect proof of characterization or assumed boron incorporation in proportional amounts to the B incorporation they used on feedstocks for CVD, targets for laser ablation or electrodes for arc-discharge.

It is necessary to keep in mind that if a sample is irradiated with a beam of electrons, the energy losses due to specific inelastic interactions can be measured, which is basically the principle of EELS spectroscopy. Nowadays, the availability of EELS in TEM has made feasible the identification of heteroatoms in various types of nanotubes. An example showing an elemental map recorded on CBx-MWNTs is shown in Fig. 8.9. Nevertheless, TEM is not the only device coupled

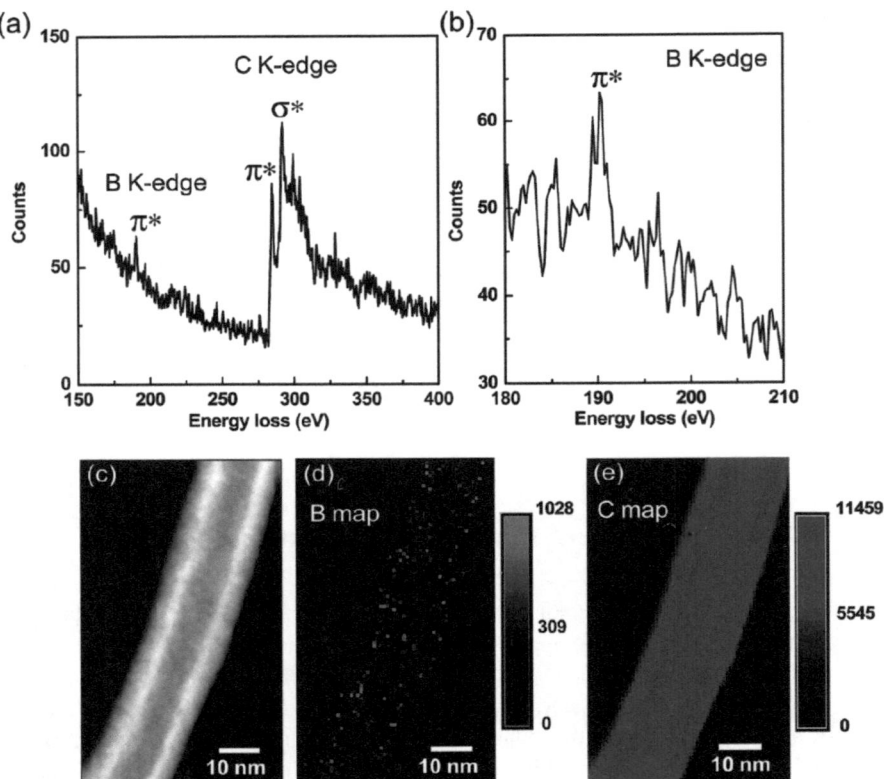

Fig. 8.9 Multiwalled B-doped nanotubes from a substitution reaction. (Image reproduced from [56] showing an EELS spectrum and high angular dark field images)

with such a system as it is nowadays known. In fact, EELS can also be performed in a purpose-built high-resolution spectrometer, which combines both, good energy and momentum resolution [92]. Unlike many electron spectroscopy techniques, this method is volume sensitive. Various types of functionalized nanotubes have been studied with this method and for such purposes the samples should ideally be prepared as films. The electronic structure of films as thin as 1 nm can be measured in transmission using high resolution(HR) EELS. A detailed description of energy and momentum resolved EELS experiments can be found in the review article in reference [92]. In contrast to optical spectroscopy which probes only dipole allowed $q = 0$ transitions, one gets here access to the momentum dependence of the dielectric function. Consequently, also non-vertical transitions can be excited. HR-EELS allows one to probe important values such as the character of excitonic excitations, the spatial extension of excitons and the degree of localization of electronic inter and intraband excitations. This applies very well especially for the case of mesoscopic systems with a very anisotropic electron density. At higher energies, the loss function describes transitions from core or shallow core levels into unoccupied states derived from orbitals. These core excitations are governed by the dipole selection rule. In this respect, this method is complementary to the TEM-EELS measurements described before. The main difference is that the high energy and momentum resolution allow to access the fine structure in the bonding environment as well as the modification in the conduction band, considering the effect of the core hole, with much better resolution. Changes in the charge transfer and bonding environments at the carbon nanotube sites. Fuentes et al. [33] reported on a spectroscopic study of the highly B doped C-SWNT using high resolution non-spatially resolved EELS and optical spectroscopy complemented by ab initio band structure calculations using FT-LDA and suggested preferential doping of semiconducting C-SWNT and the formation of BC_3 single wall nanotubes in boron substituted C-SWNT samples. However, a proof at local level is still required. The same material from substitution reactions was measured by HR-EELS and this is shown in the left panel of Fig. 8.10.

Finally, XPS, as lab-based photoemission spectroscopy method is among the most practical techniques to identify bonding environments. Here an important startpoint is that the various ionization processes have different probabilities. There is a different cross-section for each final state. Given that each element has a unique set of binding energies, XPS can be used to identify and determine the concentration of the elements and the variations in the elemental binding energies (chemical shifts). Taking care about the initial sample purity by XPS surveys is the starting point for any specific measurement and in particular for those related to the case of substitutionally doped CBx-SWNTs. Although it could seem a triviality for the expert reader, it is a step that is sometimes omitted causing the highest contribution to the initial error in most XPS measurements regarding the materials in question. The most up to date work attempting to identify bonding environments of B heteroatoms is related to graphene but spatially resolved analytic techniques still have not identified B locally in nanotubes. Gai et al. [70] used spatially resolved-EELS to study B-containing C-SWNTs samples synthesized by laser ablation,

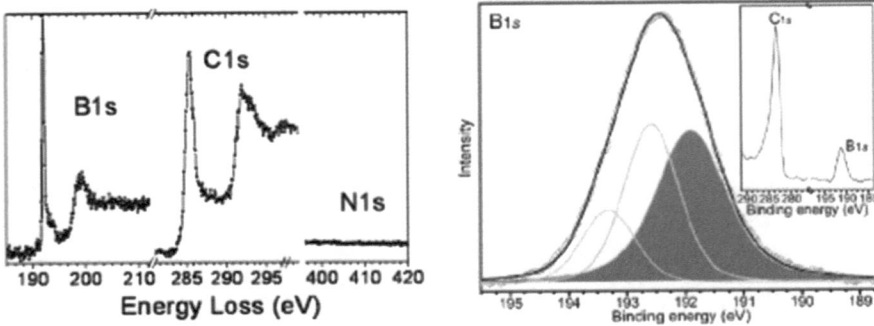

Fig. 8.10 Left: Spectra corresponding to high-resolution EELS core level excitation spectra of CBx-SWNT from [30]. This material corresponds to B- doped SWNT in the range of the B1s, C1s and N1s edges. The absence of N confirms the very successful substitution reaction, whereas the proportion of the C and B peaks gives a first overview of the high B content on the samples. Right: XPS spectrum on the B1$_s$ core level signal recorded for CBx-SWNTs synthesized from HV-CVD using a solid precursor

where they could not directly identify the presence of B within nanotube bundles produced from targets containing from 1.5% to 10% of B. The authors assumed a very low doping under the instrument detection limits and attributed the absence of B in the bundles of C-SWNTs to a preferential incorporation in nanoparticles of B_4C, which were detected in a sample synthesized from a target containing 10% of B. Another example for the B doped C-SWNT bundles are the ones reported in [30] via the substitution reaction. A TEM-EELS analysis with 10 nm beam size, which showed that a very localized substitution level could reach values up to 25 at%. Regarding B-containing thin films [93] performed an extensive XPS study on boron carbide films deposited by sputtering.

CBx-SWNTs with lower B contents have been much less explored in this context. The reason is mainly the limited availability of samples that offer the sample preparation conditions. The theoretical research related to B is vast compared to the work related to N but the experimental research is inversely proportional. The bonding environments of B associated to its presence in carbon nanotube samples are still far from understood. If B is regarded as doping agent in a C-SWNT, it has two possibilities. As explained before, the first one is to substitute the carbon atoms directly in the lattice. The second is to form nanodomains, or islands of boron atoms, which have not been substantially proved experimentally. Additionally, all different production methods give rise to different synthesis byproducts. The amount of samples and the type of measurements that have been performed so far have not been able to allow unambiguous confirmation of the possible bonding environments. If recorded with enough resolution capabilities and a proper count rate, XPS measurements are able to reveal the chemical shift that usually require a deconvolution of the core signal to discern among bonding environments. Both in laser ablation and arc discharge samples elemental B has been used for the targets and electrodes respectively in the different reports. This is why traces of elemental boron have been found in those type of materials. In CVD [65] as well as laser ablation [70] CBx-SWNTs it has been observed that B_4C is very likely to form and it is the most probable competing carbonaceous boron containing co-product. Therefore, a detailed insight of the photoemission response corresponding to the C1s is extremely important in the characterization of B doping and B content.

The majority of the works reporting the synthesis of low B-doping in single-walled nanotubes have estimated the possible B amounts solely based on the B content of the material employed for synthesis (i.e. C targets containing B compounds in the preparation mixture) which yields a strong overestimation of the actual sp^2 bonded B content. Panchakarla et al. [94] reported the presence of the B1s signal in XPS at 191.4 eV in C-DWNTs. In that study the presence of an asymmetric peak suggested at least three chemical states which have not been identified. The authors suggested that the cylindrical configuration and electron quantum confinement predict the B substitutional configuration at higher binding energy values than those for pure B ($\sim$187 eV) and B in doped carbons ($\sim$190 eV). They suggested that the signals 191.5 and 192.1 eV should be attributed to the substitutional configuration of B within the nanotubes. This range is still large and the possibility to resolve the binding energy associated to substitutional B might be resolved with the availability of purification methods that have been recently developed for doped nanotubes. Regarding additional bonding environments, other peaks have been identified for CVD tubes as well as those produced by substitution reactions suggesting that boron oxide and boric acid are chemical forms favored during the synthesis reactions. This is clearly shown in the B XPS core level spectrum on the right panel of Fig. 8.10. This corresponds to material unpublished related to the synthesis of highly B doped C-SWNTs.

Focusing briefly on XAS, it is first necessary to know that this is a tool to analyze core level excitations as well. It is a technique that requires the use of tuneable x-ray sources, like those available at synchrotron radiation centers. It is commonly applied in the total and partial electron yield , as well as the fluorescence yield mode. XAS can provide the same information as core level EELS but it has the advantage of having, in combination with third generation light sources, the utmost energy resolution of up to 10 meV at the C K-edge. Therefore this method has been very successfully applied to unravel the vHs in the conduction band of C-SWNT [95].

8.7   Electronic Structure and Transport

The literature regarding transport in CBx-SWNTs and heteronanotubes is almost restricted to theoretical calculations. Some experimental results have been shown for multi-walled structures. The electronic and transport properties of SWNTs involve a great number of considerations such as charge carrier localization, injection barriers, diffusive and ballistic regimes. In all cases, defects induce diverse physical consequences [96], and the presence of dopants must be seen in such situation as defects. Therefore, the presence of foreign atoms introduces new interesting features in each case. The experimental report by Xu et al. [97] suggested the possibility to convert metallic C-SWNTs into semiconductors by B/N co-doping. They made FET devices and later performed ab initio calculations of the structures adapting the doping levels obtained experimentally (between 3at% and 8at% considering a combined B and N total dopant concentration). Latil et al. [98] reported on the drastic modification of the electronic properties upon low incorporation (<0.5%) of boron and nitrogen in C-SWNTs. Das et al. [99] have proposed the use of Raman spectroscopy and transport measurements together to identify the doping level in C-SWNTs. Murata et al. [100] reported the observation of the Meissner effect at T_c = 12 K in thin films of CBx-SWNTs synthesized with laser ablation suggesting that the Meissner effect was very sensitive to the degree of uniformity of the samples.

8.8   Support of Contemporary Theoretical Methods

A wide range of bulk and local probes have to be applied in order to unravel the details in the modification of the structural, electronic and chemical properties of the materials here described. Understanding and controlling defects induced due to dopant incorporation is crucial if the structures should be used for nanoscale devices and engineered materials. Therefore, theoretical methods are extremely necessary to cover different aspects, from tracking the changes of structure and properties, to understanding the results of different experimental tools. Theoretical spectroscopy is in this sense extremely useful and necessary. The use of techniques with high energy, momentum and spatial resolution, does not compete with the local probing, but a complementary use represents a much better understanding of the spectroscopic response at different scales and here is where theory plays a decisive role. Regarding the theoretical analysis focused on the structures, studies are required to predict the physical behavior. An adequate description of the mechanical and electronic properties of the tubes nanotubes requires using large number of atoms and long-time scales and consequently very sophisticated techniques. The same stands for the studies on growth mechanisms that are so far not understood either. There is certainly a vast playground for theoretical research.

8.9   Application Potential and Future Perspectives

As seen in the previous sections, the presence of defects and dopants within carbon nanotubes is of great importance for new emerging technologies. Therefore, several experimental and theoretical investigations must focus towards achieving and understanding a controllable modification of the tube's properties when heteroatoms are present. So far, extensive effort has been made to synthesize the high-purity and defect-free C-SWNTs and to control their diameter and chiralities. CBx-SWNTs and heteronanotubes of C and B might encounter great potential applications such in those related to reduction reactions [101], batteries,environmental and bio-sensing [102], energy storage,or even superconductivity [103]. CBx-SWNTs could be used as tuneable semiconductors with low amounts of doping. Looking also at low doping with B as a class of weakly disordered low-dimensional systems, Latil et al. [98] reported magnetotransport features under the application of a magnetic field parallel to the tube axis. This suggested that the difference from electron conduction with much longer mean free path and positive magnetoresistance behavior. Since the boron doped tubes are hole doped this suggests that B doped nanotubes could be dilute magnetic semiconductors when incorporated into field effect transistor devices [104]. Calculations on the lithium absorption energies and electronic structures of boron-doped C-SWNTs suggests that enhanced Li absorption capability in the B-doped C-SWNT should be expected [105]. Liu et al. [106] reported results of studies on the sheet resistance and optical transmission of thin films of boron doped C-SWNTs. They suggested that B-doping did not significantly affect the optical transmittance in the visible region but it lowered the sheet resistance by 30% relative to pristine SWNT films. with this they concluded that B doping of C-SWNT as a better approach to touch-screen technology. If bulk scale applications are envisaged, the chemical modification of either sidewalls or end caps of the nanotubes has shown to be important for chemical reactivity. In this context, the heteronanotubes of B and C are very good candidates for funtionalization and the formation of polymer composites.

Most of the up to date studies regarding applications are constrained to theoretical predictions due to the still scarce availability of purified samples [107, 108]. It is clear that CBx-SWNTs have a great potential but their synthesis has only now reached a level that larger amounts of cleaner and more homogeneous samples can be prepared. Also a more precise approximation to the doping level at which a rigid band model in B-containing nanotubes is still applicable remains an open question. Considering the difficulties to develop and perform bulk scale metallicity separation of NTs sorting, the use BC_3-NTs as semiconductors showing a large chemical and thermal stability, appears very promising.

References

1. Ayala P, Arenal R, Loiseau A, Rubio A, Pichler T (2010) The physical and chemical properties of heteronanotubes. Rev Mod Phys 82:1843
2. Andreas H (2002) Functionalization of single-walled carbon nanotubes. Angew Chem Int Ed 41(11):1853
3. Ayala P, Arenal R, Rümmeli M, Rubio A, Pichler T (2010) The doping of carbon nanotubes with nitrogen and their potential applications. Carbon 48(3):575
4. Klumpp C, Kostarelos K, Prato M, Bianco A (2006) Functionalized carbon nanotubes as emerging nanovectors for the delivery of therapeutics. Biochimica et Biophysica Acta (BBA) – Biomembranes 1758(3):404. Mechanisms of carrier-mediated intracellular delivery of therapeutics

5. Jorio A, Dresselhaus M, Dresselhaus G (2008) Carbon nanotubes: advanced topics in the synthesis, structure, properties and applications. Springer, Heidelberg
6. Ayala P, Grueneis A, Gemming T, Buechner B, Ruemmeli MH, Grimm D, Schumann J, Kaltofen R, Freire FL Jr, Fonseca Filho HD, Pichler T (2007) Influence of the catalyst hydrogen pretreatment on the growth of vertically aligned nitrogen-doped carbon nanotubes. Chem Mater 19(25):6131
7. Ayala P, Grüneis A, Gemming T, Grimm D, Kramberger C, Rümmeli M, Freire FL Jr, Kuzmany H, Pfeiffer R, Barreiro A, Büchner B, Pichler T (2007) Tailoring n-doped single and double wall carbon nanotubes from a non-diluted carbon/nitrogen feedstock. J Phys Chem C 101:2879
8. Ayala P, Grüneis A, Kramberger C, Rümmeli M, Freire F Jr, Solórzano IG, Pichler T (2007) Effects of the reaction atmposphere compositon on the synthesis of single and multiwall nitrogen doped nanotubes. J Chem Phys 127:184709
9. Han W (2006) Silicon doped boron carbide nanorod growth via a solid-liquid-solid process. Appl Phys Lett 88(13)
10. Ruiz-Soria G, Susi T, Sauer M, Yanagi K, Pichler T, Ayala P (2015) On the bonding environment of phosphorus in purified doped single-walled carbon nanotubes. Carbon 81:91
11. Ruiz-Soria G, Ayala P, Puchegger S, Kataura H, Yanagi K, Pichler T (2011) On the purification of cvd grown boron doped single-walled carbon nanotubes. Phys Stat Sol B 248(11):2504
12. Chegel R (2016) Tuning electronic properties of carbon nanotubes by boron and nitrogen doping. Physica B 499:1
13. Hai-Yang S, Xin-Wei Z (2009) The effects of boron doping and boron grafts on the mechanical properties of single-walled carbon nanotubes. J Phys D 42(22)
14. Fakhrabadi MMS, Allahverdizadeh A, Norouzifard V, Dadashzadeh B (2012) Effects of boron doping on mechanical properties and thermal conductivities of carbon nanotubes. Solid State Commun 152(21):1973
15. Li YF, Wang Y, Chen SM, Wang HF, Kaneko T, Hatakeyama R (2013) Electrical transport properties of boron-doped single-walled carbon nanotubes. J Appl Phys 113(5)
16. Anand B, Podila R, Ayala P, Oliveira L, Philip R, Sai SSS, Zakhidov AA, Rao AM (2013) Nonlinear optical properties of boron doped single-walled carbon nanotubes. Nanoscale 5(16):7271
17. Pichler T, Borowiak-Palen E, Fuentes G, Knupfer M, Graff A, Fink J, Wirtz L, Rubio A (2003) Electronic structure and optical properties of boron doped single-wall arbon nanotubes. Mol Nanostruct 685(4):361
18. Krstic V, Blumentritt S, Muster J, Roth S, Rubio A (2003) Role of disorder on transport in boron-doped multiwalled carbon nanotubes. Phys Rev B 67(4)
19. Wei B, Spolenak R, Kohler-Redlich P, Ruhle M, Arzt E (1999) Electrical transport in pure and boron-doped carbon nanotubes. Appl Phys Lett 74(21):3149
20. Jalili S, Akhavan M, Schofield J (2012) Electronic and structural properties of bc3 nanotubes with defects. J Phys Chem C 116:13225
21. Cardona M, Yu P (2010) Fundamentals of semiconductors: physics and materials properties. Springer, Berlin/Heidelberg
22. Liu X, Pichler T, Knupfer M, Fink J, Kataura H (2004) Electronic properties of fecl3-intercalated single-wall carbon nanotubes. Phys Rev B 70(20):205405
23. Lee RS, Kim HJ, Fischer JE, Thess A, Smalley RE (1997) Conductivity enhancement in single-walled carbon nanotube bundles doped with k and br. Nature 388(6639):255
24. Guerini S, Souza AG, Mendes J, Alves OL, Fagan SB (2005) Electronic properties of fecl3-adsorbed single-wall carbon nanotubes. Phys Rev B 72(23)
25. Yi JY, Bernholc J (1993) Atomic structure and doping of microtubules. Phys Rev B 47(3):1708
26. Iijima S (1991) Helical microtubules of graphitic carbon. Nature 56:354
27. Iijima S, Ichihashi T (1993) Single-shell carbon nanotubes of 1-nm diameter. Nature 363:603

28. Serin V, Brydson R, Scott A, Kihn Y, Abidate O, Maquin B, Derre A (2000) Evidence for the solubility of boron in graphite by electron energy loss spectroscopy. Carbon 38(4):547
29. Terrones M, Jorio A, Endo M, Rao A, Kim Y, Hayashi T, Terrones H, Charlier JC, Dresselhaus G, Dresselhaus M (2004) New direction in nanotube science. Materialstoday, p 30
30. Borowiak-Palen E, Pichler T, Fuentes G, Graff A, Kalenczuk R, Knupfer M, Fink J (2004) Synthesis and electronic properties of b-doped single wall carbon nanotubes. Carbon 42:1123
31. Borowiak-Palen E, Pichler T, Fuentes G, Graff A, Kalenczuk R, Knupfer M, Fink J (2003) Efficient production of b-substituted single-wall carbon nanotubes. Chem Phys Lett 378(5–6):516
32. Golberg D, Bando Y, Burgeois L, Kurashima K, Sato T (2000) Large-scale synthesis and hrtem analysis of single-walled b- and n-doped carbon nanotube bundles. Carbon 38:2017
33. Fuentes G, Borowiak-Palen E, Knupfer M, Pichler T, Fink J, Wirtz L, Rubio A (2004) Formation and electronic properties of bc_3 single-wall nanotubes upon boron substitution of carbon nanotubes. Phys Rev B 69(24):245403
34. Miyamoto Y, Rubio A, Louie SG, Cohen ML (1994) Electronic properties of tubule forms of hexagonal bc_3. Phys Rev B 50(24):18360
35. Miyamoto Y, Rubio A, Cohen ML, Louie SG (1994) Chiral tubules of hexagonal bc_2n. Phys Rev B 50:4976
36. Liu AY, Wentzcovitch RM, Cohen ML (1989) Atomic arrangement and electronic structure of bc_2n. Phys Rev B 39(3):1760
37. Carroll D, Redlich P, Blase X, Charlier J, Curran S, Ajayan P, Roth S, Ruhle M (1998) Effects of nanodomain formation on the electronic structure of doped carbon nanotubes. Phys Rev Lett 81(11):2332
38. Wirtz L, Rubio A (2003) Band structure of boron doped carbon nanotubes. Mol Nanostruct 685(4):402
39. Rümmeli MH, Ayala P, Pichler T (2010) Carbon nanotubes and related structures: production and formation, chap. 1. Cambridge University Press, UK, pp 1–21
40. Kitiyanan B, Alvarez WE, Harwell JH, Resasco DE (2000) Controlled production of single-wall carbon nanotubes by catalytic decomposition of co on bimetallic co-mo catalysts. Chem Phys Lett 317(3–5):497
41. Bachilo S, Balzano L, Herrera J, Pompeo F, Resasco D, Weisman R (2003) Narrow (n,m)-distribution of single-walled carbon nanotubes grown using a solid supported catalystblase. J Am Chem Soc 125:11186
42. Tanaka T, Jin H, Miyata Y, Fujii S, Suga H, Naitoh Y, Minari T, Miyadera T, Tsukagoshi K, Kataura H (2009) Simple and scalable gel-based separation of metallic and semiconducting carbon nanotubes. Nano Lett 9(4):1497
43. Yanagi K, Iitsuka T, Fujii S, Kataura H (2008) Separations of metallic and semiconducting carbon nanotubes by using sucrose as a gradient medium. J Phys Chem C 112(48):18889
44. Arnold M, Green A, Hulvat J, Stupp S, Hersam M (2006) Sorting carbon nanotubes by electronic structure using density differentiation. Nat Nanotechnol 1:60
45. Stephan O, Ajayan P, Colliex C, Redlich P, Lambert J, Bernier P, Lefin P (1994) Doping graphitic and carbon nanotube structures with boron and nitrogen. Science 266:1863
46. Ewels C, Glerup M (2005) Nitrogen doping in carbon nanotubes. J Nanosci Nanotechnol 5(9):1345
47. Golberg D, Bando Y, Tang C, Zhi C (2007) Boron nitride nanotubes. Adv Mat 19(18):2413
48. Weng-Sieh Z, Cherrey K, Chopra NG, Blase X, Miyamoto Y, Rubio A, Cohen ML, Louie SG, Zettl A, Gronsky R (1995) Synthesis of $b_xc_yn_z$ nanotubules. Phys Rev B 51:11229
49. Redlich P, Loeffler J, Ajayan P, Bill J, Aldinger F, Ruhle M (1996) B-c-n nanotubes and boron doping of carbon nanotubes. Chem Phys Lett 260(3–4):465
50. Maultzsch J, Reich S, Thomsen C, Webster S, Czerw R, Carroll D, Vieira S, Birkett P, Rego C (2002) Raman characterization of boron-doped multiwalled carbon nanotubes. Appl Phys Lett 81(14):2647
51. Carroll D, Redlich P, Ajayan P, Curran S, Roth S, Ruhle M (1998) Spatial variations in the electronic structure of pure and b-doped nanotubes. Carbon 36(5):753

52. Babanejad SA, Malekfar R, Ashrafi F, Hosseini SMRS (2010) Production and study of boron and nitrogen-doped carbon nanotubes by arc discharge method using dispersive raman backscattering spectroscopy. Asian J Chem 22(1):245

53. Wang B, Ma Y, Wu Y, Li N, Huang Y, Chen Y (2009) Direct and large scale electric arc discharge synthesis of boron and nitrogen doped single-walled carbon nanotubes and their electronic properties. Carbon 47(8):2112

54. Deng C, Chen J, Chen X, Mao C, Nie L, Yao S (2008) Direct electrochemistry of glucose oxidase and biosensing for glucose based on boron-doped carbon nanotubes modified electrode. Biosens Bioelectron 23(8):1272

55. Han W, Bando Y, Kurashima K, Sato T (1999) Boron-doped carbon nanotubes prepared through a substitution reaction. Chem Phys Lett 299(5):368

56. Chiang WH, Chen GL, Hsieh CY, Lo SC (2015) Controllable boron doping of carbon nanotubes with tunable dopant functionalities: an effective strategy toward carbon materials with enhanced electrical properties. RSC Adv 5(118):97579

57. Golberg D, Bando Y, Kurashima K, Sato T (2001) Synthesis, hrtem and electron diffraction studies of b/n-doped c and bn nanotubes. Diamond Relat Mater 10(1):63

58. Chen C, Tsai C, Lin C (2003) The characterization of boron-doped carbon nanotube arrays. Diamond Relat Mater 12(9):1500

59. Koos AA, Dillon F, Obraztsova EA, Crossley A, Grobert N (2010) Comparison of structural changes in nitrogen and boron-doped multi-walled carbon nanotubes. Carbon 48(11):3033

60. Panchakarla LS, Govindaraj A, Rao CNR (2010) Boron- and nitrogen-doped carbon nanotubes and graphene. Inorg Chim Acta 363(15):4163

61. Lyu SC, Han JH, Shin KW, Sok JH (2011) Synthesis of boron-doped double-walled carbon nanotubes by the catalytic decomposition of tetrahydrofuran and triisopropyl borate. Carbon 49(5):1532

62. Cao Y, Yu H, Tan J, Peng F, Wang H, Li J, Zheng W, Wong NB (2013) Nitrogen-, phosphorous- and boron-doped carbon nanotubes as catalysts for the aerobic oxidation of cyclohexane. Carbon 57:433

63. Keru G, Ndungu PG, Nyamori VO (2015) Effect of boron concentration on physicochemical properties of boron-doped carbon nanotubes. Mater Chem Phys 153:323

64. Daothong S, Parjanne J, Kauppinen EI, Valkeapää M, Pichler T, Singjai P, Ayala P (2009) Study of the role of fe based catalysts on the growth of b-doped swcnts synthesized by cvd. Phys Status Solidi B 246(11–12):2518

65. Ayala P, Plank W, Grüneis A, Kauppinen E, Rümmeli MH, Kuzmany H, Pichler T (2008) A one step approach to b-doped single-walled carbon nanotubes. J Mater Chem 18:5676

66. Ayala P, Ruemmeli MH, Gemming T, Kauppinen E, Kuzmany H, Pichler T (2008) Cvd growth of single-walled b-doped carbon nanotubes. Phys Stat Sol B 245(10, Sp. Iss. SI):1935

67. Monteiro FH, Larrude DG, Maia da Costa MEH, Freire FL (2013) Estimating the boron doping level on single wall carbon nanotubes using raman spectroscopy. Mater Lett 92:224

68. Monteiro FH, Larrude DG, Maia da Costa MEH, Terrazos LA, Capaz RB, Freire FL Jr (2012) Production and characterization of boron-doped single wall carbon nanotubes. J Phys Chem C 116(5):3281

69. Ruiz-Soria G, Daothong S, Pichler T, Ayala P (2012) Spectroscopic study of the diameter distribution of b-doped single-walled carbon nanotubes. Phys Status Solidi B 249(12):2469

70. Gai P, Stephan O, McGuire K, Rao A, Dresselhaus M, Dresselhaus G, Colliex C (2004) Structural systematics in boron-doped single wall carbon nanotubes. J Mater Chem 14(4):669

71. McGuire K, Gothard N, Gai P, Dresselhaus M, Sumanasekera G, Rao A (2005) Synthesis and raman characterization of boron-doped single-walled carbon nanotubes. Carbon 43:219

72. Blackburn JL, Yan Y, Engtrakul C, Parilla PA, Jones K, Gennett T, Dillon AC, Heben MJ (2006) Synthesis and characterization of boron-doped single-wall carbon nanotubes produced by the laser vaporization technique. Chem Mater 18(10):2558

73. Ayala P, Reppert J, Grobosch M, Knupfer M, Pichler T, Rao AM (2010) Evidence for substitutional boron in doped single-walled carbon nanotubes. Appl Phys Lett 96(18)

74. Glerup M, Steinmetz J, Samaille D, Stephan O, Enouz S, Loiseau A, Roth S, Bernier P (2004) Synthesis of n-doped swnt using the arc-discharge procedure. Chem Phys Lett 387(1–3):193

75. Droppa R Jr, Hammer P, Carvalho A, Alvarez F (2002) Incorporation of nitrogen in carbon nanotubes. J Non-Cryst Solids 874:299
76. Han W, Bando Y, Kurashima K, Sato T (1998) Synthesis of boron nitride nanotubes from carbon nanotubes by a substitution reaction. Appl Phys Lett 73(21):3085
77. Golberg D, Bando Y, Han W, Kurashima K, Sato T (1999) Single-walled b-doped carbon, b/n-doped carbon and bn nanotubes synthesized from single-walled carbon nanotubes through a substitution eaction. Chem Phys Lett 308(3–4):337
78. Bystrzejewski M, Bachmatiuk A, Thomas J, Ayala P, Serwatowski J, Huebers HW, Gemming T, Borowiak-Palen E, Pichler T, Kalenczuk RJ, Buechner B, Ruemmeli MH (2009) Boron doped carbon nanotubes via ceramic catalysts. Phys Stat Sol RRL 3(6):193
79. Arenal R, Stephan O, Cochon JL, Loiseau A (2007) Root-growth mechanism for single-walled boron nitride nanotubes in laser vaporization technique. J Am Chem Soc 129(51):16183
80. Hassanien A, Tokumoto M, Shimizu T, Tokumoto H (2004) Stm on suspended single wall carbon nanotubes. Thin Solid Films 464:338. 7th international symposium on atomically controlled surfaces, interfaces and nanostructures, Nara, 16–20 Nov 2003
81. Orlikowski D, Nardelli M, Bernholc J, Roland C (2000) Theoretical stm signatures and transport properties of native defects in carbon nanotubes. Phys Rev B 61(20):14194
82. Ichimura K, Osawa M, Nomura K, Kataura H, Maniwa Y, Suziki S, Achiba Y (2002) Tunneling spectroscopy on carbon nanotubes using stm. Physica B 323(1–4):230 Tsukuba symposium on carbon nanotube in commemoration of the 10th anniversary of its discovery, Tsukuba, 03–05 Oct 2001
83. Odom T, Huang J, Lieber C (2002) Stm studies of single-walled carbon nanotubes. J Phys-Condens Matter 14(6):R145
84. Kim P, Odom T, Huang J, Lieber C (2000) Stm study of single-walled carbon nanotubes. Carbon 38(11–12):1741
85. Biro L, Gyulai J, Lambin P, Nagy J, Lazarescu S, Mark G, Fonseca A, Surjan P, Szekeres Z, Thiry P, Lucas A (1998) Scanning tunnelling microscopy (stm) imaging of carbon nanotubes. Carbon 36(5–6):689. Symposium A on fullerenes and carbon based materials, at the European-Materials-Research-Society 1997 meeting, Strasbourg, 16–20 June 1997
86. Tans SJ, Dekker C (2000) Molecular transistors: potential modulations along carbon nanotubes. Nature 404:834
87. Rubio A (1999) Spectroscopic properties and stm images of carbon nanotubes. Appl Phys A 68(3):275
88. Lambin P, Mark G, Meunier V, Biro L (2003) Computation of stm images of carbon nanotubes. Int J Quantum Chem 95(4–5):493. 43rd international symposium on theory and computations in molecular and materials sciences, biology, and pharmacology, St Augustine, 22 Feb–01 Mar 2003
89. Jorio A, Saito R, Hafner JH, Lieber CM, Hunter M, McClure T, Dresselhaus G, Dresselhaus MS (2001) Structural (n, m) determination of isolated single-wall carbon nanotubes by resonant raman scattering. Phys Rev Lett 86(6):1118
90. Caudal N, Saitta AM, Lazzeri M, Mauri F (2007) Kohn anomalies and nonadiabaticity in doped carbon nanotubes. Phys Rev B 75:115423
91. Maciel I, Anderson N, Pimenta M, Hartschuh A, Quian H, Terrones M, Terrones H, Campos-Delgado J, Rao A, Novotny L, Jorio A (2008) Electron and phonon renormalization near charged defects in carbon nanotubes. Nat Mater 7:878
92. Fink J (1989) Recent development in energy-loss spectroscopy. Adv Electron Electron Phys 75:121
93. Jacobsohn L, Schulze R, da Costa MM, Nastasi M (2004) X-ray photoelectron spectroscopy investigation of boron carbide films deposited by sputtering. Surf Sci 572:418
94. Panchakarla LS, Govindaraj A, Rao CNR (2007) Nitrogen- and boron-doped double-walled carbon nanotubes. ACS Nano 1(5):494

95. Kramberger C, Rauf H, Shiozawa H, Knupfer M, Büchner B, Pichler T (2007) Unraveling van hove singularities in the x-ray absorption response of single wall carbon nanotubes. Phys Rev B 75:235437
96. Charlier JC, Blase X, Roche S (2007) Electronic and transport properties of nanotubes. Rev Mod Phys 79(2):677
97. Xu Z, Lu W, Wang W, Gu C, Liu K, Bai X, Wang E, Dai H (2008) Converting metallic single-walled carbon nanotnbes into semiconductors by boron/nitrogen co-doping. Adv Mat 20(19):3615+
98. Latil S, Roche S, Mayou D, Charlier J (2004) Mesoscopic transport in chemically doped carbon nanotubes. Phys Rev Lett 92(25)
99. Das A, Sood AK, Govindaraj A, Saitta AM, Lazzeri M, Mauri F, Rao CNR (2007) Doping in carbon nanotubes probed by raman and transport measurements. Phys Rev Lett 99
100. Murata N, Haruyama J, Reppert J, Rao AM, Koretsune T, Saito S, Matsudaira M, Yagi Y (2008) Superconductivity in thin films of boron-doped carbon nanotubes. Phys Rev Lett 101(2):027002
101. Cheng Y, Tian Y, Fan X, Liu J, Yan C (2014) Boron doped multi-walled carbon nanotubes as catalysts for oxygen reduction reaction and oxygen evolution reactionin in alkaline media. Electrochim Acta 143:291
102. Chen X, Chen J, Deng C, Xiao C, Yang Y, Nie Z, Yao S (2008) Amperometric glucose biosensor based on boron-doped carbon nanotubes modified electrode. Talanta 76(4):763
103. Haruyama J, Matsudaira M, Reppert J, Rao A, Koretsune T, Saito S, Sano H, Iye Y (2011) Superconductivity in boron-doped carbon nanotubes. J Supercond Novel Magn 24(1–2):111
104. Owens FJ (2007) Boron and nitrogen doped single walled carbon nanotubes as possible dilute magnetic semiconductors. Nanoscale Res Lett 2(9):447
105. Zhou Z, Gao X, Yan J, Song D, Morinaga M (2004) Enhanced lithium absorption in single-walled carbon nanotubes by boron doping. J Phys Chem B 108(26):9023
106. Liu XM, Romero HE, Gutierrez HR, Adu K, Eklund PC (2008) Transparent boron-doped carbon nanotube films. Nano Lett 8(9):2613
107. Reinoso C, Berkmann C, Shi L, Debut A, Yanagi K, Pichler T, Ayala P (2019) Toward a predominant Substitutional bonding environment in B-doped single-walled carbon nanotubes. ACS Omega 4(1):1941
108. Reinoso C, Shi L, Domanov 0 RP, Pichler T, Ayala P (2018) Very high borondoping on single-w ed carbon nanotubes from a solid precursor. Carbon 140:259–264

Chapter 9
Graphene-Based Metal-Free Catalysis

Mattia Scardamaglia and Carla Bittencourt

Abstract This chapter focuses on the use of doped carbon nanomaterials in catalysis. The availability of carbon nanotubes in the '90s and graphene about 10 years later, prompted the development of fundamental research and novel nanotechnologies. We discuss this topic from a point of view that links fundamental surface science to the field of catalysis, in order to present the state of the art. We describe scientific questions that material scientists have faced during these last decades, in particular, we concentrate on the debate over the role that the different nitrogen configurations in the graphene lattice can play in certain catalytic processes.

Keywords Spectromicroscopy · Graphene · Catalyst

9.1 Introduction

Catalytic processes are at the basis of important technologies in the chemical industry in terms of present and future energy supplies, especially in the context of today's quest for renewable, clean and sustainable energy production. The oxygen reduction reaction (ORR), *in primis*, and the electrocatalytic reduction of water to molecular hydrogen via hydrogen and oxygen evolution reactions (HER and OER, respectively) are fundamental working mechanisms for the cathode of fuel cells, metal-air batteries, and dye-sensitized solar cells. Current catalysts are generally composed of expensive metals, such as platinum or its alloys, or metal oxides, which implies a high production cost. Therefore, research efforts are now being devoted to developing alternative, non-precious and highly active catalysts in order to reduce

M. Scardamaglia (✉)
MAX IV Laboratory, Lund University, Lund, Sweden
e-mail: mattia.scardamaglia@maxiv.lu.se

C. Bittencourt
Chimie des Interactions Plasma-Surface, University of Mons (UMONS), Mons, Belgium

© Springer Nature B.V. 2019
C. Bittencourt et al. (eds.), *Nanoscale Materials for Warfare Agent Detection: Nanoscience for Security*, NATO Science for Peace and Security Series A: Chemistry and Biology, https://doi.org/10.1007/978-94-024-1620-6_9

the need for expensive metals such as platinum. Potential candidates are transition metals (Co, Ni, Fe, and Mo) and their alloys – less expensive than noble metals. However, these cheaper metals suffer from inherent corrosion and oxidation susceptibility that limit their use in acidic electrolytic environments, contrary to Pt [1]. The most common poisoning agent is CO, which inhibits the dissociative adsorption of H_2 on the catalyst, lowering fuel cell performance. Research groups have reported on the use of graphene as a support for metal-based electrocatalysts that decrease the amount of platinum required. Two main features make this approach promising: the efficient charge transfer at the carbon-metal interface and the possibility to homogeneously disperse the nanoparticles without aggregation, particularly in the case of modified graphene, where the dopant atom acts as an anchoring site for their nucleation [2]. These studies focused on the use of pristine and modified graphene as a support for metal nanoparticle catalysts, taking advantage of the improved performances to reduce the amount of platinum, substituting it with less expensive platinum alloys (such as PtPd core-shell nanorings [3], FePt and PtCo nanoparticles [4]) or non Pt-based metal catalysts (Au, CuPd). Research on low-cost and metal-free alternative catalysts has been conducted for decades. Back in 1926, research on the catalytic properties of carbon alloys with nitrogen was first published: Rideal and Wright reported their studies on the oxidation of oxalic acid in charcoal containing nitrogen and iron [5]. In 1966, an enhanced activity for the electrochemical reduction of oxygen in fuel cells was attained by using carbon that had been activated through heating treatments at high temperatures in the presence of ammonia [6]. However, the possibility of having a metal-free catalyst for ORR was not considered feasible [7, 8] until two milestones raised interest for carbon-based catalysts: 1) the prediction of the remarkable conducting properties of carbon nanotubes (CNTs) in 1993 [9–11] and 2) the findings on graphene reported in 2004 by Geim and Novoselov [12]. These reports triggered an explosion in fundamental research in many fields of application, including catalysis [13–16].

In 2006, Matter and Ozkan reported on a metal-free ORR catalyst containing nitrogen-doped carbon fibres. The purified fibres showed significant ORR activity, although less than the non-purified iron-containing fibres [17]. Afterwards, in 2009, Gong and colleagues reported on the first metal-free catalyst that showed ORR activity superior to commercial Pt in alkaline fuel cells [18]; it was comprised of nitrogen-doped vertically-aligned carbon nanotubes (N-vCNTs). Using a rotating ring-disk electrode (RRDE) steady-state voltammogram in an alkaline solution, the authors showed that the ORR activity of N-vCNTs is superior to that of undoped vCNTs, to random oriented N-CNTs, and, in particular, to commercially available platinum-loaded carbon (Vulcan XC-72R). The electrochemical mechanism for the ORR was reported to be the same for aligned and disordered CNTs. However, the electrocatalytic performance was shown to improve for v-CNTs; this improvement was associated with the better-defined surface area of the aligned tips at the interface with the electrolyte solution, which facilitates the electrolyte/reactant diffusion. A significant strengthening of the electric conduction was observed for N-CNTs as opposed to undoped nanotubes. Furthermore, the Gong team demonstrated that the glass-supported N-vCNTs electrode was more stable in ORR than the platinum-loaded one, which showed deterioration after repeated cycles due to the migration and aggregation of the Pt nanoparticles. The electrocatalytic activity of nitrogen-doped carbon nanotubes was reported to be unaffected by CO poisoning, in contrast with commercial electrodes. Gong et al. showed that the typical RRDE voltammogram of pristine vCNTs involves a two-step process with an onset potential of about -0.2 V and -0.7 V, corresponding to a two-electron pathway that involves the formation of hydrogen peroxide ions. And, on the other hand, nitrogen-doped (N-CNTs) and commercial Pt-C electrodes have a one-step process for the ORR with a half-wave potential of -0.1 V, suggesting a more efficient four-electron pathway that directly produces water as an end product by combining oxygen with electrons and protons. Furthermore, these authors showed that the steady-state diffusion current was greater for N-vCNTs and lower for Pt-C electrodes.

In 2010, Qu et al. reached similar conclusions when, for the first time, they substituted a nitrogen-doped graphene electrode for a commercial Pt one [19]. Similar to the un-doped CNT [18], the pristine graphene electrode was characterised by a two-step, two-electron process for ORR, however with onset potentials at -0.45 V and -0.7 V, while for N graphene it is a one-step, four-electron pathway with higher steady-state current than a commercial Pt-C electrode. These authors discovered long-term operation stability and tolerance to poison effects, which they had tested using methanol and CO (Fig. 9.1). The N-graphene response was unaffected by the presence of the poisoning gas, while the Pt-C electrode current abruptly decreased when the gas was introduced.

More recently, in 2013, Zhao and co-authors described the synthesis and operation of the first metal-free catalyst that exhibited OER activity that was comparable to non-precious metal catalysts [20]. In their work, the authors used nitrogen-doped graphite nano-materials that had been synthesized from a nitrogen-rich polymer (melamine/formaldehyde). In 2015, using nitrogen- and phosphorous-co-doped carbon foam, Zhang and colleagues obtained the first bifunctional electrocatalyst for both ORR and OER [21] for high-performance rechargeable zinc-air batteries. The same material was used by Xue et al. as a counter electrode in dye-sensitized solar cells, yielding high performances [22].

Nitrogen-doped carbon nanostructures have proved their overall superiority and greater efficiency as metal-free catalysts when compared to commercially available Pt-based electrodes. Functionalized graphene and CNTs were also reported to have good gas-sensing properties [23–25] for gases such as NO_2, CO, H_2, avoiding the need for expensive metal atoms. Nitrogen atoms can be incorporated in the hexagonal carbon lattice in different configurations, according to the number of bonding C atoms and to the hybridization, as we will describe later. The role played in the catalytic mechanism by the nitrogen atom in different configurations is still under debate and no clear conclusion has been reached so far.

The examples reported above show the importance of carbon materials for non-metal-based catalysis. In the next section, we will take a photoelectron spectroscopic look at the fundamental electronic properties of nitrogen-doped graphene and connect them to applications in chemical sensing and catalysis, focusing on the inherent ORR activity. The main properties that justify promoting graphene and carbon nanotubes are examined, as well as the most common nitrogen-doping techniques.

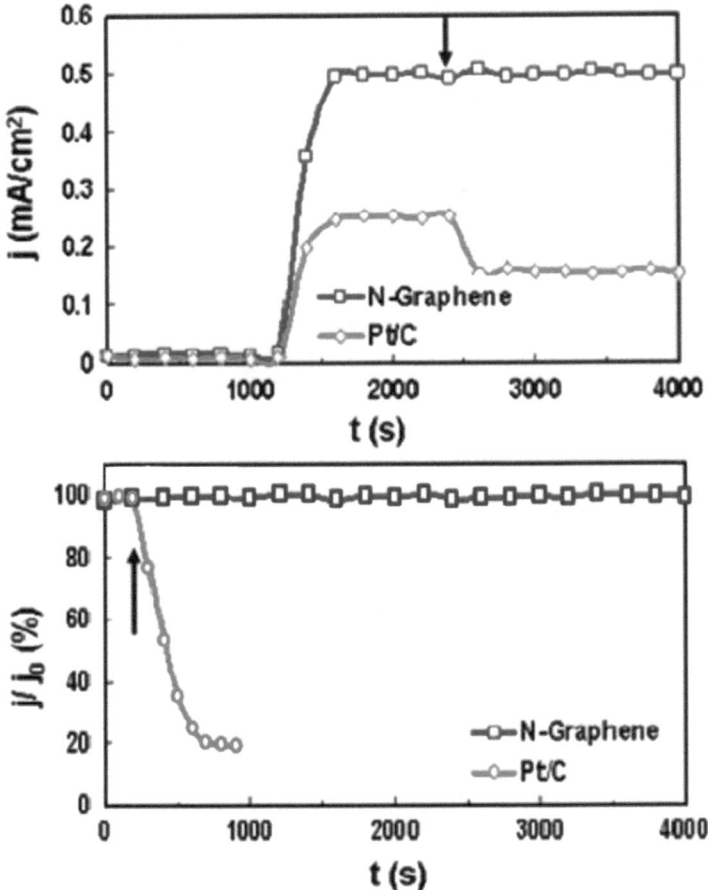

Fig. 9.1 (top) Current density (j) – time (t) chronoamperometric responses obtained at the Pt-C (circle) and N-graphene (square) electrodes at −0.4 V in air saturated 0.1 M KOH. The arrow indicates the addition of 2% (w/w) methanol into the air-saturated electrochemical cell. (bottom) Current (j) - time (t) chronoamperometric responses of Pt-C (circle) and N-graphene (square) electrodes to CO. The arrow indicates the addition of 10% (v/v) CO into air saturated 0.1 M KOH at −0.4 V; j_0 defines the initial current. (Adapted with permission from [19]. Copyright (2010) American Chemical Society)

9.2 Graphene and Its Modification

Carbon nanotubes and graphene share many characteristics in terms of structure and other properties, such as high surface area to volume ratio, good mechanical properties and extraordinary electronic properties. The one-atom-thick bidimensional planar structure of graphene sheets makes it superior to CNTs for certain applications because it facilitates electron transport [26], a key feature of a working electrode. Therefore, graphene may be used in many applications where CNTs have already proved viable.

The excellent mechanical properties, low reactivity and high stability [27] of graphene ensure long-lasting use as a catalyst in working conditions. Another very important property of graphene is its high thermal and electric conductivity [28], allowing good heat diffusion and conduction during catalytic reactions.

Most of the peculiar properties of graphene arise from its honeycomb lattice of carbon atoms arranged at the vertices of hexagons. These atoms are bound together in the basal plane by strong covalent bonds in sp^2 hybridization. Using standard tight-binding calculation, the electronic band structure and the well-known linear dispersion of the energy near the K point of the Brillouin Zone can be obtained. The Fermi velocity $v_F \approx 1 \times 10^6$ m/s is independent from the momentum, unlike usual crystals where a quadratic behaviour is found in momentum, making graphene a special material for different applications. Through geometrical considerations, due to the symmetry of the honeycomb lattice, the energy dispersion typical of ultra-relativistic massless particles is obtained. The low-energy quasiparticles can be described by a Dirac-like Hamiltonian rather than the Schrödinger equation [29]. The Dirac point for pure graphene is found exactly at the Fermi level, therefore graphene is defined as a semi-metal (or a zero-gap semiconductor). This peculiarity represents one of the biggest challenges for its use in concrete applications because it results in a very low density of states (DOS) at the Fermi level for typical doping, so it is intrinsically inert for sensing and catalysis. It has even been reported that for gas-sensing devices, pristine graphene without proper surface modification shows poor selectivity and slow recovery [30].

To overcome these issues, several strategies have been employed to tailor the properties of graphene. Since it is very sensitive to local perturbations, any modification of the lattice or adsorption of foreign atoms or molecules produces an obvious change in the density of states (DOS) demonstrated by a shift of the Dirac cone or by a change in the resistance caused by the tuning of the density of charge carriers [30–33]. To meet specific requirements demanded by a given application, chemical modification of pristine carbon nanomaterials is then essential to make possible, for example, improved sensitivity to atoms and molecules that interact with the modified carbon nanostructure surface.

The activation of the surface with the creation of active sites is a key step for obtaining an optimal working catalyst as well as for gas- or bio-sensors. These sites make possible the dissociation of molecules and the bond-formation of products, such as the molecular adsorption of oxygen and its reduction. This situation is maximised on graphene due to its 2D nature, a surface without bulk, in which every atom takes part in the interaction mechanism, thus providing the highest specific surface area ever (>2600 m^2 g^{-1}) [32], and thus allowing a high density of active surface sites. In contrast to many metal catalysts, graphene has no subsurface [34, 35], consequently, the chemical complexity is confined to the bi-dimensional interface between the graphene and the reactants, without the complications of intercalation chemistry.

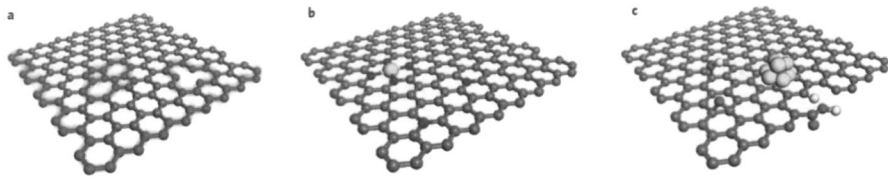

Fig. 9.2 (**a**) Active sites from defects and edges. (**b**) Active sites from doped heteroatoms. (**c**) Active sites from functional groups and metal clusters. (Reprinted by permission from Springer Nature [32])

In addition to the properties discussed above, their ease in tuning makes nanocarbon materials very adaptable. The graphitic network can be easily modified by introducing heteroatoms, functional groups or defects, such as dislocations, vacancies or edges [31, 36, 37], as reported in Fig. 9.2. They themselves act as active sites for electron localisation or they are useful to anchor metal clusters or foreign molecules that will be active sites. The change in reactivity can then be obtained in a controllable way for systematic optimisation. Changing the structure of current metal catalysts in order to study their dynamic behaviour is a daunting task and has rarely been investigated [38], while it is routinely done for graphene and carbon nanotubes.

Alternatively, the tuning of the electronic states of graphene can be obtained by reducing its dimensions to obtain graphene nanoribbons [39]. In nanoribbons, a bandgap is opened due to the quantum confinement effect, with the bandgap energy depending on the size of the nanoribbon itself and on its edge termination, since armchair or zigzag edges strongly enhance the DOS locally, with different electronic structures [40–42]. It is important to mention that the edges, in particularly zigzag ones, are chemically active, as demonstrated through DFT calculations by Jiang et al. [43], in contrast with armchair edges and the graphene basal plane, which are inactive [44, 45]. The curvature of the graphene sheet, hence the radius of a carbon nanotube, is also an important parameter which plays a role in the modification of the electronic structure [46, 47] and in lowering the activation energy for oxygen dissociation [48].

9.2.1 Heteroatom-Doped Carbon Nanostructures

Among the strategies to dope graphene, heteroatom doping is a promising way to improve graphene's reactivity by introducing active sites that facilitate interaction with foreign gases, already at room temperature [23]. Heteroatoms change the electronic properties of graphene, modify its morphology and, in addition, activate its surface. Heteroatom, or substitutional, doping is obtained when a foreign atom replaces a carbon atom in the hexagonal lattice; typical heteroatoms are nitrogen [49–53], boron [54–56], sulphur [51, 57, 58], and phosphorus [21, 59, 60]. The first two have a similar atomic radius to that of the carbon atom but one electron difference (seven in N, five in B, so a different electronegativity), so they also constitute a straightforward model to achieve n- and p-type charge-carrier doping of graphene, respectively. Like nitrogen, phosphorus has five valence electrons, but, being on the third shell, its atomic radius is bigger; its incorporation in the carbon lattice will then cause a protrusion out of the graphitic plane [59]. At the same time, other functional groups containing oxygen [61], hydrogen [62, 63] or fluorine [64–67], also contribute to the modification of graphene, affecting its electronic states. In general, point defects generate localised states at the Fermi level, easily identifiable as protrusions in scanning tunnelling microscopy (STM) [68], while carbon vacancies are responsible for a gap opening [69]. The increase in the DOS around the Fermi level after doping [70] is usually responsible for the enhancement of the catalytic activity of the material. The heteroatoms in the graphitic network make the catalyst non-electron neutral. In the particular case of nitrogen doping, the carbon atom that is the nearest neighbour of an N atom is considered to be the active site [18, 32] because of the net positive charge accumulating on it due to the higher electronegativity of the N atom [71, 72], so it will attract electrons from the anode more easily and facilitate the ORR. Yu et al. [71], using DFT calculations, showed an increase in the DOS at the Fermi Level of the C atom neighbour to the substituting N atom, as shown in Fig. 9.3a. This is due to the combination of two effects: the higher electronegativity of nitrogen than carbon and the back-donation of the lone-pair electrons from N to C. In this context, Scardamaglia et al. evaluated the increase in the DOS at the Fermi level of nitrogen-doped vertically-aligned carbon nanotubes (N-vCNTs) by measuring valence band (VB) spectra in high resolution [70]. They correlated the change in the DOS to the different types of nitrogen functionalities incorporated in the carbon matrix, by using XPS and in-vacuum annealing. In particular, by increasing the temperature up to 950 °C the amount of pure substitutional (graphitic) nitrogen increases followed by the DOS, as reported in Fig. 9.3b. As mentioned, the low-energy electronic excitation and the selectivity for oxygen dissociation strongly depend on the different configurations of nitrogen dopant in the sample, as we will discuss in the next paragraph. The increasing interest in new functionalization strategies of carbon nanomaterials has been triggered by their wide range of potential applications.

9.2.2 Co-doping

After the reports on the catalytic performance of nitrogen-doped carbon nanomaterials for ORR, many groups found that co-doping with nitrogen and another element could further enhance those properties in a synergetic effect. The first report showing that the co-doping effect enhanced the metal-free catalytic activities of carbon-based catalysts dates back to 2011 and is based on vertically-aligned BCN (boron-carbon-nitrogen) nanotubes [73]. Besides boron [73–75], other elements that have been used in combination with nitrogen are sulphur [76] and phosphorus [77]. Similar to

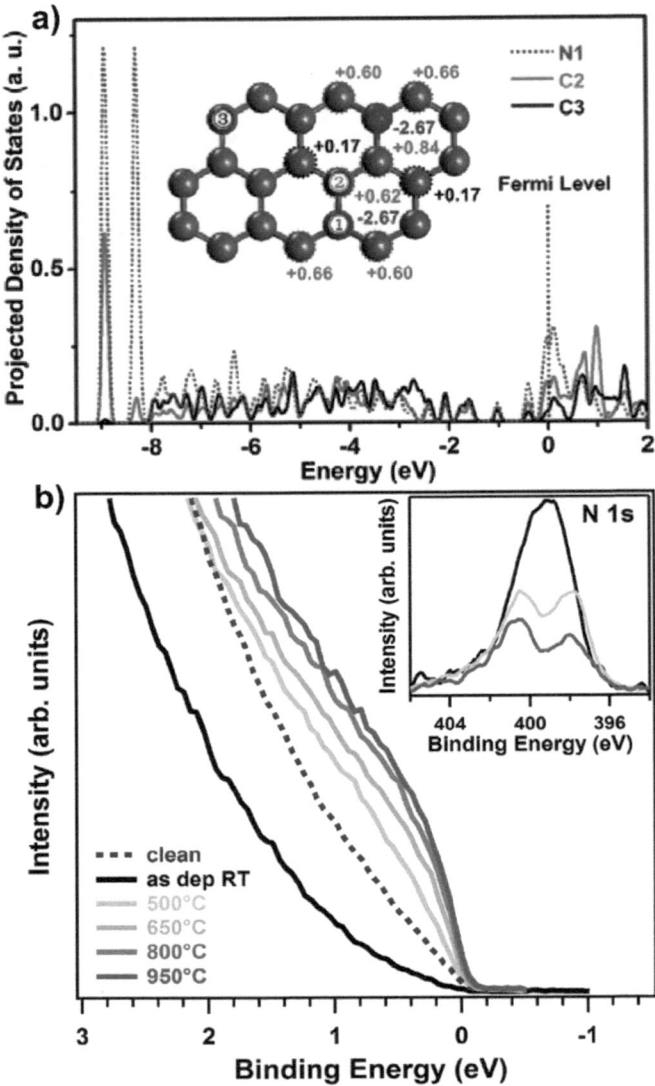

Fig. 9.3 (a) Projected DOS on p_z orbital of the N and C atoms denoted in the inset. The inset shows the net charges on N and its adjacent C atoms. Reprinted by permission from Nature [32]. (b) Valence band spectra near the Fermi energy level for pristine, nitrogen-functionalized (as dep RT) and annealed nitrogen-functionalized v-CNTs. The integrated spectra were recorded normal to the tip of vertically aligned CNT, using a photon excitation of 31 eV. In the inset are the XPS N 1 s core levels recorded after the nitrogen functionalization (black) and after two heating treatments: 500 °C (yellow) and 950 °C (red). Colors code for graphics in the original references. (Reprinted from [70] with permission from Elsevier)

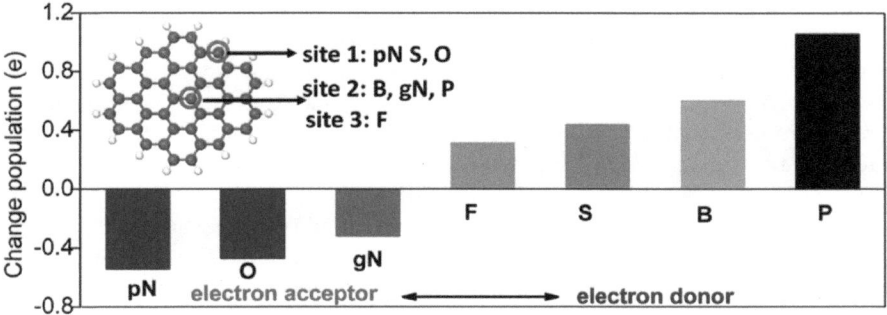

Fig. 9.4 Natural bond orbital (NBO) population analysis of six different non-metallic heteroatoms in a graphene matrix. pN and gN represent pyridinic and graphitic types of N, respectively. The inset shows the proposed doping sites for different elements, sites 1 and 2 are the edge and centre in-plane sites, respectively, and site 3 is an out-of-plane centre site in graphene. (Reproduced from [77], copyright 2014 American Chemical Society, published under a CC-BY 4.0 International license, https://creativecommons.org/licenses/by/4.0/)

N, boron can take a substitutional position with three-fold coordination or form a vacancy complex changing the bond lengths in the carbon lattice, and consequently these configurations have different responses to a target gas [23]. Wu et al. explained the interesting behaviour that takes place when oxygen interacts with boron-containing carbon nanomaterials: due to its accepting nature, B easily oxidizes in a site-dependent way [78]. In nitrogen-doped materials, dual doping promotes long-term stability and resistance to poisoning agents for working catalysts. In general, the principle of dual doping is the introduction in the hexagonal carbon lattice of two elements with reverse electronegativity in relation to that of carbon ($\chi = 2.55$), such as N ($\chi = 3.04$) from one side and B, P or S ($\chi = 2.04, 2.19, 2.58$, respectively) on the other side, as sketched in Fig. 9.4. In this way, the coupling of the donor and acceptor activity of both dopants generates an increase in the ORR activity, with better performance than single-doped materials. The reason for the enhanced catalytic activity cannot simply be associated with an increase in the available active sites, but the synergistic effect of both dopants needs to be considered as well. A high-performance anode material for lithium-ion batteries was obtained using nitrogen- and fluorine-co-doped graphene, prepared through hydrothermal reaction of graphene oxide in aqueous dispersion with trimethylamine tri(hydrofluoride) [79].

9.2.3 Synthesis and Properties of Nitrogen-Doped Graphene

Heteroatom (in particular nitrogen) doping of carbon nanostructures can be performed either during synthesis or in post-synthetic treatments. The most common technique for doping during synthesis is chemical vapour deposition (CVD), similar to the synthesis of the pristine material [38, 80, 81], though using nitrogen-containing precursors such as benzylamine [82], acetonitrile [83, 84], phthalocyanines [85, 86] or ammonia [87]. Nevertheless, post-treatments are particularly effective for achieving surface functionalization, while the use of nitrogen-containing precursors along with a carbon source is more suitable for obtaining a homogeneous incorporation of nitrogen in 3D graphene foam [88], multiwall CNTs [89] or porous carbon [90].

Post-synthesis treatments such as cold plasmas and ion implantation are emerging as versatile options to engineer nanostructured carbon materials [91]. These techniques make it possible to reach high dopant concentrations (order of 10 at.%) and design different architectures adjusting the ion kinetic energy, plasma parameters and/or controlled annealing treatments [92, 93]. Recently, in accordance with the theoretical report by Krasheninnikov et al. [94], Scardamaglia et al. showed that for the doping of carbon nanostructures in a suspended geometry, the use of energetic ions, with kinetic energy from hundreds eV to keV, does not lead to the damage or sputtering of the material [49, 69, 95, 96]. This effect was explained taking into consideration the difference in the dissipation of the ion energy in nanostructured materials with respect to bulk material. In the first case, many ions traverse the nanostructures without inelastic scattering, while for bulk or graphene supported on heavy metallic substrates, the energy of the incident ion species is dissipated at the surface, generating a cascade process of backscattering, leading to the sputtering of the sample's surface (Fig. 9.5a). Using molecular dynamics combined with the analytical potential and density functional theory methods, Lehtinen et al. reported on the influence of the ion kinetic energy and mass in the probability of defect formation during irradiation of suspended graphene sheets and single-walled CNTs [97]. Being proportional to the ion mass, the aforementioned probability is lower for nitrogen than for argon. The energy required to displace a C atom from the hexagonal network is about 22 eV [98], which creates carbon vacancies that are then filled by N atoms. It was shown that at higher kinetic energies, the cross section for vacancy creation decreases and more complex defect configurations are created, in the form of di-vacancies plus distortions (Fig. 9.5b) [94, 97, 98]. The effect of different ion kinetic energies on supported graphene has been highlighted in the case of CF_4 plasma: for energies close to the threshold for carbon substitution, a functionalization takes place, while at higher energies the main electronic signatures of graphene are compromised, indicating the destruction of the hexagonal lattice due to the metallic atoms backscattering from the substrate [66, 99]. On the other hand, in suspended graphene, the main effect of the variable ion energies was the creation of different defect configurations without compromising the properties of the original material [49, 66, 67].

Other advantages of the post-synthesis functionalization methods include the possibility of performing the doping after integrating the nanostructures into a device and the possibility of creating different nano-patterns using appropriate masks [100]. In this context, plasma-based methods (radio frequency or microwave) have proved to be effective for grafting different atomic species, such as oxygen [93, 101], fluorine [65, 66], nitrogen [69, 102, 103], and boron [104] on both

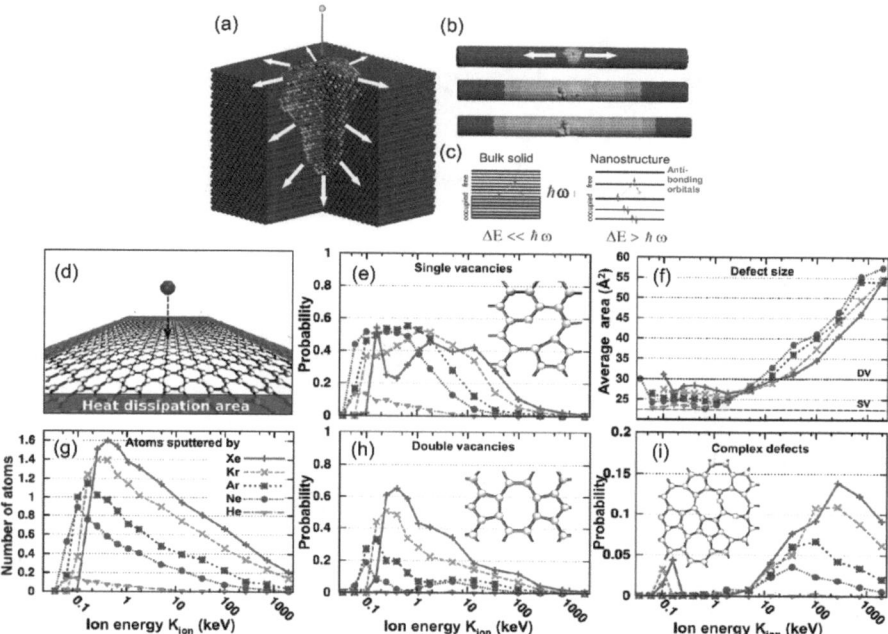

Fig. 9.5 (**a-c**) Conversion of the projectile initial kinetic energy into thermal energy in bulk and nanosystems. (**a**) Impact of an energetic ion onto a bulk metal target. The ion kinetic energy is transferred ballistically to the target atom, which results in temperature rise. The excess energy is dissipated in an essentially 3D system. The atoms are coloured according to their kinetic energy from zero energy (blue) to high energies (red). A quarter of the target was cut out for better visualisation. (**b**) Impact of an ion on a carbon nanotube, a quasi-1D system. The excess energy dissipates in only two directions, which may affect the temperature profile and cause additional defects. (**c**) The sketch of the electronic structure of bulk and nanoscale objects, illustrating the so-called "phonon bottleneck" problem. The excitation relaxation time is enhanced when the spacing between the size-quantized energy levels ΔE is larger than the vibrational energy $\hbar\omega$. This mechanism is discussed for illustration purposes only. There are many other non-radiative relaxation channels in nanosystems that affect the excitation lifetimes. (Reproduced with permission from [94], copyright 2010 American Institute of Physics". (**d-i**) Production of defects in graphene under ion irradiation as revealed by analytical potential molecular dynamics. (**d**) Simulation setup. "(**e, h**) Probability for the formation of single and double vacancies as a function of the ion energy. The insets show the atomic structures of the reconstructed vacancies. (**f**) Average area of defects as a function of the ion energy. The areas corresponding to single vacancies (SV) and double vacancies (DV) are marked. (**g**) Number of sputtered atoms per ion impact as a function of the ion energy. (**i**) Probability for creating defects other than SV/DV (except Frenkel pairs or Stone–Wales defecs), see the inset for an example. Reprinted figure with permission from [97], copyright 2010 American Physical Society)

carbon nanotubes and graphene. However, to avoid contamination with undesired elements, these methods require a UHV environment and typical surface preparation techniques should be considered. Controlled doping using plasma causes a change in the electronic properties without affecting the morphology. In particular, in the

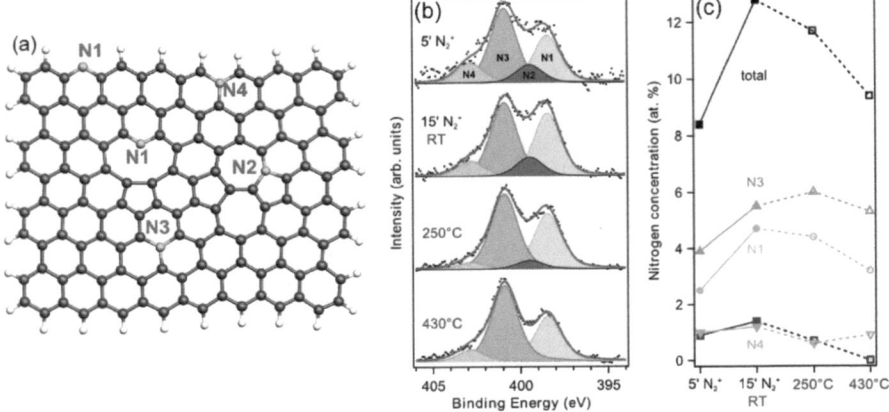

Fig. 9.6 (**a**) Schematic representation of proposed nitrogen configurations in graphene. Nitrogen atoms appear in blue, sites are labelled for (N1) pyridinic N, (N2) pyrrolic N, (N3) graphitic N, and (N4) graphitic "valley" N. Note that these are only indicative of potential structures and do not represent an exhaustive list of possibilities. (**b**) N 1 s core level spectra corresponding to 5 and 15 min nitrogen implantation of CVD few-layer graphene at 250 °C and 430 °C annealing from top to bottom, respectively; experimental data (dotted line), peaks resulting from a least-square fitting procedure (continuous red line). A Shirley-type background was subtracted. (**c**) Nitrogen content in the sample for each component and total amount (black squares) as a function of implantation time (full symbols connected by full line) and annealing temperature (open symbols connected by dotted line). The second and third points represent respectively nitrogen concentration after 15 min of implantation (second point) and after annealing to 250 °C (third point). Colors code for graphics in the original reference. (Adapted from [49] with permission from Elsevier)

case of nitrogen doping, Scardamaglia et al. [49] showed that the sp^2 character of the material is maintained or easily recovered through thermal annealing (Fig. 9.6). A nitrogen atom can be hosted in the hexagonal carbon network in many forms: the three most common configurations are pyridinic (N1), pyrrolic (N2) and graphitic (N3, N4). While the pyrrolic N is a pentagonal structure, the pyridinic and the graphitic are sp^2 hybridised and, in both configurations, the N atom takes the place of a C atom. It is important to note that the graphitic configuration is purely substitutional, unlike the pyridinic, which has a vacancy as a neighbour.

The different nitrogen species are easily distinguished according to their binding energy thanks to XPS: pyridinic nitrogen is usually found at 398–399 eV, pyrrolic and other defective components at 399.5–400.5 eV, graphitic at 400–401 eV [15, 105, 106]. At a slightly higher binding energy with respect to graphitic N, the component was assigned to graphitic-valley [107] – a graphitic that is close to a vacancy or edge (N4 in Fig. 9.6a). Nitrogen oxide is usually found at binding energies higher than 402 eV [108]. It was reported that at the binding energy corresponding to pyrrolic nitrogen, many other defective components may be found, such as single C-N bonds, nitrogen adatoms or Stone-Wales defects. The latter, in particular, are three-fold coordinated N atoms incorporated between two pairs of five-membered and seven-membered rings.

The nitrogen content and the nitrogen functionalities on graphene and carbon nanotubes can be tuned through thermal annealing or through interaction with different substrates. Temperature treatments were reported to promote the conversion of pyridinic N into graphitic N and the fine-tuning of the DOS at the Fermi level, which increases proportionally with graphitic N content in the carbon lattice. The five valence electrons of the N atom are distributed differently, therefore graphitic and pyridinic nitrogen atoms have different effects on the electronic properties of graphene. While the first one contributes with three σ bonds and two p_z orbitals, with the extra electron available for conduction in a partially-occupied π^* band, the latter forms only two σ bonds with the neighbouring carbon atoms, one electron occupies the p_z orbital, and the other two form a lone electron pair. Since there is no occupation of the π^* band in the pyridinic nitrogen, it does not behave as an electron dopant (whereas graphitic N does). Furthermore, due to the proximity of the carbon vacancy, the pyridinic vacancy complex has a hole-doping effect. This behaviour is reflected in the DOS at the Fermi level, and it can be clearly observed in Fig. 9.3b where the VB spectra of N-doped CNTs are reported together with the respective N 1 s core level spectra [70]. After plasma exposure, the amount of defective nitrogen components (pyrrolic, pyridinic) was higher than for graphitic N: this is reflected in a decrease of the DOS with respect to the pristine sample. When UHV annealing was performed, the N 1 s lineshape (onset of Fig. 9.3b) evolved due to the desorption of pyrrolic N and the transformation of pyridinic into graphitic N; both these effects were reported to be responsible for the increase in the DOS, which exceeds that of the pristine sample. Recently, these predictions were experimentally confirmed following the Dirac cone shift upon modification of the graphitic-pyridinic ratio through thermal treatment of graphene [69] and by intercalating different metal atoms [109]. The authors reported the calculated and experimental Dirac point positions for different pyridinic-graphitic ratios (see Fig. 9.7). There is a counterbalancing effect between the two doping actions that obscures the evaluation of the effective charge transfer to/from the dopant atoms.

The amount of nitrogen is also an important parameter for preserving the stability and the electronic properties of graphene: it has been shown that there is an upper limit for local substitutional nitrogen concentration in an sp^2 layered carbon system [105]. For bilayer graphene, with a nitrogen concentration lower than 15–20 at. % the system remains fully sp^2-coordinated; above this threshold, cross-linking between the layers starts to occur and carbon begins to be sp^3-coordinated [105]. At intermediate concentrations the system is stabilised by nitrogen saturation of edges and vacancies, implying an increase in pyridinic N content with respect to pure substitutional, as was observed experimentally in ion-bombarded suspended graphene [49].

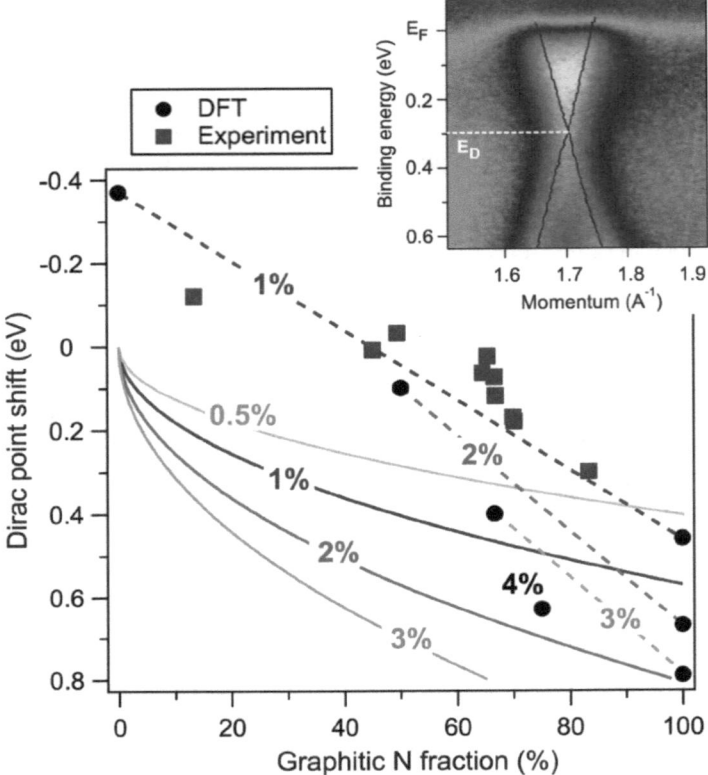

Fig. 9.7 Measured Dirac point position (blue symbols) at different concentrations of pyridinic and graphitic N, compared to the DFT calculation results (black circles connected by dashed lines). Numbers denote the total concentration of nitrogen (NP + NG). Solid lines demonstrate the hypothetical situation when pyridinic N provides no charge transfer, while graphitic N donates one electron. The inset shows the ARPES intensity of an N-graphene/Au/Ni(111) system at the K-point of the BZ, measured after nitrogen conversion in the ΓK direction at a temperature of 40 K. Colors code for graphics in the original references (Reprinted with permission from [109]. Copyright (2014) American Chemical Society)

9.3   The Role of the Active Sites

As reported in the previous section, nitrogen doping of carbon nanomaterials improves their catalytic activity, which in some cases overcomes the performances of commercially available metal-based catalysts. Nitrogen atoms can have different configurations in the hexagonal carbon lattice, and the role of each of these structures in the catalytic or sensing process has not yet been clarified. In the work by Qu et al. [19], a first report on the electrocatalytic performance of N-doped graphene, the authors found a threefold steady-state current density with respect to commercial Pt-C electrodes. This was the first report where N-doped graphene was employed as an electrode for ORR with performances better than the commercial electrodes. Although the role of the active sites was not addressed in detail, the electrode performance was already remarkable. The N/C ratio was approximately 4% but the energy resolution of the reported XPS data was too poor to correctly distinguish the different N species, and the N 1 s peak was deconvolved into two components: pyridinic at 398.3 eV and pyrrolic at 400.5 eV. The absence of a graphitic component in the spectrum analysis may be due to poor data interpretation rather than due to an actual physical phenomenon.

The correlation between catalytic activity and different nitrogen functionalities is investigated by Rao et al. [110]. Evaluating the ORR activity of v-CNTs with different amounts of incorporated nitrogen (from 4 to 11 at.%), they found that the v-CNTs with the highest amount of pyridine showed the highest onset potential (the potential where the current density increases to 10 μA/cm^2) and the most positive reduction peak (0.35 V). The results obtained using nitrogen-doped v-CNTs were superior to when un-doped nanotubes were used but less remarkable than commercial Pt-C electrodes. Nevertheless, they attributed the improved performances to the conjugation effect of the nitrogen lone pair electrons found on the pyridinic and graphene π-system, as already reported for carbon nano-fibres by Maldonado and Stevenson [111]. A more detailed study was performed by Kundu et al. [112] on N-CNTs prepared via pyrolysis of acetonitrile using cobalt catalysts. Samples were synthesized at two different temperatures, 550 and 750 °C, allowing for approximately the same amount of nitrogen (around 7 at.%) but a different graphitic/pyridinic ratio, 0.7 and 2.3 respectively. The obtained N 1 s core level spectra are shown in Fig. 9.8a, b.

The reported catalytic activity measured using CV and RDE in oxygen-saturated H$_2$SO$_4$ at 900 rpm is shown in Fig. 9.8c. The results indicated that the catalytic activity for ORR of nitrogen-doped CNTs, with onset potential at about 0.5 V, is much more positive than pristine CNT, although 0.1 V lower than the Pt-C catalyst. In particular, the performance of the sample with the higher amount of pyridine (NCNT-L) is superior to the one with higher graphitic content.

The reports of Rao and Kundu, mentioned above, suggested that pyridinic nitrogen is responsible for the high ORR catalytic performance of nitrogen-doped carbon nanostructures. Remarkably, Luo and colleagues succeeded to synthesize pure pyridinic-N-doped graphene through CVD of hydrogen, ethylene and ammonia on Cu foils [113], with nitrogen content up to 16 at.%. A graphene with only one type of N dopant is an optimum platform to study the role of this particular site for ORR activity. However, contrary to the previous reports that attributed the high ORR performance to the pyridinic N, in their work, Luo et al. showed that all the graphene samples had poor ORR activity with 0.3 V difference in the onset potential with respect to the Pt disk. The low performance was associated with the mutual repulsion between the lone pair electrons in both pyridinic N and O$_2$, which produces a higher energy barrier to the molecular oxygen activation.

Along the same lines, two other groups reported that catalysts with a higher amount of graphitic N exhibit a higher ORR activity [114, 115]. Niwa et al. [114]

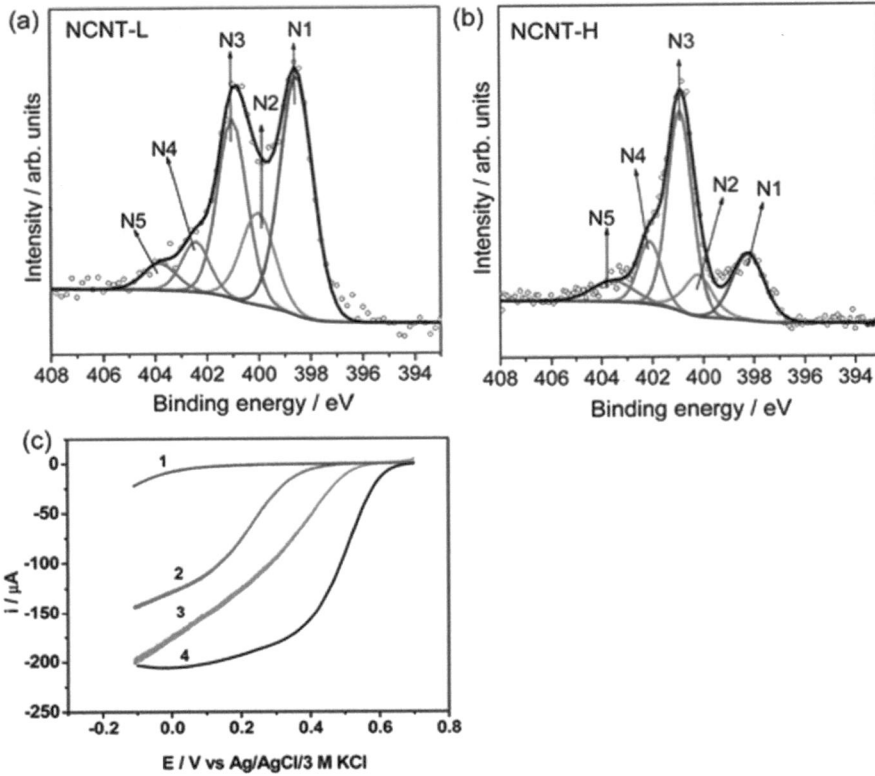

Fig. 9.8 N 1 s core level spectra for the N-CNT samples grown at 550 °C (**a**) and 750 °C (**b**). N1 pyridinic, N2 pyrrolic, N3 graphitic, N4 pyridine-N-oxide, and N5 chemisorbed nitrogen oxide. (**c**) RDE linear sweep voltammograms for ORR in oxygen-saturated 0.5 M H2SO4 at a rotation speed of 900 rpm and a scan rate of 5 mV/s with (1) un-doped CNT, (2) NCNT-H, (3) NCNT-L, and (4) E-TEK Pt-C (20 wt %) catalysts. (Adapted with permission from [112]. Copyright (2009) American Chemical Society)

studied the introduction of nitrogen into various carbon-based cathode catalysts for the polymer electrolyte fuel cell (PEFC) by using x-ray absorption spectroscopy (XAS) and ORR activity measurements. The material used in their work was carbon nanostructure alloys, a precursor of graphene used as a replacement for the conventional support, carbon black (Vulcanic XC-72) [116]. Different preparation methods were used: nitrogen doping with ammonia resulted in a higher concentration of pyridinic N, while the amount of graphitic N was higher using pyrolysis of nitrogen-containing precursors. Using x-ray absorption spectroscopy, (Fig. 9.9), the authors were able to identify different nitrogen species: three characteristic peaks are individuated at 399.1 eV, 400.1 eV and 401.5 eV and were assigned to pyridinic, cyanide (triple C-N bond) and graphitic N, respectively, in agreement with previous

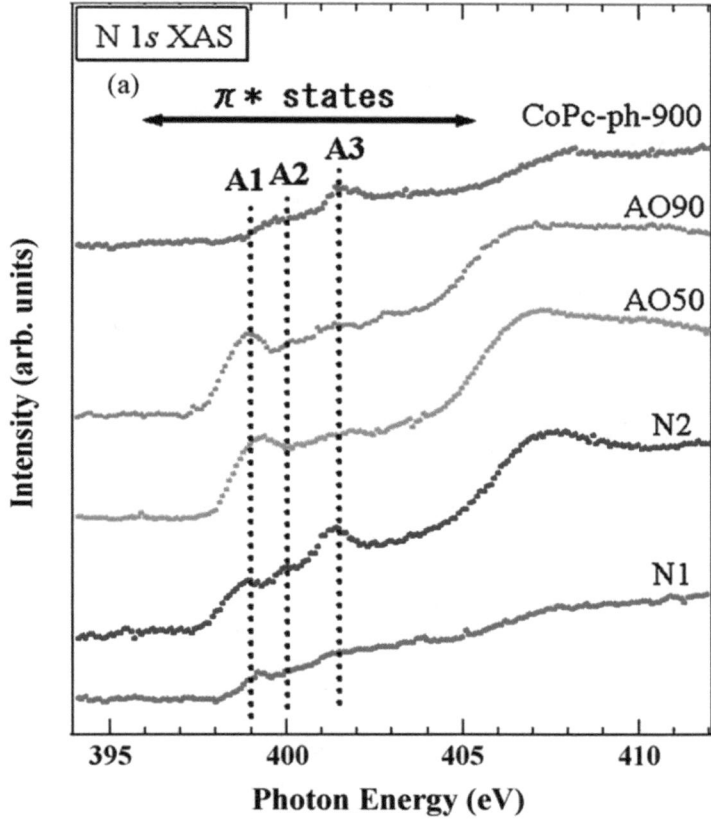

Fig. 9.9 X-ray absorption spectroscopy N 1 s → π* spectra for carbon alloy catalysts. Pyridinic (A1), cyanide (A2) and graphitic (A3) nitrogen atoms in graphene. The spectra correspond to different samples: nanoshell carbon derived from pyrolyzed cobalt phthalocyanine and nitrogen-containing polymers followed by acid washing to remove cobalt (CoPc-ph -900); nitrogen-doped carbon alloys prepared by ammo-oxidation (AO50 and AO90); nitrogen-doped carbon materials prepared by nitrogen-containing precursors (N1 and N2). (Reprinted from [114] with permission from Elsevier)

reports that correlate XAS and XPS [117, 118]. By correlating XAS and RRDE, they assumed that the π* profile is an indicator of ORR activity.

Nagaiah and colleagues arrived at similar results in nitrogen-doped CNTs [115]. In their work, post-synthesis heating treatments were used to modify the surface composition, resulting in different nitrogen functionalities. The material with the best catalytic performance for ORR was found to be the one with the highest graphitic/pyridinic ratio.

Different groups calculated the energy barriers for oxygen molecule adsorption and dissociation on both pristine and N-doped SWCNTs and graphene using DFT [72, 119–121]. Ni and co-workers [120], in particular, studied the energy barriers depending on different nitrogen configurations and found that they decrease with all types of nitrogen- doping configurations. Graphitic and Stone-Wales defects are more efficient than pyridinic nitrogen in lowering the energy barrier for an oxygen reaction. In fact, pyridinic nitrogen has a negligible occupation of π^* anti-bonding orbital, due to its two-fold coordination and the lone pair electrons, contrary to both graphitic N and Stone-Wales defects, which are three-fold coordinated. The dissociation of the oxygen molecule was reported to take place on the carbon atom neighbouring the substitutional nitrogen; the two dissociated O atoms are then adsorbed onto an adjacent C-C bridge site, while the C-N bridge site is not available for O adsorption. This mechanism was recently observed by Scardamaglia et al. who carried out high-resolution synchrotron measurements in a fully *in situ* experiment [95] in order to avoid contamination. They functionalized single-layer graphene grown by CVD on iridium (111) using nitrogen plasma, then they studied the interaction with molecular oxygen (O_2). This led to oxygen dissociation and the formation of single carbon-oxygen bonds on graphene. The change of the N 1 s core level spectrum indicates that graphitic nitrogen is involved in the observed mechanism: the adsorbed oxygen molecule is dissociated and the two O atoms chemisorb with epoxy bonds to the nearest carbon neighbours of the graphitic nitrogen, as schematized in Fig. 9.10.

Theoretical studies on graphene nanoribbons made it possible to distinguish the catalytic activity of different nitrogen atom positions with respect to the graphitic lattice, with atomistic precision. Oxygen adsorption at edge sites has a much lower energy barrier than "planar bulk" sites, independently from any possible functionalization due to their intrinsic reactivity [45]. Nitrogen doping not only increases the activity on the edges but also enhances the electron transfer rate because the energy difference between the Fermi level and the unoccupied $2p$ orbital state in the adsorbed oxygen molecule is reduced. Furthermore, the ORR pathway in the presence of nitrogen changes (for nanoribbons) to a most efficient four-electron reduction process, instead of two-electron. Kim et al. studied the localization of different graphitic -N sites with respect to the edge and found that the most catalytically active site is the near-edge one, the outermost graphitic, or graphitic-valley [121]. Their results are in agreement with Ikeda and colleagues, who stated that an O_2 molecule is preferentially adsorbed associatively at C sites on graphene-like zigzag edges if a graphite-like N is located nearby [119]. Additionally, Kim et al. suggested that the graphitic N catalyses the reaction through a ring opening of the cyclic C-N bond, which results in the formation of pyridinic N, as shown in Fig. 9.11. In this way, the controversy over the role of graphitic and pyridinic nitrogen in the catalytic process is partially solved.

In agreement with the theoretical findings of Ikeda and Kim [119, 121], in 2012, Sharifi et al. investigated the formation of different active sites through heat treatment of nitrogen-doped multiwalled CNTs and subsequent ORR measurements [107]. The nanotubes were synthesized using a pyridine precursor resulting in N contents in the range 3–6 at.%. However, the starting material had an O content comparable to that of the N, due to the synthesis condition, that can be translated into a relatively high concentration of nitrogen oxide groups, as shown by the N

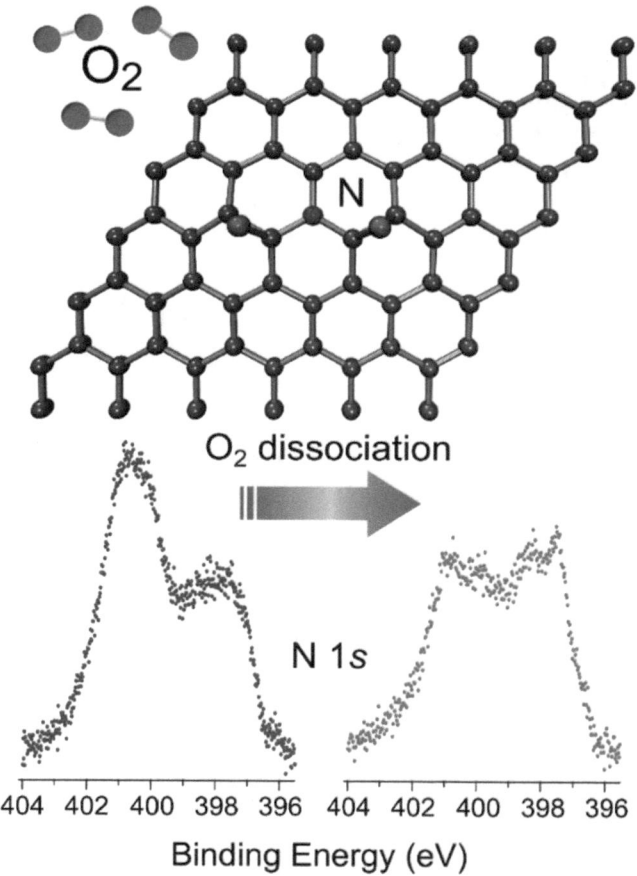

Fig. 9.10 Schematic representation of the molecular oxygen dissociation on nitrogen-doped graphene. Colors code for graphics in the original reference. (Adapted from [95] under copyright 2017 the authors, published under a CC-BY 4.0 International license, https://creativecommons.org/licenses/by/4.0/)

1 *s* core level spectra. Annealing the samples in argon atmosphere at temperatures ranging from 500 to 1000 °C, the changes induced in the amount and the surface rearrangement of the nitrogen species were studied: a decrease in pyrrolic N and a general transformation into graphitic N, consistent with other works, was observed [70]. The electrocatalytic performances were measured using cyclic voltammetry in oxygen-saturated 1 M KOH electrolyte. By correlating XPS and ORR, the samples with the highest amount of graphitic-valley N sites were identified as the most active ones with a four-electron process, as reported in Fig. 9.12, in agreement with theoretical calculations [119, 121].

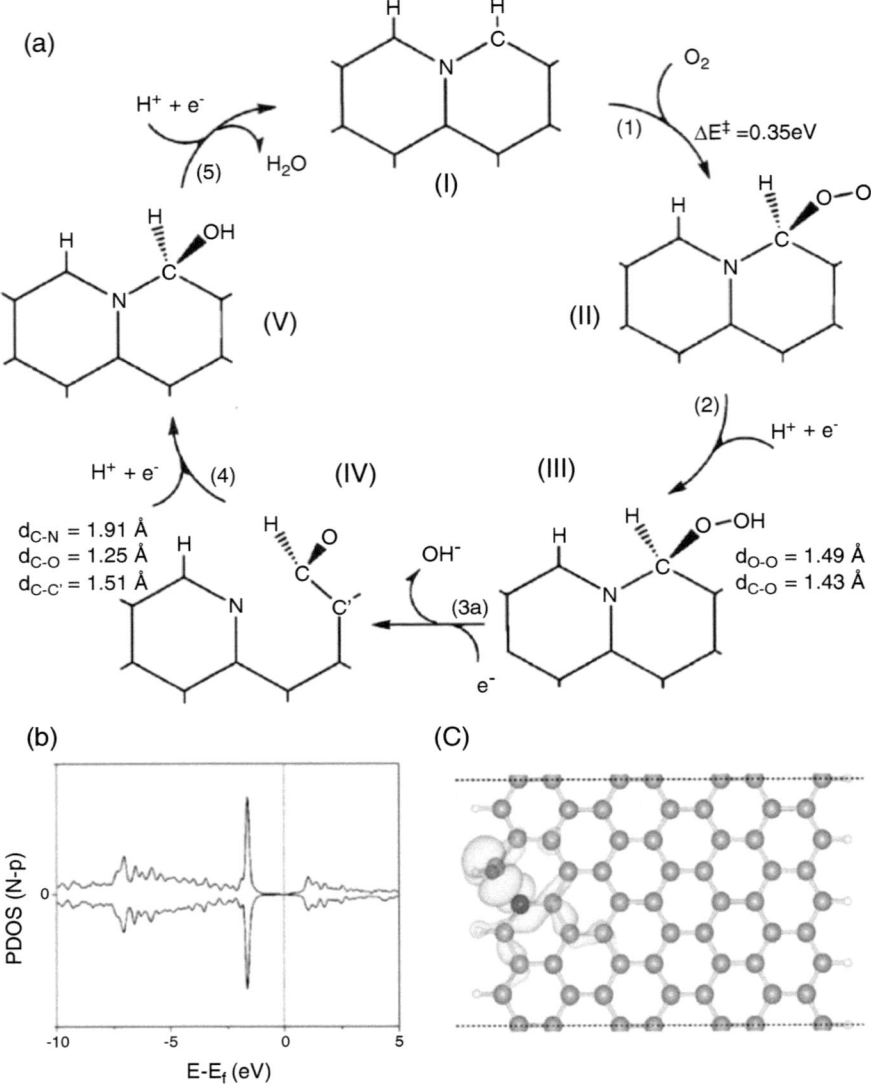

Fig. 9.11 (a) The proposed ORR catalytic cycle for the N0 structure. (b) The partial density of state (PDOS) for the p-orbital of the nitrogen atom in the N0 structure in stage (IV) of (a). (c) The isosurface of electron density for the band of peak position around −2 eV. The blue and red atoms are nitrogen and oxygen, respectively. Colors code for atoms in the original reference. (Reproduced from Ref. [121] with permission from the Royal Society of Chemistry)

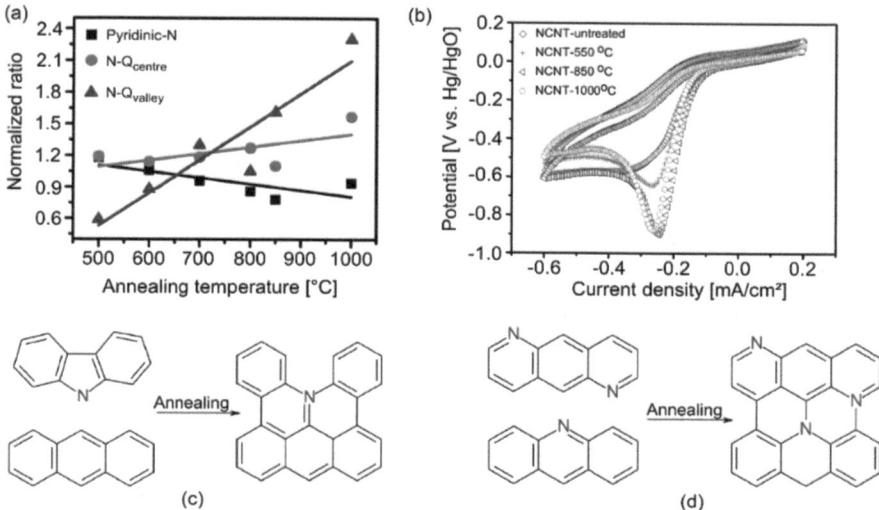

Fig. 9.12 (a) Normalized ratio of nitrogen functionalities vs annealing temperature. (b) Background-corrected cyclic voltammograms (CV) in oxygen-saturated 1 M KOH electrolyte for nitrogen-doped CNTs annealed at high temperatures. (c) Condensation reactions by annealing from pyrrolic to graphitic valley and (d) from pyridinic to graphitic (center and valley). (Adapted with permission from [107]. Copyright (2012) American Chemical Society)

Simultaneously, Lai et al. studied the catalytic activity of two distinct N-doped graphene samples that had undergone different temperature treatments to tune the surface composition of the different nitrogen functionalities, as observed through XPS [122]. They produced N-graphene in two ways: (1) by annealing GO in NH_3 and (2) by annealing rGO and a composite of N-containing polymer (polyaniline or polypyrrole). In the first case, they obtained mostly graphitic and pyridinic N, in the second one, pyridinic and pyrrolic N. The total concentration of N they have is between 3 and 7 atomic %. Correlating XPS with electrochemical tests carried out in a standard three-electrode cell with a Pt plate as the counter electrode, they showed that the total N content is not a fundamental parameter in the ORR process, while graphitic and pyridinic nitrogen have different roles. Graphitic nitrogen provides the maximum current density, while the pyridinic configuration improves the onset potential for ORR and might convert the ORR reaction mechanism from a 2-electron process to a 4-electron process.

More recently, Guo and colleagues published a detailed work on engineered HOPG (highly oriented pyrolytic graphite) with well-defined and controlled N species [123]. They achieved the synthesis of model HOPG catalysts with just one nitrogen dopant configuration – pyridinic or graphitic – with different concentrations up to 11 at. %. By measuring the ORR activity using cyclic voltammograms (CV) in an acidic electrolyte (0.1 M H_2SO_4), they individuated the pyridinic-HOPG as the best catalysts for ORR, with higher activity at high voltages, with a linear relation between current density and pyridinic concentration, without any dependence on the amount of graphene. Using *ex situ* post-ORR XPS, they observed a transformation of pyridinic N into pyridonic N after the ORR: this is consistent with the adsorption of an O atom on the C atom next to the pyridinic. Similar results were previously obtained on multilayer graphene by performing post-ORR XPS measurements [124]. This behaviour was ascribed to the Lewis basicity of the C atom due to the localized DOS near the Fermi level, as observed through STM [125].

9.4 Conclusions and Future Perspectives

As we have seen, in all the experiments reported in this Chapter, the complete picture of the nitrogen active sites in carbon nanomaterials for ORR activity is far from being clear, since many results are contradictory in nonetheless similarly designed experiments. In general, it can be concluded that both species have a role: graphitic nitrogen determines the maximum current density that facilitates electron transfer from graphene to the antibonding orbitals of the oxygen molecule [113–115, 122], while pyridinic nitrogen improves the onset potential for the ORR by weakening the O-O bond [110, 111, 123, 126]. From a theoretical point-of-view, it has been shown that the activation barrier for O_2 adsorption is much higher on both the pyridinic nitrogen and the nearest carbon atom [72, 119–121] if compared to the adsorption on graphitic nitrogen; the proximity and the structure of the edge's flake is another important parameter in lowering such a barrier. The reason this debate has lasted so long is primarily due to experimental limitations. The low doping level and the lack of control of the nature of the grafted heteroatom functionalities, as they strongly depend on synthesis conditions, causes an inhomogeneous mix of different types of nitrogen species. Furthermore, reported experimental studies of the active sites on heteroatom-doped carbon nanomaterials are often indirect. One way to make important progress in the field of catalysis is to make links with our understanding of the fundamental surface science behind the catalytic process [127]. In particular, from an experimental perspective, recently developed techniques such as (near) ambient-pressure XPS may be a promising tool to explore this field since it can probe the materials in *operando* conditions [128–131]. In this way, it is possible to directly identify the different nitrogen configurations possible in a carbon network using XPS during the chemical reactions or to probe the mechanism of interaction between doped graphene and a target gas. Of course, together with advanced measurement techniques, the engineering of the catalyst should be improved in order to obtain the optimum concentration of active sites without inactive components.

References

1. Conway BE, Tilak BV (2002) Electrochim Acta 47:3571–3594. https://doi.org/10.1016/S0013-4686(02)00329-8
2. Perivoliotis DK, Tagmatarchis N (2017) Carbon 118:493–510. https://doi.org/10.1016/j.carbon.2017.03.073
3. Li S-S, Lv J-J, Teng L-N, Wang A-J, Chen J-R, Feng J-J (2014) ACS Appl Mater Interfaces 6:10549–10555. https://doi.org/10.1021/am502148z
4. Ma Y, Liu Z, Wang B, Zhu L, Yang J, LI X (2012) New Carbon Mater 27:250–257. https://doi.org/10.1016/S1872-5805(12)60016-X
5. Rideal EK, Wright WM (1926) J Chem Soc 129:1813–1821. https://doi.org/10.1039/JR9262901813
6. Mrha J (1966) Collect Czech Chem Commun 31:715–734. https://doi.org/10.1135/cccc19660715
7. Lalande G, Côté R, Guay D, Dodelet JP, Weng LT, Bertrand P (1997) Electrochim Acta 42:1379–1388. https://doi.org/10.1016/S0013-4686(96)00361-1
8. Strelko VV, Kuts VS, Thrower PA (2000) Carbon 38:1499–1503. https://doi.org/10.1016/S0008-6223(00)00121-4
9. Iijima S, Ichihashi T (1993) Nature 363:603–605. https://doi.org/10.1038/363603a0
10. Bethune DS, Klang CH, de Vries MS, Gorman G, Savoy R, Vazquez J, Beyers R (1993) Nature 363:605–607. https://doi.org/10.1038/363605a0
11. Monthioux M, Kuznetsov VL (2006) Carbon 44:1621–1623. https://doi.org/10.1016/j.carbon.2006.03.019
12. Novoselov KS (2004) Science 306:666–669. https://doi.org/10.1126/science.1102896
13. Liu X, Dai L (2016) Nat Rev Mater 1:16064. https://doi.org/10.1038/natrevmats.2016.64
14. Liu J, Song P, Ning Z, Xu W (2015) Electrocatalysis 6:132–147. https://doi.org/10.1007/s12678-014-0243-9
15. Wang D-W, Su D (2014) Energy Environ Sci 7:576. https://doi.org/10.1039/c3ee43463j
16. Wu Z, Iqbal Z, Wang X (2015) Front. Chem Sci Eng 9:280–294. https://doi.org/10.1007/s11705-015-1524-4
17. Matter PH, Ozkan US (2006) Catal Letters 109:115–123. https://doi.org/10.1007/s10562-006-0067-1
18. Gong K, Du F, Xia Z, Durstock M, Dai L (2009) Science 323:760–764. https://doi.org/10.1126/science.1168049
19. Qu L, Liu Y, Baek J-B, Dai L (2010) ACS Nano 4:1321–1326. https://doi.org/10.1021/nn901850u
20. Zhao Y, Nakamura R, Kamiya K, Nakanishi S, Hashimoto K (2013) Nat Commun 4:2390. https://doi.org/10.1038/ncomms3390
21. Zhang J, Zhao Z, Xia Z, Dai L (2015) Nat Nanotechnol 10:444–452. https://doi.org/10.1038/nnano.2015.48
22. Xue Y, Liu J, Chen H, Wang R, Li D, Qu J, Dai L (2012) Angew Chemie Int Ed 51:12124–12127. https://doi.org/10.1002/anie.201207277
23. Adjizian J-J, Leghrib R, Koos A a, Suarez-Martinez I, Crossley A, Wagner P, Grobert N, Llobet E, Ewels CP (2014) Carbon 66:662–673. https://doi.org/10.1016/j.carbon.2013.09.064
24. Koós A a, Nicholls RJ, Dillon F, Kertész K, Biró LP, Crossley A, Grobert N (2012) Carbon 50:2816–2823. https://doi.org/10.1016/j.carbon.2012.02.047
25. Dai J, Yuan J, Giannozzi P (2009) Appl Phys Lett 95:232105. https://doi.org/10.1063/1.3272008
26. Wu J, Pisula W, Müllen K (2007) Chem Rev 107:718–747. https://doi.org/10.1021/cr068010r
27. Frank IW, Tanenbaum DM, van der Zande AM, McEuen PL (2007) J. Vac. Sci Technol B Microelectron Nanom Struct 25:2558. https://doi.org/10.1116/1.2789446
28. Balandin AA, Ghosh S, Bao W, Calizo I, Teweldebrhan D, Miao F, Lau CN (2008) Nano Lett 8:902–907. https://doi.org/10.1021/nl0731872

29. Castro Neto AH, Guinea F, Peres NMR, Novoselov KS, Geim AK (2009) Rev Mod Phys 81:109–162. https://doi.org/10.1103/RevModPhys.81.109
30. Schedin F, Geim AK, Morozov SV, Hill EW, Blake P, Katsnelson MI, Novoselov KS (2007) Nat Mater 6:652–655. https://doi.org/10.1038/nmat1967
31. Gierz I, Riedl C, Starke U, Ast CR, Kern K (2008) Nano Lett 8:4603–4607. https://doi.org/10.1021/nl802996s
32. Deng D, Novoselov KS, Fu Q, Zheng N, Tian Z, Bao X (2016) Nat Nanotechnol 11:218–230. https://doi.org/10.1038/nnano.2015.340
33. Lv R, Chen G, Li Q, McCreary A, Botello-Méndez A, Morozov SV, Liang L, Declerck X, Perea-López N, Cullen DA, Feng S, Elías AL, Cruz-Silva R, Fujisawa K, Endo M, Kang F, Charlier J-C, Meunier V, Pan M, Harutyunyan AR, Novoselov KS, Terrones M (2015) Proc. Natl. Acad. Sci 112:14527–14532. https://doi.org/10.1073/ pnas.1505993112
34. Vass EM, Hävecker M, Zafeiratos S, Teschner D, Knop-Gericke A, Schlögl R (2008) J Phys Condens Matter 20:184016. https://doi.org/10.1088/0953-8984/20/18/184016
35. Fu Q, Bao X (2017) Chem Soc Rev 46:1842–1874. https://doi.org/10.1039/C6CS00424E
36. Georgakilas V, Otyepka M, Bourlinos AB, Chandra V, Kim N, Kemp KC, Hobza P, Zboril R, Kim KS (2012) Chem Rev 112:6156–6214. https://doi.org/10.1021/cr3000412
37. Boukhvalov   DW,   Katsnelson   MI   (2008)   Nano   Lett   8:4373–4379. https://doi.org/10.1021/nl802098g
38. Su DS, Zhang J, Frank B, Thomas A, Wang X, Paraknowitsch J, Schlög R (2010) Chem Sus Chem 3:169–180. https://doi.org/10.1002/cssc.200900180
39. Son Y-W, Cohen ML, Louie SG (2006) Nature 444:347–349. https://doi.org/10.1038/ nature05180
40. Nakada K, Fujita M, Dresselhaus G, Dresselhaus MS (1996) Phys Rev B 54:17954–17961. https://doi.org/10.1103/PhysRevB.54.17954
41. Enoki   T,   Kobayashi   Y,   Fukui   K-I   (2007)   Int   Rev   Phys   Chem   26:609–645. https://doi.org/10.1080/ 01442350701611991
42. Fujii S, Enoki T (2013) Acc Chem Res 46:2202–2210. https://doi.org/10.1021/ ar300120y
43. Jiang D, Sumpter BG, Dai S (2007) J Chem Phys 126:134701. https://doi.org/10.1063/ 1.2715558
44. Deng D, Yu L, Pan X, Wang S, Chen X, Hu P, Sun L, Bao X, Magalhaes-Paniago R, Pimenta MA, Geim AK, Bao XH (2011) Chem Commun 47:10016. https://doi.org/10.1039/c1cc13033a
45. Shen A, Zou Y, Wang Q, Dryfe RAW, Huang X, Dou S, Dai L, Wang S (2014) Angew Chemie – Int Ed 53:10804–10808. https://doi.org/10.1002/anie.201406695
46. Li Z, Cheng Z, Wang R, Li Q, Fang Y (2009) Nano Lett 9:3599–3602. https://doi.org/10.1021/ nl901815u
47. Chae SJ, Güneş F, Kim KK, Kim ES, Han GH, Kim SM, Shin H-J, Yoon S-M, Choi J-Y, Park MH, Yang CW, Pribat D, Lee YH (2009) Adv Mater 21:2328–2333. https://doi.org/10.1002/adma.200803016
48. Srivastava D, Susi T, Borghei M, Kari L (2014) RSC Adv 4:15225. https://doi.org/10.1039/ c3ra47784c
49. Scardamaglia M, Aleman B, Amati M, Ewels CP, Pochet P, Reckinger N, Colomer J-F, Skaltsas T, Tagmatarchis N, Snyders R, Gregoratti L, Bittencourt C (2014) Carbon 73:371–381. https://doi.org/10.1016/ j.carbon.2014.02.078
50. Susi T, Ayala P (2015) Carbon nanomaterials for advanced energy systems. Wiley, Hoboken, pp 133–161. https://doi.org/10.1002/9781118980989.ch4
51. Su Y, Zhang Y, Zhuang X, Li S, Wu D, Zhang F, Feng X (2013) Carbon 62:296. https://doi.org/10.1016/j.carbon.2013.05.067
52. Podyacheva OY, Ismagilov ZR (2015) Catal Today 249:12–22. https://doi.org/10.1016/ j.cattod.2014.10.033
53. Wang   H,   Maiyalagan   T,   Wang   X   (2012)   ACS   Catal   2:781–794. https://doi.org/10.1021/cs200652y

54. Gebhardt J, Koch RJ, Zhao W, Höfert O, Gotterbarm K, Mammadov S, Papp C, Görling A, Steinrück H-P, Seyller T (2013) Phys Rev B 87:155437. https://doi.org/10.1103/PhysRevB.87.155437
55. Agnoli S, Favaro M (2016) J Mater Chem A 00:1–24. https://doi.org/10.1039/C5TA10599D
56. Ferrighi L, Trioni MI, Di Valentin C (2015) J Phys Chem C 119:150305145047002. https://doi.org/10.1021/jp512522m
57. Huang H, Zhu J, Zhang W, Tiwary CS, Zhang J, Zhang X, Jiang Q, He H, Wu Y, Huang W, Ajayan PM, Yan Q (2016) Chem Mater 28:1737–1745. https://doi.org/10.1021/acs.chemmater.5b04654
58. Yang Z, Yao Z, Li G, Fang G, Nie H, Liu Z, Zhou X, Chen X, Huang S (2012) ACS Nano 6:205–211. https://doi.org/10.1021/nn203393d
59. Susi T, Hardcastle TP, Hofsäss H, Mittelberger A, Pennycook TJ, Mangler C, Drummond-Brydson R, Scott AJ, Meyer JC, Kotakoski J (2017) 2D Mater 4:021013. https://doi.org/10.1088/2053-1583/aa5e78
60. Xue Y, Wu B, Liu H, Tan J, Hu W, Liu Y (2014) Phys Chem Chem Phys 16:20392–20397. https://doi.org/10.1039/c4cp02935f
61. Yan J-A, Chou MY (2010) Phys Rev B 82:125403. https://doi.org/10.1103/PhysRevB.82.125403
62. Balog R, Andersen M, Jørgensen B, Sljivancanin Z, Hammer B, Baraldi A, Larciprete R, Hofmann P, Hornekær L, Lizzit S (2013) ACS Nano 7:3823–3832. https://doi.org/10.1021/nn400780x
63. Scheffler M, Haberer D, Petaccia L, Farjam M, Schlegel R, Baumann D, Hänke T, Grüneis A, Knupfer M, Hess C, Büchner B (2012) ACS Nano 6:10590–10597. https://doi.org/10.1021/nn303485c
64. Poh HL, Pumera M (2015) Chem Electro Chem 2:190–199. https://doi.org/10.1002/celc.201402307
65. Struzzi C, Scardamaglia M, Hemberg A, Petaccia L, Colomer J-F, Snyders R, Bittencourt C (2015) Beilstein J Nanotechnol 6:2263–2271. https://doi.org/10.3762/bjnano.6.232
66. Struzzi C, Scardamaglia M, Reckinger N, Colomer J-F, Sezen H, Amati M, Gregoratti L, Snyders R, Bittencourt C (2017) Nano Res 10:3151–3163. https://doi.org/10.1007/s12274-017-1532-4
67. Struzzi C, Sezen H, Amati M, Gregoratti L, Reckinger N, Colomer J, Snyders R, Bittencourt C, Scardamaglia M (2017) Appl Surf Sci 422:104–110. https://doi.org/10.1016/j.apsusc.2017.05.258
68. Rasool HI, Song EB, Mecklenburg M, Regan BC, Wang KL, Weiller BH, Gimzewski JK (2011) J Am Chem Soc 133:12536–12543. https://doi.org/10.1021/ja200245p
69. Scardamaglia M, Struzzi C, Osella S, Reckinger N, Colomer J-F, Petaccia L, Snyders R, Beljonne D, Bittencourt C (2016) 2D Mater. 3:011001. https://doi.org/10.1088/2053-1583/3/1/011001
70. Scardamaglia M, Struzzi C, Aparicio Rebollo FJ, De Marco P, Mudimela PR, Colomer J-F, Amati M, Gregoratti L, Petaccia L, Snyders R, Bittencourt C (2015) Carbon 83:118–127. https://doi.org/10.1016/j.carbon.2014.11.009
71. Yu L, Pan X, Cao X, Hu P, Bao XJ (2011) Catalogue 282:183–190. https://doi.org/10.1016/j.jcat.2011.06.015
72. Huang S-F, Terakura K, Ozaki T, Ikeda T, Boero M, Oshima M, Ozaki J, Miyata S (2009) Phys Rev B 80:235410. https://doi.org/10.1103/PhysRevB.80.235410
73. Wang S, Iyyamperumal E, Roy A, Xue Y, Yu D, Dai L (2011) Angew. Chemie – Int. Ed. 50:11756–11760. https://doi.org/10.1002/anie.201105204
74. Zheng Y, Jiao Y, Ge L, Jaroniec M, Qiao SZ (2013) Angew Chemie - Int Ed 52:3110–3116. https://doi.org/10.1002/anie.201209548
75. Wang S, Zhang L, Xia Z, Roy A, Chang DW, Baek JB, Dai L (2012) Angew. Chemie - Int. Ed. 51:4209–4212. https://doi.org/10.1002/anie.201109257
76. Liang J, Jiao Y, Jaroniec M, Qiao SZ (2012) Angew. Chemie - Int. Ed 51:11496–11500. https://doi.org/10.1002/anie.201206720

77. Zheng Y, Jiao Y, Li LH, Xing T, Chen Y, Jaroniec M, Qiao SZ (2014) ACS Nano 8:5290–5296. https://doi.org/10.1021/nn501434a
78. Wu   X,   Radovic   LR   (2004)   J   Phys   Chem   A   108:9180–9187. https://doi.org/10.1021/jp048212w
79. Huang S, Li Y, Feng Y, An H, Long P, Qin C, Feng W (2015) J Mater Chem A 3:23095–23105. https://doi.org/10.1039/C5TA06012E
80. Seah   C-M,   Chai   S-P,   Mohamed   AR   (2014)   Carbon   70:1–21. https://doi.org/10.1016/j.carbon.2013.12.073
81. Wang H, Xie M, Thia L, Fisher A, Wang X (2014) J Phys Chem Lett 5:119–125. https://doi.org/10.1021/jz402416a
82. Terrones M, Kamalakaran R, Seeger T, Rühle M, Zhu YQ, Kroto HW, Walton DRM, Kohler-Redlich P, Rühle M, Zhang JP, Cheetham AK, Walton DRM (2000) Chem Commun 59:2335–2336. https://doi.org/10.1039/b008253h
83. Keskar G, Rao R, Luo J, Hudson J, Chen J, Rao AM (2005) Chem Phys Lett 412:269–273. https://doi.org/10.1016/j.cplett.2005.07.007
84. Kudashov AG, Okotrub AV, Bulusheva LG, Asanov IP, Shubin YV, Yudanov NF, Yudanova LI, Danilovich VS, Abrosimov OG (2004) J Phys Chem B 108:9048–9053. https://doi.org/10.1021/jp048736w
85. Khene   S,   Nyokong   T   (2012)   J   Porphyr   Phthalocyanines   16:130–139. https://doi.org/10.1142/S1088424611004439
86. Zhi L, Gorelik T, Friedlein R, Wu J, Kolb U, Salaneck WR, Müllen K (2005) Small 1:798–801. https://doi.org/10.1002/smll.200500150
87. Wei D, Liu Y, Wang Y, Zhang H, Huang L, Yu G (2009) Nano Lett 9:1752–1758. https://doi.org/10.1021/nl803279t
88. Wu J, Liu M, Sharma PP, Yadav RM, Ma L, Yang Y, Zou X, Zhou X-D, Vajtai R, Yakobson BI, Lou J, Ajayan PM (2016) Nano Lett 16:466–470. https://doi.org/10.1021/acs.nanolett.5b04123
89. Terrones   M,   Filho   AGS,   Rao   AM   (2007)   Carbon   nanotubes.   Topics in   applied   physics,   vol   111.   Springer,   Berlin/Heidelberg,   pp   531–566. https://doi.org/10.1007/978-3-540-72865-8_17
90. Shen W, Fan W (2013) J Mater Chem A 1:999. https://doi.org/10.1039/c2ta00028h
91. Bangert U, Pierce W, Kepaptsoglou DM, Ramasse Q, Zan R, Gass MH, Van den Berg JA, Boothroyd CB, Amani J, Hofsäss H (2013) Nano Lett 13:4902–4907. https://doi.org/10.1021/nl402812y
92. Krasheninnikov   AV,   Banhart   F   (2007)   Nat   Mater   6:723–733. https://doi.org/10.1038/nmat1996
93. Felten A, Bittencourt C, Pireaux JJ, Van Lier G, Charlier JC (2005) J Appl Phys 98:074308. https://doi.org/10.1063/1.2071455
94. Krasheninnikov   A,   V;   Nordlund   K   (2010)   J   Appl   Phys   107:071301. https://doi.org/10.1063/1.3318261
95. Scardamaglia M, Susi T, Struzzi C, Snyders R, Di Santo G, Petaccia L, Bittencourt C (2017) Sci Rep 7:7960. https://doi.org/10.1038/s41598-017-08651-1
96. Scardamaglia M, Amati M, Llorente B, Mudimela PR, Colomer J-F, Ghijsen J, Ewels CP, Snyders R, Gregoratti L, Bittencourt C (2014) Carbon 77:319–328. https://doi.org/10.1016/j.carbon.2014.05.035
97. Lehtinen O, Kotakoski J, Krasheninnikov AV, Tolvanen A, Nordlund K, Keinonen J (2010) Phys Rev B 81:153401. https://doi.org/10.1103/PhysRevB.81.153401
98. Åhlgren EH, Kotakoski J, Krasheninnikov AV (2011) Phys Rev B 83:115424. https://doi.org/10.1103/PhysRevB.83.115424
99. Struzzi C, Scardamaglia M, Reckinger N, Sezen H, Amati M, Gregoratti L, Colomer J-F, Ewels C, Snyders R, Bittencourt C (2017) Phys Chem Chem Phys 19:31418–31428. https://doi.org/10.1039/C7CP05305C

100. Struzzi C, Erbahar D, Scardamaglia M, Amati M, Gregoratti L, Lagos MJ, Van Tendeloo G, Snyders R, Ewels C, Bittencourt C (2015) J Mater Chem C 3:2518–2527. https://doi.org/10.1039/C4TC02478H
101. Felten A, Eckmann A, Pireaux J-J, Krupke R, Casiraghi C (2013) Nanotechnology 24:355705. https://doi.org/10.1088/0957-4484/24/35/355705
102. Orlando F, Lacovig P, Dalmiglio M, Baraldi A, Larciprete R, Lizzit S (2016) Surf Sci 643:214–221. https://doi.org/10.1016/j.susc.2015.06.017
103. Lin Y, Ksari Y, Aubel D, Hajjar-Garreau S, Borvon G (2016) Carbon 100:337–344. https://doi.org/10.1016/j.carbon.2015.12.094
104. Tang Y-B, Yin L-C, Yang Y, Bo X-H, Cao Y-L, Wang H-E, Zhang W-J, Bello I, Lee S-T, Cheng H-M, Lee C-S (2012) ACS Nano 6:1970–1978. https://doi.org/10.1021/nn3005262
105. Ewels CP, Erbahar D, Wagner P, Rocquefelte X, Arenal R, Pochet P, Rayson M, Scardamaglia M, Bittencourt C, Briddon P (2014) Faraday Discuss 173:215–232. https://doi.org/10.1039/C4FD00111G
106. Susi T, Pichler T, Ayala P (2015) Beilstein J. Nanotechnol. 6:177–192. https://doi.org/10.3762/bjnano.6.17
107. Sharifi T, Hu G, Jia X, Wågberg T (2012) ACS Nano 6:8904–8912. https://doi.org/10.1021/nn302906r
108. Kumar B, Asadi M, Pisasale D, Sinha-Ray S, Rosen B a, Haasch R, Abiade J, Yarin AL, Salehi-Khojin A (2013) Nat Commun 4:2819. https://doi.org/10.1038/ncomms3819
109. Usachov D, Fedorov A, Vilkov O, Senkovskiy B, Adamchuk VK, Yashina LV, Volykhov AA, Farjam M, Verbitskiy NI, Grüneis A, Laubschat C, Vyalikh DV (2014) Nano Lett 14:4982–4988. https://doi.org/10.1021/nl501389h
110. Rao CV, Cabrera CR, Ishikawa Y (2010) J Phys Chem Lett 1:2622–2627. https://doi.org/10.1021/jz100971v
111. Maldonado S, Stevenson KJ (2005) J Phys Chem B 109:4707–4716. https://doi.org/10.1021/jp044442z
112. Kundu S, Nagaiah TC, Xia W, Wang Y, Dommele S, Van; Bitter JH, Santa M, Grundmeier G, Bron M, Schuhmann W, Muhler M (2009) J Phys Chem C 113:14302–14310. https://doi.org/10.1021/jp811320d
113. Luo Z, Lim S, Tian Z, Shang J, Lai L, MacDonald B, Fu C, Shen Z, Yu T, Lin J (2011) J Mater Chem 21:8038. https://doi.org/10.1039/c1jm10845j
114. Niwa H, Horiba K, Harada Y, Oshima M, Ikeda T, Terakura K, Ozaki J, Miyata S (2009) J Power Sources 187:93–97. https://doi.org/10.1016/j.jpowsour.2008.10.064
115. Nagaiah TC, Kundu S, Bron M, Muhler M, Schuhmann W (2010) Electrochem Commun 12:338–341. https://doi.org/10.1016/j.elecom.2009.12.021
116. Shao Y, Sui J, Yin G, Gao Y (2008) Appl Catal B Environ 79:89–99. https://doi.org/10.1016/j.apcatb.2007.09.047
117. Ripalda JM, Román E, Díaz N, Galán L, Montero I, Comelli G, Baraldi A, Lizzit S, Goldoni A, Paolucci G (1999) Phys Rev B 60:R3705–R3708. https://doi.org/10.1103/PhysRevB.60.R3705
118. Hellgren N, Guo J, Luo Y, Såthe C, Agui A, Kashtanov S, Nordgren J, Ågren H, Sundgren J-E (2005) Thin Solid Films 471:19–34. https://doi.org/10.1016/j.tsf.2004.03.027
119. Ikeda T, Boero M, Huang S-F, Terakura K, Oshima M, Ozaki J (2008) J Phys Chem C 112:14706–14709. https://doi.org/10.1021/jp806084d
120. Ni S, Li Z, Yang J (2012) Nanoscale 4:1184–1189. https://doi.org/10.1039/c1nr11086a
121. Kim H, Lee K, Woo SI, Jung Y (2011) Phys Chem Chem Phys 13:17505. https://doi.org/10.1039/c1cp21665a
122. Lai L, Potts JR, Zhan D, Wang L, Poh CK, Tang C, Gong H, Shen Z, Lin J, Ruoff RS (2012) Energy Environ Sci 5:7936. https://doi.org/10.1039/c2ee21802j
123. Guo D, Shibuya R, Akiba C, Saji S, Kondo T, Nakamura J (2016) Science 351:361–365. https://doi.org/10.1126/science.aad0832.
124. Xing T, Zheng Y, Li LH, Cowie BCC, Gunzelmann D, Qiao SZ, Huang S, Chen Y (2014) ACS Nano 8:6856–6862. https://doi.org/10.1021/nn501506p

125. Kondo T, Casolo S, Suzuki T, Shikano T, Sakurai M, Harada Y, Saito M, Oshima M, Trioni MI, Tantardini GF, Nakamura J (2012) Phys Rev B 86:035436. https://doi.org/10.1103/PhysRevB.86.035436
126. Feng L, Yang L, Huang Z, Luo J, Li M, Wang D, Chen Y (2013) Sci Rep 3:3306. https://doi.org/10.1038/srep03306
127. Ertl G, Freund H-J (1999) Phys Today 52:32–38. https://doi.org/10.1063/1.882569
128. Zhang C, Grass ME, McDaniel AH, Decaluwe SC, El Gabaly F, Liu Z, McCarty KF, Farrow RL, Linne M a, Hussain Z, Jackson GS, Bluhm H, Eichhorn BW (2010) Nat Mater 9:944–949. https://doi.org/10.1038/nmat2851
129. Salmeron M, Schlögl R (2008) Surf Sci Rep 63:169–199. https://doi.org/10.1016/j.surfrep.2008.01.001
130. Amati M, Abyaneh MK, Gregoratti LJ (2013) Instrumentalist 8:T05001–T05001. https://doi.org/10.1088/1748-0221/8/05/T05001
131. Toyoshima R, Kondoh H (2015) J Phys Condens Matter 27:083003. https://doi.org/10.1088/0953-8984/27/8/083003

Chapter 10
Graphene for Photodynamic Therapy

Selene Acosta and Mildred Quintana

Abstract Photodynamic therapy is a medical tool to treat diverse diseases such as cancer and infections caused from bacteria and fungi. Photodynamic therapy involves the use of light sources and photosensitive molecules to produce reactive oxygen species responsible for killing ill cell, bacteria and fungi. Recent advances in nanotechnology and nanoparticle fabrication open new strategies to improve the efficiency and selectively of photodynamic therapy. In this chapter, we summarize the principles that govern photodynamic process and advantages in using graphene in conjugation with photodynamic agents. It is expected that graphene and graphene-based material enhance the photosensitivity of common agents improving the therapy results.

Keywords Graphene · Phototherapy

10.1 Photodynamic Therapy

Photodynamic therapy (PDT) is a medical technology that uses visible light sources to activate photosensitive molecules that works as anti-cancer drugs or drugs to fight against bacteria and fungi. The PDT was discovered more than 100 years ago (in the year 1900) where it was observed that *Paramecia* microorganisms died when they were exposed to a photosensitizer (Ps) and sunlight at the same time [1].

Since then PDT has been studied and developed as a therapy against cancer and as antifungal and antibacterial therapy. This therapy results from the interaction of photons of visible light with a Ps that is located inside the cell or anchored to its membrane, this Ps can produce highly reactive oxygen species that are responsible for killing cancer cells, bacteria or fungi.

S. Acosta · M. Quintana (✉)
Universidad Autónoma de San Luis Potosí, San Luis Potosí, Mexico
e-mail: mildred@ifisica.uaslp.mx

© Springer Nature B.V. 2019
C. Bittencourt et al. (eds.), *Nanoscale Materials for Warfare Agent Detection: Nanoscience for Security*, NATO Science for Peace and Security Series A: Chemistry and Biology, https://doi.org/10.1007/978-94-024-1620-6_10

The use of PDT in clinics consists in administrating to the patients the Ps, or an inducer of it by a systemic (oral or intravenous) or topical, route, and wait the time in which the Ps selectively accumulates in altered tissue to later irradiate with light of the wavelength at which it absorbs, to activate it [2].

10.1.1 Mechanism of Action

The mechanism of action of PDT involves the absorption of a photon of light at a wavelength that overlaps with the absorption band of the light-sensitive molecule. This photon absorbed by the Ps is responsible for exciting an electron of the valence band which passes into an excited state, the excited molecule has a very short lifetime (of the order of nanoseconds) returning to its basal state by emitting the energy absorbed as fluorescence or by internal conversion. However, this excited state may undergo an electronic transition leading to a relatively more stable state with a longer lifetime (of the order of microseconds), by a process known as intersystem crossing, known as the triplet excited state. The longer life of this molecule in the triplet state allows it to react with the oxygen of the environment or with molecules that have oxygen-oxygen bonds, this reaction can happen in two different ways called Type I and Type II, shown in Fig. 10.1. At the end of these reactions the Ps is degraded by light, what is known as quenching [2].

The Type I and Type II reactions occur in constant competition. If one reactions take place more than the other depends on the type of photosensitizer and the nature

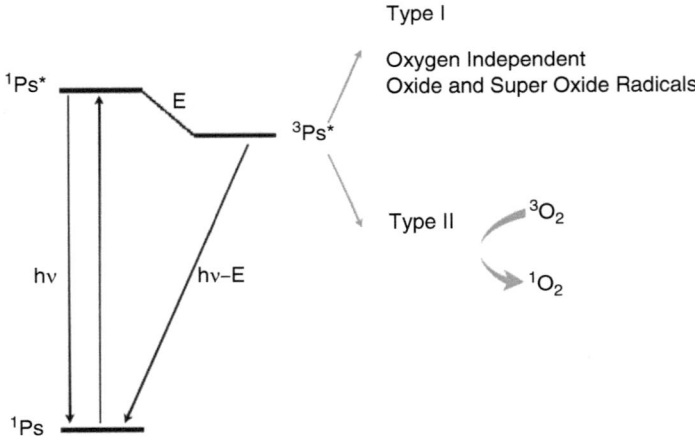

Fig. 10.1 Modified Jablonski's diagram. Photochemical pathways of Photodynamic Therapy. (hv) absorbed wavelength. (^{1}Ps) basal state. (^{1}Ps*) excited state. (E) electronic transition. (^{3}Ps*) triplet state. ($^3O_2$) oxygen in the triplet state. ($^1O_2$) oxygen in singlet state. Orange arrows indicate the reactions of the molecule in the triplet state with molecular oxygen (Type II) and independent of oxygen (Type I) for the formation of free radicals

of the molecules with which the products of reactions, I and II react [2]. The free radicals produced in PDT are highly reactive oxygen species (ROS) that can damage all types of biomolecules that are near them, such as proteins, lipids and nucleic acids, causing cellular oxidative stress and in this way, killing the cell [3].

10.1.2 Type I Reactions

Type I reactions do not necessarily require oxygen to cause cell damage. This type of reactions can cause damage through the action of free radicals by transferring an electron to the substrate or subtracting a hydrogen atom from it to produce peroxide and superoxide radicals [2].

10.1.3 Type II Reactions

Type II reaction involves the transfer of energy from the Ps in its triplet state (^{3}Ps*) to molecular oxygen in its normal triplet state ($^3O_2$) to produce singlet oxygen ($^1O_2$) and Ps in its basal state (^{1}Ps). In addition to being highly reactive this oxygen radical has a short lifetime (microseconds), which limits its cell diffusion with an action ratio of 0.05 μm [4, 5].

10.1.4 Light Sources

There is a high spectrum of light sources that can be used in PDT, including light sources at a continuous wavelength (red, blue, white or green), lasers, incoherent light sources, among others. However, red light has been reported to have more tissue penetration than blue light [6]. The advantage of using lasers is the short exposure time compared to inconsistent light sources. On the other hand, the use of a LED systems is a promising option as a light source for PDT because they have many advantages over incandescent and fluorescent light sources, such as low power consumption, longer life, reduced size, resistance to vibration, reduced heat emission and economic prices.

10.1.5 Photosensitizers

The Ps used in PDT must meet certain characteristics. First, they must be non-toxic before they are irradiated with light. In addition, they must produce a high amount of ROS and must have a high absorption coefficient at a wavelength that can penetrate

Fig. 10.2 Absorption spectrum of porphyrins and penetration of visible light into the tissue. The darker pink region represents epidermis and the lighter pink region represents dermis. Porphyrins have an absorption band between 600–700 nm which coincides with wavelengths with greater penetration in tissues

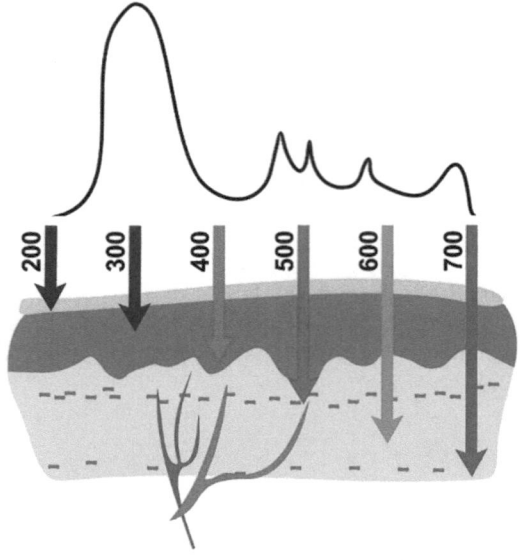

the cellular tissue, as schematically shown in Fig. 10.2. Molecules possessing these characteristics typically have rigid planar structures and possess a high degree of electron conjugation. The classes of molecules most used in PDT are porphyrins, phenothiazines and phthalocyanines.

10.1.6 Porphyrins Based Photosensitizers

Porphyrins are heterocyclic molecules derived from four pyrrole-like subunits interconnected by methine bonds. The absorption spectrum of the porphyrins exhibits a maximum in the region of the visible spectrum between 360 and 400 nm, known as the Soret band, followed by four small peaks between 500 and 700 nm, the so-called Q bands [7] observed in Fig. 10.2.

10.1.7 Photosensitizer Penetration in Cell Membranes

The penetration that a Ps has in the cells on which PDT is performed is very important for a positive effect to be achieved with this therapy. The degree of amphiphilicity or hydrophobicity of the Ps plays a very important role in the penetration, besides other factors such as distribution of the cellular membrane and

Fig. 10.3 Scheme of a lipid
vesicle and a phospholipid

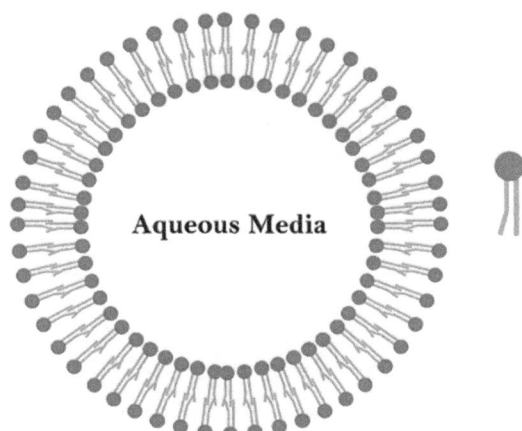

Aqueous Media

pathways of cellular uptake. In the case of Ps based on porphyrins, the degree of
self-aggregation that these molecules present is a crucial factor for penetration, this
aggregation can vary with minimal changes in its structure or pH values [8]. Due
to these factors, the behavior about penetration into cell membranes of any Ps is
difficult to predict. For this reason, there is currently great interest in understanding
the components that modulate the interactions between Ps and membranes.

The lipid bilayer is the basic component of cell membranes which gives structure
to cells and compartmentalize subcellular entities such as nucleus and mitochondria
[9]. However, the cell membrane is not solely composed of the lipid bilayer, also
possesses sugars, cholesterol and a great variety of proteins, this complexity in
composition and organization has motivated the development of simpler systems
that serve as models for the understanding of biological membranes. One example
of these models, is the unilamellar vesicles, which are models of a lipid bilayer
composed by phospholipids, whose structure consists of a polar head and a
hydrophobic tail. A schematic representation is shown in Fig. 10.3. The formation
of these vesicles occurs due to the self-assembly that phospholipids present in water
thanks to their amphiphilic character [9]. The characteristics of this vesicles such as
phospholipids composition, size and geometry can be adapted with high precision
[10].

In the report by Vermathen et al. the penetration of porphyrin Ce6, a common
Ps, into lipid bilayers was evaluated by nuclear magnetic resonance. These authors
show that Ce6 is capable of intercalating in the membrane at a pH ranging from 6.5
to 8, as schematically shown in Fig. 10.4 [8]. This intercalation takes place because
at this pH the carboxylic groups of Ce6 are deprotonated leaving a negative charge
that interacts with the positive charge of the polar head of the phospholipids.

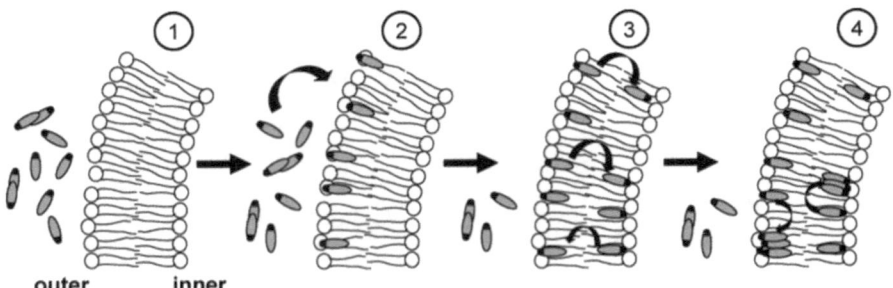

Fig. 10.4 Steps of chlorin interaction with the phospholipid bilayer (PL). (*1*) External addition of chlorin to chlorin-free vesicles. (*2*) Adsorption of chlorin molecules to the outer PL monolayer. (*3*) Chlorin distribution across the bilayer by flip-flop processes. (*4*) Proposed secondary kinetic steps that may comprise intramembrane clustering of chlorin molecules [8]

10.1.8 Cell Death Mechanisms Activated by PDT

As mentioned above, PDT generates reactive oxygen species which, due to their high reactivity, react with biomolecules that are within their reach causing oxidative stress in the cells. It is known that when cells are subjected to stress they can respond to damage by initiating a rescue mechanism and leading to cell death either by apoptosis, necrosis or autophagy, although the first is the most frequent and principal effect in the case of cells damaged by PDT [2]. This response of cells to stress generally involves changes in gene expression, expressing the stress proteins that allow the cell to cope with the damage induced by the stress. The type of cellular response caused by PDT depends on the intracellular location of Ps. Also, other determinants for cell death occurring either by necrosis or apoptosis, are the cell type and the dose of PDT [11, 12].

Apoptosis known as programmed cell death, is a process that requires energy in which the cell responds to the presence or absence of external stimuli or internal structural alterations, to destroy itself [2]. During apoptosis, the cell shrinks, the nuclear chromatin is grouped and condenses with the nuclear membrane, and finally the cytoplasm and nucleus are broken to form the apoptotic bodies. Although apoptosis is the major cell death pathway activated by PDT, it has been shown that autophagy can also be activated by it. Autophagy is a type of cell death accompanied by the formation of vacuoles in the cytoplasm, without chromatin condensation [2, 13].

Finally, necrosis is another of the cell death pathways caused by PDT. The necrosis consists of an irreversible state of the cell, with rapid loss of the plasma membrane integrity and escape from cytoplasmic components, absence of caspase signaling and denaturation of proteins. Necrosis occurs in response to excessive photodynamic injury to cellular components [13].

10.1.9 Photodynamic Therapy against Fungi

The frequency of fungal infections has notably increased in the last 20 years. Among the reasons why the incidence of these infections has increased, are: increasing use of antineoplastic and immunosuppressive drugs, broad-spectrum antibiotics, prosthetic devices and increased aggressive surgery [14]. On the other hand, people with a compromised immune system such as those with HIV, or massive burns are predisposed to fungal infection [15]. Antifungal treatments are very restricted to a certain number of drugs, longer treatments are required and sometimes are ineffective or present several serious side effects [14]. Consequently, new antifungal therapies such as PDT have been proposed to combat these infections. In a study reported by Dawei Hana et al. in 2016, PDT was used to combat mycosis fungoides (MF) using aminolovulinic acid (ALA) as Ps. Three patients were treated with ALA-PDT, in two of them the skin lesions caused by MF were completely eradicated, while in the remaining patient the lesions disappeared in 75% [16]. On the other hand, the antifungal properties of the porphyrins have been investigated extensively proving that after the irradiation, these molecules can effectively kill yeast cells [17–19].

10.1.10 Candida albicans

A common example of fungal infections is those caused by *Candida albicans*. A dimorphic fungus, this means, *Candida albicans* develops differently depending on the temperature of growth. As yeast, normally at 37 °C in the host, and as a filamentous fungus, at 25 °C in nature. It belongs to the *Ascomycota* phylum and reproduces asexually by budding. In the yeast form, it has an appearance of round or oval cells of 2–7 microns in size. Grouped into small groups, whereas, in the form of filamentous fungus, the cells lengthen and diversify taking the appearance of filaments, hyphal and pseudohyphal morphologies. Macroscopically, on agar it grows into white, soft, creamy and smooth colonies. A variety of *Candida* species are harmless and are commonly found in the gastrointestinal and genitourinary tract. However, there are species that are pathogenic that cause superficial infections such as vaginitis and other more severe in the mouth and esophagus of patients with compromised immune system [20].

Candida albicans is one of the most used organisms to test the effectiveness of different Ps in PDT against fungi. In 2016, Arash and collaborators used indocyanine green as Ps and a light source of 808 nm to fight *C. albicans*, this treatment caused a significant reduction in the ability to form colonies of this fungus [21]. On the other hand, methylene blue was used as Ps in PDT to combat *C. albicans*-induced vaginal candidiasis in murine, obtaining positive effects where methylene blue/PDT significantly reduces the inflammation caused by vaginitis and also reduce the presence of *C. albicans* in murine [22].

10.2   Carbon Materials

Carbon is a very versatile element present in a great variety of chemical compounds, is an essential element in organic molecules and it is considered as the fundamental basis of all living organisms. Carbon has unique properties, including the ability to form large chains of atoms and multiple bonding facilitating the formation of double and triple bonds with itself or with other elements [23].

This element can be observed in nature in different allotropic forms, allotropes are molecules formed by a single element that possess different molecular structure, the best known allotropic forms of carbon are diamond and graphite [24]. However, there are other allotropic forms with nanometer sizes, these forms have very interesting and unique physicochemical properties, such as superconductivity, high flexibility and hardness, among many others. Due to these properties carbon nanostructures have become a very important source of study in science, to be used in fields such as power generation, electronics and nanomedicine, to mention a few. Some of the allotropic nanometric forms of carbon are graphene, nanodiamonds, fullerenes, and carbon nanotubes [25, 26].

10.2.1   Graphite

Graphite is one of the most well-known allotropic carbon forms. It is reported as the softest material, with a laminar structure, each sheet is composed by only carbon atoms with a sp^2 hybridization, connected in a hexagonal arrangement. Graphite can be reduced in smaller allotropes. A single sheet of graphite was defined as graphene [27]. A schematic representation of a graphene sheet in graphite is observed in Fig. 10.5.

10.2.2   Graphene

Graphene is a carbon allotrope that has dimensions on the nanometer scale, which consists of a sheet of an atom of thickness, so it is considered a two-dimensional (2D) material. A schematic representation of a graphene sheet and a TEM image of stabilized graphene are showed in Fig. 10.6.

This material exhibits a variety of unique mechanical, optical and electronic properties, which will be described in more detail in later sections. Due to these properties graphene has been proposed as a material that could have a wide variety of applications in different fields such as electronics, bioengineering, renewable energy, biomedicine, among many others.

Graphite Structure

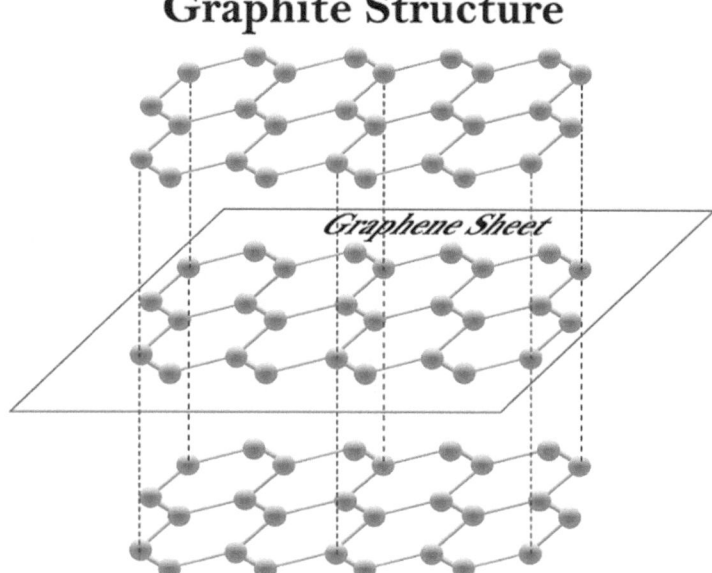

Fig. 10.5 Laminar structure of graphite

(a)   (b)

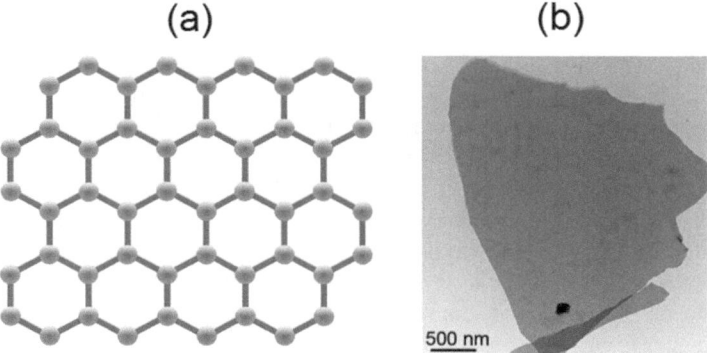

Fig. 10.6 Structure of graphene. (**a**) schematic representation of a graphene sheet, (**b**) TEM images of stabilized graphene in NMP [28]

To understand the formation of the graphene allotrope and its diversity in behavior and properties, it is necessary to know the electronic configuration of the carbon atom and the way in which it is linked to other atoms.

Carbon is the first element in the IVA series of the periodic table. This atom has 6 electrons that are distributed in atomic orbitals according to the following configuration: $1s^2 2s^2 2p_x^1 2p_y^1$. The electrons located in the outermost layer of the atom are those that are available to intervene in the formation of bonds, and are

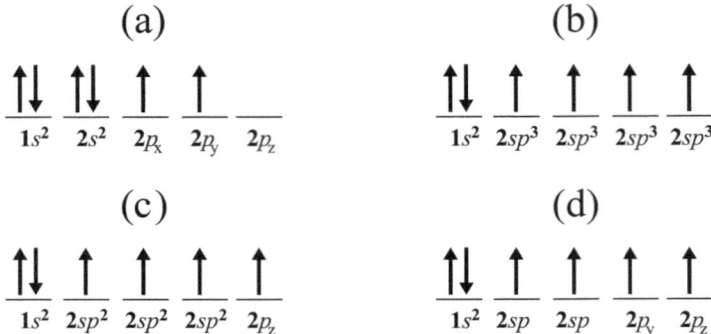

Fig. 10.7 Carbon electronic (the relative energy of the different electronic levels is not shown). (**a**) basal carbon state. (**b**) hybridization sp³. (**c**) hybridization sp². (**d**) hybridization sp

called valence electrons. In carbon, valence electrons are the two unpaired electrons of the $2p$ orbital and give rise to divalent carbon although in the form of a highly unstable intermediate. However, most of the compounds formed by carbon and its allotropes are tetravalent which is explained by the promotion of an electron from the $2s$ orbital to the empty orbital $2p_z$, generating four unpaired electrons available for bonding. The promotion of an electron from the orbital $2s$ to the orbital $2p_z$ clarifies the tetravalence of the carbon and allows to observe that the orbitals can interact to form new hybrid orbitals, in this case are generated four hybrid orbitals sp, of the four orbitals formed, one is from the carbon orbital s ($2s$) and three from the orbitals p ($2p_x$, $2p_y$ and $2p_z$) giving rise to a sp^3 type hybridization.

When hybridization is given only by one orbital s ($2s$) and two p ($2p_x$ and $2p_y$) three hybrid orbitals sp are formed generating sp^2 type hybridization. The atoms that form sp^2 hybridizations can form double bond compounds, the single bonds are known as sigma bonds (σ), and the double bonds are composed of a sigma (σ) bond and a pi (π). Hybridization of one orbital s ($2s$) and one orbital p ($2p_x$) generates two hybrid orbitals sp and the hybridization known as sp. Hybridized typed sp can form triple bonds of which one is a σ link and two are π [29]. The basal state, and the sp, sp^2, and sp^3 electronic configurations of C are shown in Fig. 10.7.

Previously, it was thought that graphene was only a theoretical prediction that could never be a physical reality because the instability of 2D materials would make it become a more stable 3D allotrope.

Because graphene is found naturally in graphite, it seemed logical to try to isolate from it. This idea culminated favorably with the exfoliation of a single sheet of graphene by Andre Geim and Kostya Novoselov at the University of Manchester in the year 2004 making use of the technique known as micromechanical exfoliation or the commonly known method of adhesive tape [30].

Although graphene is described as a two-dimensional material, it has been discovered that it is not completely flat, there are corrugations in the sheet that distribute the surface energy stabilizing the 2D structure, as shown in Fig. 10.8.

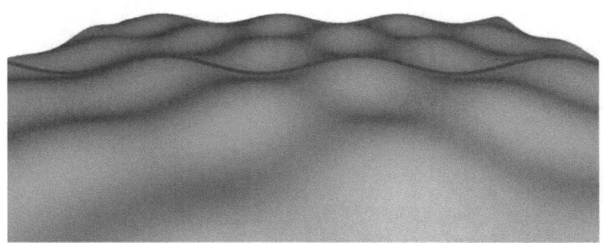

Fig. 10.8 Model of undulated graphene [31]

10.2.3   Methods of Synthesis

Nanostructures can be synthesized by different methods. These methods can be classified into two types: top-down and bottom-up methods. Graphene can be manufactured by various methodologies including mechanical exfoliation, chemical vapor deposition, carbide decomposition and solvent exfoliation.

Top-down methods start from a macroscopic material and involve the removal or rearrangement of atoms to obtain the desired structure on the nanometer scale. The main disadvantage of this method is the presence of defects in the final structure and the presence of impurities that can become very difficult to remove from the material. These imperfections and impurities may have a significant effect on the properties of the synthesized nanostructures [32].

In methods that go from bottom to top atoms, molecules or atoms are used as raw material to create more complex nanostructures. This methodology has the advantage that structures with less defects and impurities and a homogeneous chemical composition can be produced. Unfortunately, these methodologies include processes that involve highly controlled expensive and complex chemical synthesis [32].

Despite the number of methods by which graphene can be produced, large-scale production of graphene remains a challenge to be overcome.

10.2.4   Liquid Phase Exfoliation of Graphite

The most commonly used methodology to produce graphene is a top-down method. Graphite exfoliation is a technique that involves the breakdown of van der Waals interactions between graphene sheets. Ultrasonication in solvents has been used to break these forces and in this way separate sheets of graphene from graphite.

Graphite has been exfoliated in solvents such as N-methyl-2-pyrrolidone (NMP) in an ultra-sonication bath. In this method, macroscopic particles are precipitated by centrifugation, in the resulting solution are found single layer graphene and few layers graphene (FLG) [33].

The main disadvantage of this process is that ultrasonic waves break graphene into very small fragments. Using this methodology sheets are obtained with lateral dimensions of 100 nm in average, in addition to that, graphene sheets tend to aggregate with the time evolution.

10.2.5 Graphite Exfoliation Assisted by Molecules

Graphite exfoliation assisted by molecules emerges as a graphene-obtaining technique that seeks to counteract the previously mentioned aggregation problems. The objective of these molecules is to intercalate between the graphene and in this way, screen the attraction that exists between them. Thanks to this intercalation, the graphene solution is stabilized and aggregation is prevented. In 2016 Hernández et al. reported the graphene exfoliation from graphite using water or phosphate buffer saline (PBS) as solvents, and the porphyrin chlorine e6 as stabilizing molecule [34]. This exfoliation was made possible by the chlorine e6 molecules being intercalated between the graphene sheets during the exfoliation allowing these sheets to be stable in water and PBS, both of them considered biocompatible solvents, preventing graphene sheets from being added to form graphite again. This intercalation is possible because both, graphene and chlorine e6, have a conjugated system of π electrons that allows them to interact through π interactions.

10.2.6 Graphene Properties

10.2.6.1 Physical and Structural Properties

Carbon atoms present in graphene layers have a sp^2 hybridization, the carbon-carbon bonds have a length of 1.42 Å. The interplanar space between one sheet and another is 3.45 Å. As mentioned above, the stability of graphene has been attributed to microscopic undulations which are formed in the sheets, these curvatures disturb the electronic structure of graphene, many chemical and electronic properties of graphene, such as superconductivity, have been attributed to these undulations [35, 36].

10.2.6.2 Mechanical Properties

The mechanical behavior of a material is expressed as the ability to resist actions of loads or forces. A material can be subjected basically to three forces: tension, compression and cutting. The tension test is the one that gives us more information about the mechanical properties of a material, from this test we obtain parameters such as modulus of elasticity, elastic limit, tensile strength, elongation percentage, among others [37]. Graphene has an elastic modulus or Young's modulus of approximately 1 TPa and an intrinsic force of 130 GPa [38], so it is a very rigid material, this means that it can withstand very large stresses before breaking. Despite this, the graphene is also highly flexible, which is evidenced in the formation of folds in its leaves. These values indicate that graphene is the strongest known material.

10.2.6.3 Thermal Properties

Thermal conductivity indicates the ability of a material to transfer heat. In 2008 the thermal conductivity of a graphene sheet was measured using a non-contact optical technique, the value obtained was 5000 W/mK at room temperature [39]. The thermal conductivity of graphene is greater than any other known material.

10.2.6.4 Electronic Properties

Because graphene has only sp^2 hybridization carbons, an unhybridized p_z orbital remains in its electron configuration. The orbital $2p_z$ can be represented as a pair of lobes oriented perpendicular to the plane at an angle of 90° with the plane containing the three hybrid orbitals, resulting in a delocalized electronic state called the system π, as indicated in the schematic representation shown in Fig. 10.9a.

The electron found in these orbitals is weakly linked, allowed to jump from orbital to orbital taking form of a continuous electronic distribution above and below the plane of graphene. See Fig. 10.9b.

Electrons in graphene can be propagated for relatively long distances in the order of microns through the graphene sheet, the great mobility of the delocalized electrons is what determines the electronic properties of this π material.

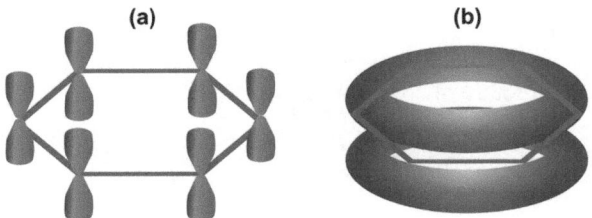

Fig. 10.9 Non-bonded graphene electrons. (**a**) Each carbon atom has a $2p_z$ orbital perpendicular to the plane of the carbon-carbon bonds, (**b**) the orbitals of the π bond take the form of a continuous electronic distribution above and below the plane of the ring

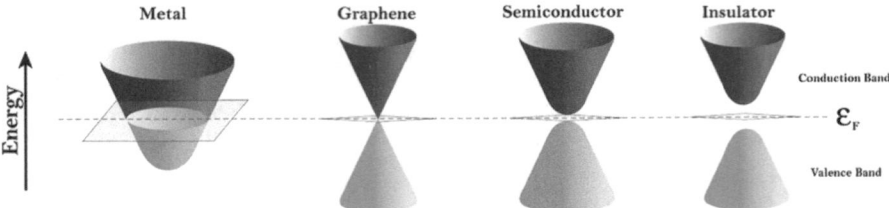

Fig. 10.10 Electronic energy bands around the Fermi level for metals, graphene, semiconductor materials and insulator

Monolayers, bilayers and three layers graphene are electronically distinct materials. In some aspects clusters of up to 10 layers may exhibit electronic properties different from those of graphite and may be called graphene, above 10 layers all graphene is electronically indistinguishable from graphite.

The electrical properties of a solid material depend in detail on the arrangement of the outermost electronic bands and on the way in which they are occupied with electrons [40]. Based on this, there are three main classes of bands structures in solids: the structure corresponding to metallic materials, insulators and semiconductors.

In Fig. 10.10, the structures of the electronic bands of materials are described. The purple shapes represent the conduction bands and the yellow forms are the valence band. In the metallic materials, the conduction band and the valence band overlap. In insulator materials appears a prohibited band that hinders the free movement of the electrons of a band to another one. The band structure of semiconductors is similar to the band structure of insulators except that the energy gap is relatively narrow making possible to overpass it [40].

Graphene shows an unconventional behavior. Defined as a semiconductor with zero forbidden band, it shares with metals the fact that the separation band is null. This is due to the unusual cone shape of their bands, where the conduction band joins in a single point with the top of the valence band. As for the semiconductor materials, it shares with them both types of load conveyors: electrons and holes. This hybrid material between metal and semiconductor owes many of its properties to this electronic structure of bands [41].

10.2.6.5 Chemical Properties

It has been described that single layer graphene is at least 10 times more reactive than two-layer graphene and multi-layer graphene. Likewise, the reactivity of the edges is at least twice as high as in the rest of the sheet [42]. Graphene can be functionalized through radical reactions such as nitrene or diazonium, which introduce reactive species covalently bound to graphene [43]. These groups can serve as template to bond graphene to other functional groups. The chemical functionalization of graphene can be monitored through the effects on conductivity when it is forming bonds with different molecules. It is also possible to perform the graphene functionalization selectively on the edges or at the surfaces, [44] opening the controlled modification of graphene properties by chemical functionalization.

10.2.7 Graphene Applications

Due to the large number of properties that graphene presents and even though experiments are still being done to better understand this material, the applications that graphene may have are very evident. However, the commercial viability of most of these applications is severely restricted by the limited production of graphene at large scales.

Despite these constraints, graphene has now been integrated into gas sensor devices for NH_3 and NO_2, to cite some examples, in which the conductance of graphene increases linearly with the increase in the sample of the gas [45]. The advantages of these sensors are high sensitivity, fast response time and low power consumption.

The reliability and speed of the optoelectronic devices depend on the temperature. Therefore, it is necessary to have materials with very high thermal conductivities to efficiently dissipate the heat generated in those devices. In this sense, graphene has a very high thermal conductivity which makes it a conductor of better quality compared to silicon, this feature allows to profile applications in transistors or circuits that work faster [39]. For this reason, it is thought that graphene can replace silicon in this type of devices in the following years.

10.2.8 Graphene Bioapplications

In addition to the applications of graphene in the field of energies and electronics, its unique properties have increasingly attracted attention for their applications in biotechnology, mainly due to its conductivity, flexibility and its ability to form covalent and non-covalent bonds with biomolecules. It is thought that graphene could play a significant role in the transport of drugs, nucleotides and peptides, antiviral and antibacterial materials, biosensors, cancer treatments, electrical stimulation for tissue and cell engineering, biological markers for the diagnosis of cancer, cell imaging and others [46].

10.2.9   Graphene Derivatives

10.2.9.1   Graphane

Graphane is a highly saturated hydrocarbon derived from graphene, with formula CH. Graphane was first described theoretically as a stable structure in 2007 [47]. Two years later, it was experimentally demonstrated that graphene can be hydrogenated and converted to graphane using a low pressure (0.1 mbar) and a mixture of hydrogen and argon [48].

10.2.9.2   Graphene Oxide (GO)

GO consists of a sheet of graphene with a random distribution of oxygen-rich functionalities on its surface. The presence of oxygenated functional groups makes the graphene sheets highly hydrophilic and causes the van der Waals interactions between sheets to weaken, allowing the introduction of water molecules between them. On the other hand, the increase in the distance between sheets and their hydrophilic property allows GO to be easily exfoliated in water and in various polar organic solvents by the application of external energy such as ultrasonic vibration [49]. The dispersions obtained are composed of sheets stacked in packets of only a few overlapping GO layers stabilized by electrostatic repulsion from the negative charge that they acquire in dispersion due to the ionization of the hydroxyl and carboxyl groups located on the planes and at its edges [50].

The GO structure has been the subject of several studies, however, the Lerf-Klinowski model is the one that is believed best describes it. In this model, GO is represented as a material constructed of non-oxidizing aromatic segments of varying sizes that are separated from each other by oxidized regions containing epoxides and hydroxyl groups on the surface of their planes and carbonyl and carboxyl groups, presumably located at the ends of the sheets. Recent studies have proposed that this model also contains ketone, lactone groups and tertiary alcohols [51].

10.2.9.3   Reduced Graphene Oxide (rGO)

Graphene due to its properties, such as high thermal conductivity [39], a Young's modulus of 1TPa and an intrinsic force of 130Gpa [38], is thought to have several applications in different fields such as biotechnology, electronics, power generation, among others. The most commonly used methods of graphene production are mechanical exfoliation, chemical vapor deposition, carbide decomposition and solvent exfoliation. Notwithstanding, these methods produce graphene with few imperfections, large-scale production has not been achieved. In graphene oxide production, however, GO is synthesized on a larger scale and at a lower cost, because of that its application in different fields is now possible. The introduction of oxygenated groups in the GO structure breaks the conjugated system of π electrons in the graphene sheets, with this change in its structure, the great mobility of the π electrons is lost and with it some graphene's properties such as superconductivity and optical absorbance [52].

Due to GO reduction, the oxygen groups on the sheets structure are removed and the conjugated system of π electrons is restored thus obtaining graphene on a larger scale [53]. GO reduction has been carried out in different ways. The most used is thermal reduction, where the rapid increase in temperature causes the oxygen groups to decompose into gases. During this method, small and wrinkled rGO sheets are produced [54]. Another methodology to obtaining rGO is chemical reduction of GO, the most commonly used reducing agent is hydrazine and its derivatives [55, 56]. Chemical reduction can be carried out at room temperature or with a slight heating and is much less expensive than the thermal reduction [53]. On each synthesis method rGO is obtained with different properties. Although the main objective of reducing GO is to generate graphene having the same properties and structure as graphene produced by methods such as liquid phase exfoliation, this is not possible, rGO always possesses residues of oxygenated groups in its structure in addition to severe imperfections that do not allow it to reach the unique and interesting properties of graphene [53]. However, work groups around the world are working to obtain a protocol that allows the synthesis of rGO with properties that are more like those of graphene.

10.3 Graphene in PDT

Graphene family nanomaterials (GFN) includes: single layered graphene, few layer graphene (FLG), graphene oxide (GO), and the reduced form of graphene oxide (rGO) [52]. GFN due to its unique and interesting properties have been proposed as starting material that could enhance different applications and photodynamic therapy is not the exception. Different types of graphene have been used in photodynamic therapy to improve the therapeutic effects of this medical technique.

In the work reported by Peng et al. in 2011, folic acid (FA)-GO conjugated was designed, prepared and used as a carrier of the photosensitizer Chlorin e6 (Ce6) being this the first report of controlled and specific release of Ce6 using GO as a carrier [57].

Most porphyrin-based molecules used as Ps, including Ce6, have limited use because of their poor solubility in water and because they are not selective with specific cells. GO is used to try to correct these disadvantages due to its solubility in water. It is expected that GO provides biocompatibility, while due to its high surface area and its abundant oxygenated functional groups its functionalization with the drugs would relatively easy bring selectivity [57]. During this work GO was conjugated with FA molecules (FA-GO), FA was used for targeting specific cells with folate receptors, after that, FA-GO was conjugated to Ce6 (FA-GO-Ce6). Tumorigenic cells MGC803 were incubated with FA-GO-Ce6, and then the composite was introduced into cells by endocytosis. When endosomes were

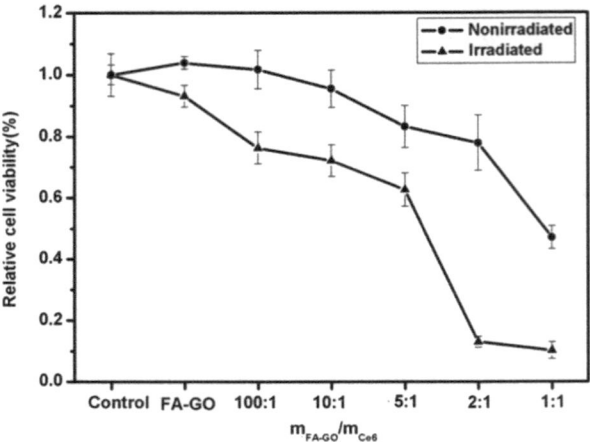

Fig. 10.11 Photodynamic effect of FA-GO-Ce6 in MGC803 cells. The FA-GO-Ce6 concentration was varied from 0-100 μM. Irradiation was carried out for 10 min with 632.8 nm laser [57]

converted into liposomes Ce6 was released into the cytoplasm due to pH change. FA-GO-Ce6 had no effect on MGC803 cells without exposure to the light source, indicating that it has no toxicity alone, on the contrary, in cells that had FA-GO-Ce6 and were irradiated with light from a 632.8 nm laser, the viability of MGC803 was 90% reduced, Fig. 10.11 shows a graph with these results, it can be observed that FA-GO increases the accumulation of Ce6 in tumor cells [57].

GO was also used to carry Indiocyanin Green (ICG), a Ps that absorbs in the near infrared that could have application in the PDT against Gram-positive and Gram-negative bacteria from the oral cavity. However, this molecule possesses a negative charge which is thought reduce the interaction with the negatively charged surface of microorganisms which disturbs its effect on PDT [58]. In a study reported by Akabari et al. In 2017, GO was conjugated with ICG (GO-ICG) and used to combat *Enterococcus faecalis*, a predominant bacterium in endodontic infections. This methodology improved the binding of ICG to microorganisms and increased the production of ROS by ICG. It had been previously reported that conjugation of ICG with GO allows to increase the production of singlet oxygen [59]. GO was treated with 3-aminopropyltriethoxysilane (APTES) and L-carnitine to add positive charge to its surface followed by the conjugation of ICG by means of π-π stacking interactions. *Enterococcus faecalis* cells were exposed to ICG and GO-ICG at ICG concentrations of 200 μg/mL and 1000 μg/mL, respectively. These cells were then irradiated for 60 s with an 810 nm laser. The growth of *E. faecalis* was 99.4% reduced with GO-ICG-PDT and 78.2% with ICG-PDT, results are show in Fig. 10.12. These results prove that GO increases the capacity as Ps of ICG in PDT [58].

Nanocrystals such as CdTe and CdSe can generate reactive oxygen species upon light irradiation and convert energy from light to heat, for these reasons it is

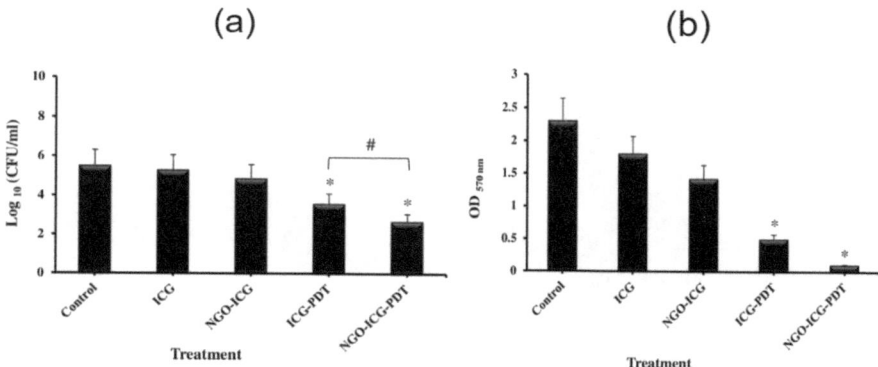

Fig. 10.12 Effects of ICG-PDT and NGO-ICG-PDT in *Enterococcus faecalis*. (**a**) Reduction of colony forming units per millimeter (CFU/mL). (**b**) Inhibition of biofilm formation ability [58]

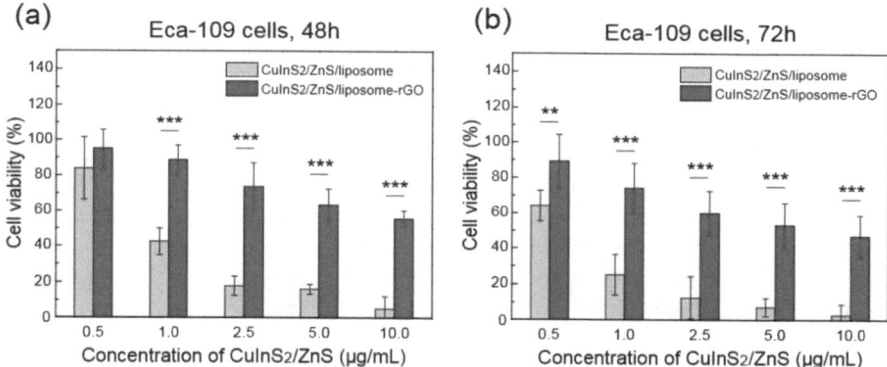

Fig. 10.13 Toxicity of the CuInS$_2$/ZnS/liposome and CuInS$_2$/ZnS/liposome-rGO to Eca-109 cells. (**a**) 48 h of incubation. (**b**) 72 h of incubation [60]

thought that nanocrystals may function as Ps in PDT and in photothermal therapy. Unfortunately, these nanocrystals are composed of heavy metals presenting toxic side effects and their use is limited [60]. In 2016, Qiang Wu et al. used rGO to improve the biocompatibility of CuInS$_2$/ZnS nanocrystals wrapped in liposomes, to improving their efficiency in PDT. GO was synthesized by the modified Hummers method [61] and then reduced with hydrazine. The rGO incorporation reduced the release of heavy metals (Cu^{2+}) from nanocrystals due to rGO barriers improving the biocompatibility of this nanocrystals as we can see in Fig. 10.13. In addition, ROS generation (in vivo and in vitro experiments) was higher in CuInS$_2$/ZnS/liposome-rGO than in CuInS$_2$/ZnS/liposome and rGO separately. Also, after 671 nm laser irradiation CuInS$_2$/ZnS/liposome-rGO nanocomposites efficiently induced apoptosis in carcinoma cells and a tumor size reduction of this type of cells in mouse was also observed [60].

In 2016, Fan et al. reported the behavior of a silver and rGO nanocomposite (Ar/rGO) as Ps in PDT against breast cancer cells using as a light source a 532 nm laser. Ar/rGO was prepared using silver nitrate and GO as precursors. Ar/rGO showed little toxicity in cells that were used in this study, whereas Ar/GO/532 nm laser had a large cytotoxic effect, indicating that Ag/GO is a promising Ps in PDT [62].

10.4   Conclusions and Perspectives

The study of graphene and its derivatives is a research with many possibilities, both, in the development of future applications or as new materials with better properties. New results respected to graphene are frequently published in journals from different branches of science such as biology, physics, chemistry, materials, among many others. Many potential applications are still under investigation, among these applications are those related to biomedicine, such as PDT and Ps, although it has been seen that due to its properties graphene and its derivatives can improve biotechnological applications its use is limited by the toxicity that could present in the human body. Experiments using graphene derivatives, such as GO and rGO, with cell cultures or in some cases in animals like mice, show very preliminary results. While non-significant toxicity is observed in many of these systems, it is required to generate much more knowledge in this area before thinking about a real application in humans. On the other hand, the use of graphene in cellular systems has been conditioned due to the limitations of graphene production, including the use of solvents or precursors that are not biocompatible or cytotoxic and very difficult to eliminate. The production of graphene on a larger scale and in a biocompatible media is of crucial importance for biomedicine applications. Although graphene derivatives (GO and rGO) can be produced on a larger scale and have been found to be biocompatible, the properties of graphene such as thermal and electrical conductivity necessary for PDT are considerably reduced in these compounds. Pristine graphene synthesis in biocompatible media might open new possibilities for the generation of highly stable materials for PDT.

Acknowledgments The present work was financially supported by CONACYT through the project PN-1767. SA is thankful to CONACYT for the masters' scholarship number 486938. Thanks are given to Marcos Moshinsky Foundation for finantial support. Authors are grateful to M.C. Jonathan S. de Lira Escobedo for help with images.

References

1. Zeyd I, Hamblin MR (2015) Photodynamic therapy of infectious disease mediated by functionalized fullerenes. In: Rai M, Kon K (eds) Nanotechnology in diagnosis, treatment and prophylaxis of infectious diseases. Elservier Inc, London, pp 69–70

2. Eva Ramón G (2015) Terapia fotodinámica: teoría y práctica. Instituto Politécnico Nacional, Mexico
3. Broekgaarden M, Weijer R, van Gulik TM, Hamblin MR, Heger M (2015) Tumor cell survival pathways activated by photodynamic therapy: a molecular basis for pharmacological inhibition strategies. Cancer Metastasis Rev 34(4):643–690
4. Dougherty TJ, Gomer CJ, Henderson BW, Jori G, Kessel D, Korbelik M et al (1998) Photodynamic therapy. JNCI: J Natl Cancer Inst 90(12):889–905
5. Moan J (1990) On the diffusion length of singlet oxygen in cells and tissues. J Photochem Photobiol B Biol 6(3):343–344
6. Brown SB (2003) The role of light in the treatment of non-melanoma skin cancer using methyl aminolevulinate. J Dermatol Treat 14(sup3):11–14
7. Kalka K, Merk H, Mukhtar H (2000) Photodynamic therapy in dermatology. J Am Acad Dermatol 42(3):389–413
8. Vermathen M, Marzorati M, Vermathen P, Bigler P (2010) pH-dependent distribution of chlorin e6 derivatives across phospholipid bilayers probed by NMR spectroscopy. Langmuir 26(13):11085–11094
9. Bhatia T, Husen P, Brewer J, Bagatolli LA, Hansen PL, Ipsen JH, Mouritsen OG (2015) Preparing giant unilamellar vesicles (GUVs) of complex lipid mixtures on demand: mixing small unilamellar vesicles of compositionally heterogeneous mixtures. Biochim Biophys Acta (BBA)-Biomembr 1848(12):3175–3180
10. van Swaay D (2013) Microfluidic methods for forming liposomes. Lab Chip 13(5):752–767
11. Moor AC (2000) Signaling pathways in cell death and survival after photodynamic therapy. J Photochem Photobiol B Biol 57(1):1–13
12. Agostinis P, Buytaert E, Breyssens H, Hendrickx N (2004) Regulatory pathways in photodynamic therapy induced apoptosis. Photochem Photobiol Sci 3(8):721–729
13. Dewaele M, Verfaillie T, Martinet W, Agostinis P (2010) Death and survival signals in photodynamic therapy. In: Photodynamic therapy: methods and protocols. Humana Press, New York, pp 7–33
14. Donnelly RF, McCarron PA, Tunney MM (2008) Antifungal photodynamic therapy. Microbiol Res 163(1):1–12
15. Eggimann P, Garbino J, Pittet D (2003) Epidemiology of Candida species infections in critically ill non-immunosuppressed patients. Lancet Infect Dis 3(11):685–702
16. Han D, Xue J, Wang T, Liu Y (2016) Observation of clinical efficacy of photodynamic therapy in 3 patients with refractory plaque-stage mycosis fungoides. Photodiagn Photodyn Ther 16:9–11
17. Quiroga ED, Mora SJ, Alvarez MG, Durantini EN (2016) Photodynamic inactivation of Candida albicans by a tetracationic tentacle porphyrin and its analogue without intrinsic charges in presence of fluconazole. Photodiagn Photodyn Ther 13:334–340
18. Mora SJ, Cormick MP, Milanesio ME, Durantini EN (2010) The photodynamic activity of a novel porphyrin derivative bearing a fluconazole structure in different media and against Candida albicans. Dyes Pigments 87(3):234–240
19. Cormick MP, Alvarez MG, Rovera M, Durantini EN (2009) Photodynamic inactivation of Candida albicans sensitized by tri-and tetra-cationic porphyrin derivatives. Eur J Med Chem 44(4):1592–1599
20. Sudbery P, Gow N, Berman J (2004) The distinct morphogenic states of Candida albicans. Trends Microbiol 12(7):317–324
21. Azizi A, Amirzadeh Z, Rezai M, Lawaf S, Rahimi A (2016) Effect of photodynamic therapy with two photosensitizers on Candida albicans. J Photochem Photobiol B Biol 158:267–273
22. Machado-De-Sena RM, Correa L, Kato IT, Prates RA, Senna AM, Santos CC et al (2014) Photodynamic therapy has antifungal effect and reduces inflammatory signals in Candida albicans-induced murine vaginitis. Photodiagn Photodyn Ther 11(3):275–282
23. Rayner-Canham G, García RLE, Garcés SB (2000) Química inorgánica descriptiva. Pearson educación, México

24. Yasuda EI (ed) (2003) Carbon alloys: novel concepts to develop carbon science and technology. Gulf Professional Publishing, Elsevier Science Ltd, Kidlington, Oxford, UK
25. Shenderova OA, Zhirnov VV, Brenner DW (2002) Carbon nanostructures. Crit Rev Solid State Mater Sci 27(3–4):227–356
26. Neto AC, Guinea F, Peres NM (2006) Drawing conclusions from graphene. Phys World 19(11):33
27. Vajtai R (2013) Handbook springer of nanomaterials. Editorial Springer Dordrecht Heidelberg (New York, EUA)
28. Quintana M, Spyrou K, Grzelczak M, Browne WR, Rudolf P, Prato M (2010) Functionalization of graphene via 1, 3-dipolar cycloaddition. ACS Nano 4(6):3527–3533
29. Morrison RT, Boyd RN (1998) Química Orgánica, 5th edn. Addison Wesley Longman de México, México
30. Novoselov KS, Geim AK, Morozov SV, Jiang D, Zhang Y, Dubonos SV et al (2004) Electric field effect in atomically thin carbon films. Science 306(5696):666–669
31. Kocman M, Pykal M, Jurečka P (2014) Electric quadrupole moment of graphene and its effect on intermolecular interactions. Phys Chem Chem Phys 16(7):3144–3152
32. Cao G (2006) Nanoestructures and nanomaterials synthesis, properties and applications. Imperial College Press, London, pp 7–10
33. Hernandez Y, Nicolosi V, Lotya M, Blighe FM, Sun Z, De S et al (2008) High-yield production of graphene by liquid-phase exfoliation of graphite. Nat Nanotechnol 3(9):563–568
34. Hernández-Sánchez D, Scardamaglia M, Saucedo-Anaya S, Bittencourt C, Quintana M (2016) Exfoliation of graphite and graphite oxide in water by chlorin e 6. RSC Adv 6(71):66634–66640
35. Meyer JC, Geim AK, Katsnelson MI, Novoselov KS, Booth TJ, Roth S (2007) The structure of suspended graphene sheets. Nature 446(7131):60–63
36. Fasolino A, Los JH, Katsnelson MI (2007) Intrinsic ripples in graphene. Nat Mater 6(11):858–861
37. Ballesteros MNS (2009) Tecnología de proceso y transformación de materiales, 2nd edn. Ediciones UPC, Barcelona
38. Lee C, Wei X, Kysar JW, Hone J (2008) Measurement of the elastic properties and intrinsic strength of monolayer graphene. Science 321(5887):385–388
39. Balandin AA, Ghosh S, Bao W, Calizo I, Teweldebrhan D, Miao F, Lau CN (2008) Superior thermal conductivity of single-layer graphene. Nano Lett 8(3):902–907
40. Callister WD (2007) Materials science and engineering: an introduction, 7th edn. Wiley, Hoboken, pp 668–671
41. Neto AHC (2010) The carbon new age. Mater Today 13(3):12–17
42. Sharma R, Baik JH, Perera CJ, Strano MS (2010) Anomalously large reactivity of single graphene layers and edges toward electron transfer chemistries. Nano Lett 10(2):398–405
43. He H, Gao C (2010) General approach to individually dispersed, highly soluble, and conductive graphene nanosheets functionalized by nitrene chemistry. Chem Mater 22(17):5054–5064
44. Quintana M, Montellano A, Del Rio-Castillo E, Van Tendeloo G, Bittencourt C, Prato M (2011) Selective organic functionalization of graphene bulk or graphene edges. Chem Commun 47(11):9330–9332
45. Fowler JD, Allen MJ, Tung VC, Yang Y, Kaner RB, Weiller BH (2009) Practical chemical sensors from chemically derived graphene. ACS Nano 3(2):301–306
46. Wang Y, Li Z, Wang J, Li J, Lin Y (2011) Graphene and graphene oxide: biofunctionalization and applications in biotechnology. Trends Biotechnol 29(5):205–212
47. Sofo JO, Chaudhari AS, Barber GD (2007) Graphane: a two-dimensional hydrocarbon. Phys Rev B 75(15):153401
48. Elias DC, Nair RR, Mohiuddin TMG, Morozov SV, Blake P, Halsall MP et al (2009) Control of graphene's properties by reversible hydrogenation: evidence for graphane. Science 323(5914):610–613
49. Stankovich S, Piner RD, Nguyen ST, Ruoff RS (2006) Synthesis and exfoliation of isocyanate-treated graphene oxide nanoplatelets. Carbon 44(15):3342–3347

50. Li D, Müller MB, Gilje S, Kaner RB, Wallace GG (2008) Processable aqueous dispersions of graphene nanosheets. Nat Nanotechnol 3(2):101–105
51. Dreyer DR, Park S, Bielawski CW, Ruoff RS (2010) The chemistry of graphene oxide. Chem Soc Rev 39(1):228–240
52. Jaleel JA, Sruthi S, Pramod K (2017) Reinforcing nanomedicine using graphene family nanomaterials. J Control Release 255:218–230
53. Pei S, Cheng HM (2012) The reduction of graphene oxide. Carbon 50(9):3210–3228
54. Schniepp HC, Li JL, McAllister MJ, Sai H, Herrera-Alonso M, Adamson DH et al (2006) Functionalized single graphene sheets derived from splitting graphite oxide. J Phys Chem B 110(17):8535–8539
55. Stankovich S, Piner RD, Chen X, Wu N, Nguyen ST, Ruoff RS (2006) Stable aqueous dispersions of graphitic nanoplatelets via the reduction of exfoliated graphite oxide in the presence of poly (sodium 4-styrenesulfonate). J Mater Chem 16(2):155–158
56. Zhu Y, Cai W, Piner RD, Velamakanni A, Ruoff RS (2009) Transparent self-assembled films of reduced graphene oxide platelets. Appl Phys Lett 95(10):103104
57. Huang P, Xu C, Lin J, Wang C, Wang X, Zhang C, Zhou X, Guo S, Cui D (2011) Folic acid-conjugated graphene oxide loaded with photosensitizers for targeting photodynamic therapy. Theranostics 1:240–250
58. Akbari T, Pourhajibagher M, Hosseini F, Chiniforush N, Gholibegloo E, Khoobi M et al (2017) The effect of indocyanine green loaded on a novel nano-graphene oxide for high performance of photodynamic therapy against enterococcus faecalis. Photodiagn Photodyn Ther 20:148–153
59. Li Y, Dong H, Li Y, Shi D (2015) Graphene-based nanovehicles for photodynamic medical therapy. Int J Nanomedicine 10:2451
60. Wu Q, Chu M, Shao Y, Wo F, Shi D (2016) Reduced graphene oxide conjugated with CuInS 2/ZnS nanocrystals with low toxicity for enhanced photothermal and photodynamic cancer therapies. Carbon 108:21–37
61. Hummers WS Jr, Offeman RE (1958) Preparation of graphitic oxide. J Am Chem Soc 80(6):1339–1339
62. Fan B, Guo H, Shi J, Shi C, Jia Y, Wang H et al (2016) Facile one-pot preparation of silver/reduced graphene oxide nanocomposite for cancer photodynamic and photothermal therapy. J Nanosci Nanotechnol 16(7):7049–7054

Chapter 11
Novel Supported Nanostructured Sensors for Chemical Warfare Agents (CWAs) Detection

Gabriela S. García-Briones, Miguel Olvera-Sosa, and Gabriela Palestino

Abstract Recently, the use of chemical warfare agents (CWAs) during terrorist attacks has been intensified affecting mainly civilian population around the world. Since these events are impossible to predict or prevent, the only plausible solution is to design and synthesize novel materials that allow developing more effective portable on-site sensors which at the same time could be produced at low cost in industrial scale. Nanomaterials for their outstanding properties have become ideal candidates for developing emergent platforms applied to the detection of toxic agents and biological threats. The goal of this chapter is to provide an updated overview of the latest research focused on the use of nanotechnology for developing CWAs sensors.

Keywords Nerve agent simulants · Nanoparticles · Graphene · Chemiresistor · Colorimetric sensor

11.1 Introduction

Use of chemical warfare agents (CWAs) can be retracted as early as 1000 B.C. when ancient civilizations employed arsenical smokes and mandrake roots to desecrate the enemies' wine looking for an advantage into battlefield [1]. In the following centuries, as technology and scientific discoveries marched on, CWAs formulations become more sophisticated leading the creation of highly toxic synthetic chemical weapons. These compounds can be released to the environment in form of gas, liquid, aerosol or powder and its main characteristic is that they can incapacitate, injure or kill human beings [2].

Modern history of these heinous chemicals goes back to the First World War when Germany released large quantities of chlorine-containing compounds in 1915

G. S. García-Briones · M. Olvera-Sosa · G. Palestino (✉)
Universidad Autónoma de San Luis Potosí, San Luis Potosí, Mexico
e-mail: palestinogabriela@uaslp.mx

© Springer Nature B.V. 2019
C. Bittencourt et al. (eds.), *Nanoscale Materials for Warfare Agent Detection: Nanoscience for Security*, NATO Science for Peace and Security Series A: Chemistry and Biology, https://doi.org/10.1007/978-94-024-1620-6_11

at Ypres, Belgium [3]. Later in the Second World War, chemical and scientific knowledge allowed the sophistication of chlorine (Cl_2) through its replacement by phosgene ($COCl_2$) and then by the mustard gas. Many years later, Iraq used nerve agents (sarin, tabun, VX) and mustard gas against Kurdish civilian population in Halabja, during the Iraq-Iran War in 1988, causing around 5000 fatalities [4]. However, CWAs were not only used in the battlefield, but also in a domestic terrorist attacks such as the occurred in Japan where members of "Aum Shinrikyo" cult, released sarin gas in Matsumoto (1994) and Tokyo subway system (1995), causing 12 casualties and 5500 injuries. More recently during the Syrian Civil War, sarin gas was employed against Ghouta, in the suburbs of Damascus in 2013, killing 1400 civilians and severely affecting health of thousands more. On April 4, 2017, another chemical attack was perpetrated over Khan Sheikhoun town, in northwestern region of Syria. The release of toxic gas which included sarin or sarin-like substance, killed at least 74 people and injured more than 557 [5]. Use of CWAs in these incidents illustrates why these compounds are considered massive destruction weapons and the importance to develop novel, useful tools to deal with them. The discussion in this chapter will be focused on the latest research of matrix-supported sensors for in-situ and real-time detection of simulant nerve agents (NA) and some chemical warfare agents (CWAs) with particular emphasis in the key role played by nanostructured materials to enhance functionality of mentioned sensors.

11.2   Classification of CWAs

11.2.1   Nerve Agents

Chemical warfare agents possess different features and can be classified in terms of their unique physicochemical, physiological and chemical properties. Based on their chemical structure they can be grouped by organophosphorus (OP), organosulfur and organofluorine compounds and arsenicals. For decades, the most used classification system has been based on the physiological effects produced on humans. Thereby, CWAs are still classified as nerve agents, vesicants, blood and blister agents and toxins [6]. Nerve agents (NA) are organophosphorus compounds which inhibit the enzyme acetylcholinesterase, are the deadliest of CWA's and can be classified in two major groups: G-agents (letter "G" represents the country of origin, "Germany") and V-agents (letter "V" possibly denotes "venom or venomous"). Tabun (GA), sarin (GB) and soman (GD) belong to the first group (Fig. 11.1a), while the second group contains five well known members: VE, VG, VM, VR and VX, being the VX agent the most studied member of V-series (Fig. 11.1b). The main difference between them, besides chemical structure, is their persistence on the environment. In general, due to their oily consistency and low volatility, V-series are more persistent on surfaces and soil than G-series, making them more dangerous at prolonged times [7]. As mentioned before, nerve agents possess an

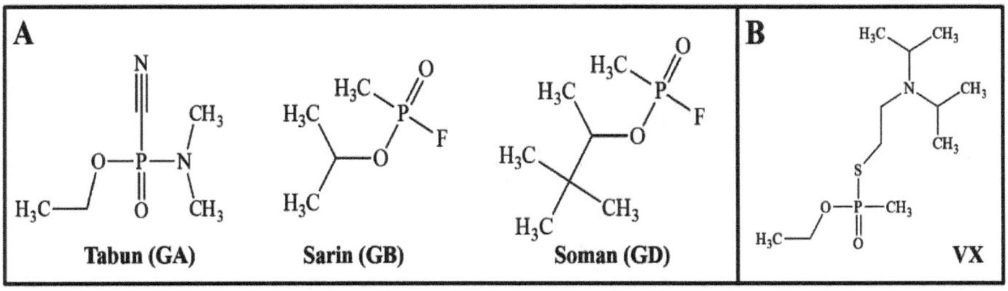

Fig. 11.1 Molecular structures of most common nerve agents; (**a**) G-series and (**b**) V-series

organophosphoric acid ester or an organophosphorus structure which is an acetylcholinesterase (AChE) enzyme inhibitor [8]. AChE is an enzyme whose primary function is to hydrolyze the neuromediator acetylcholine to avoid its accumulation into neurosynaptic clefts. Clinical effects from AChE inhibition by nervous agents include end-organ overstimulation and death within few minutes to several hours after exposure, depending on the chemical compound and concentration (e.g. VX LD50 is estimated to be 0.14 mg/kg) [9].

11.2.2 Blood Agents

These agents usually enter the body by inhalation and are distributed through the bloodstream. Nevertheless, these compounds do not affect the blood *per se* but interrupt the electron transport chain in the mitochondria exerting a massive toxic effect at the cellular level. Examples of blood agents include hydrogen cyanide (HCN) and arsine (SA), being this last one the most lethal with LD_{50} value of 2.5 ppm [10].

11.2.3 Toxins

Toxins are poisonous compounds produced in nature by living organisms such as bacteria, algae, terrestrial or marine animals. Protein toxins consist of long, folded amino acid chains while non-protein toxins are complex small molecules. Based on their mechanism of action they can be classified as cardiotoxins, hepatotoxins, neurotoxins, among others. Botulinum toxin (BoNT) is the most toxic agent ever known and is produced by bacteria *Clostridium botulinum* which grows on poorly preserved food. BoNTs have been categorized into seven highly virulent distinct serotypes (from A to G), being BoNT/A one of the most important bioterrorist agent due to extreme toxicity and easy production. Toxin works by blocking the release of AChE from the cholinergic nerves. Lethal amounts reported for crystalline BoNT/A have been estimated to be 0.09–0.15 μg by injection, 0.70–0.90 μg by inhalation, or 70 μg orally for a 70-kg human [11].

Considering the effectiveness of these agents at very low doses and easy access to reagents, combined with relatively simple synthesis methods is not difficult to see what makes them a highly attractive massive destruction powerful weapon [12]. After having pointed out lethality of these compounds, it is understandable that a better alternative for conducting a safe research in the laboratories worldwide is the widespread use of simulant compounds.

11.3 Simulants

These agents are innocuous molecules which imitate all relevant chemical and physical properties of NA avoiding its extreme associated toxicity. Simulants, also named surrogates or mimics, are less expensive than real one compounds and are also readily available around the world. Another advantage is that they can be manipulated in less equipped laboratories, avoiding the installation of expensive facilities and more specialized training for its personnel [13]. For these reasons, laboratory-scale study of detection, adsorption and degradation mechanisms of NA is generally carried out using simulants or imitators. A significant number of surrogates for G and V series have been used, and some of the most well know are shown in Fig. 11.2a and b, respectively. For sensing applications, DMMP and paraoxon are the best choices to simulate G-series sorption-desorption behavior (governed by log K_{ow}) due their log K_{ow} values are closely related to original CWAs as well as its chemical structures. Following same criteria, malathion is the best

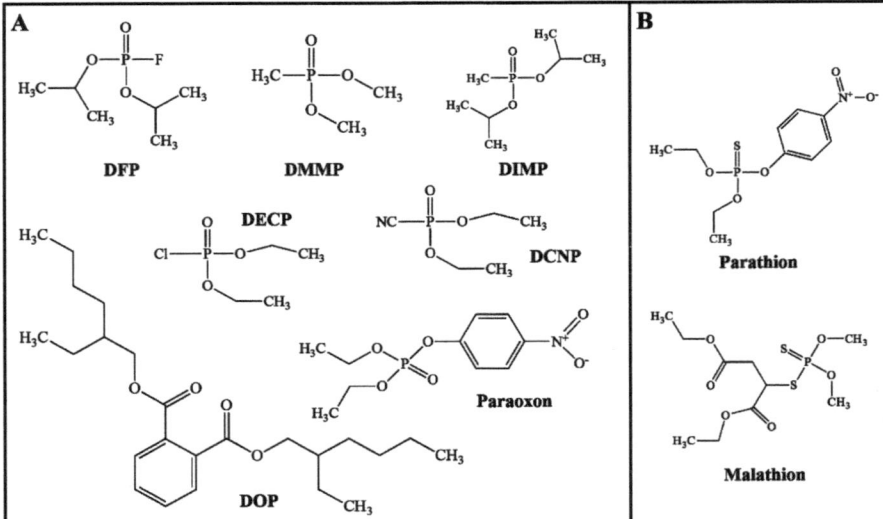

Fig. 11.2 Chemical structures of CWAs surrogates; (**a**) G-series and (**b**) V-series

simulant to represent VX sorption although is important to notice that in terms of their chemical structure, none surrogate gives "the unique intramolecular amino-nitrogen effect of VX" [14]. Low toxicity is ideally another relevant criterion for choosing a simulant, being malathion and DMMP the best options for each series. In fact, DMMP is by far the "workhorse" for nerve-agent like research, as can be seen in the next sections where newest detection/sensing studies are described [15].

11.4 Supported Sensors for the Detection of CWA and NA Simulants

Considering the overall background exposed above, designing of more selective, fast, low-cost and easy in-field sensing systems to enhance detection of compounds as toxins, explosives and chemical warfare agents (CWAs) are currently needed [16]. Albeit existent sensors have enough technology to detect these substances as low as ppb level in a selective fashion [17], they are also expensive, uncomfortable and usually require specialized training before its use. In this context, employing nanomaterials to develop novel sensing platforms to mitigate these undesirable features seems to be an advantageous alternative [15].

Nanomaterials or nanostructured materials, which usually have dimensions smaller than 100 nm, are a new class of materials with unique properties such as large surface-to-volume ratio, quantum size effect, high adsorption, and reactive capacity. These features give nanomaterials the opportunity to play a crucial role in the development of nanotechnology, especially in important fields like biomedicine [18] and technology associated to safety in food industry [19]. Such progress in nanoscience has led the incorporation of these novel materials to create powerful and more specialized detection systems. In the next paragraphs, we will provide a brief review of most currently used nanomaterials as sensing platforms.

11.4.1 Carbon-Based Materials

Nanoscale or nanostructured carbon-based materials include single and multiwalled carbon nanotubes (SWNTs and MWNTs, respectively), graphene, nanodiamonds and fullerenes among many others. These materials have been recently used for sensors development, particularly electric and electrochemical, due to their unique chemical behavior and electronic properties [20]. Sensing mechanisms used by these materials can be classified in four types, but here only will be listed the three related with applications explained in next sections. First, carbon materials can work as transducers since interaction between them and adsorbates usually generates a measurable electronic perturbation, which can be easily monitored by convenient methods [21]. Secondly, they can be used as electrodes for electrochemical sensors; in this case, carbon materials work as a bridge for transducing the electric signals generated from electrochemical reactions on the electrodes surface. Finally, carbon materials can be incorporated as nanofillers into polymer matrices to create composites with improved dispersion and mechanical stability for enhancing some critical properties like redox behavior, biocompatibility, selectivity and sensitivity [22]. Obviously, among the universe of these materials only some of them can be considered optimal for using certain sensing mechanisms mainly due its electronic hybridation and physicochemical features. Nevertheless, an exhaustive review of each material and its application for detection purposes is beyond the scope of this chapter, so in here, only will be examined carbon nanotubes and graphene, as well as some carbon hybrid-polymer composites. At this respect, some polymers (e.g. PVA, polystyrene, polypropylene) have been used as suitable matrices for sensing devices due its unique synergistic effects, like highly enhancing conductive properties at low cost and mechanical reinforcement [23]. As a novel example, M. Facure and coworkers proposed an impedimetric electronic tongue to analyze and discriminate mixtures of organophosphate (OPs) pesticides. The sensor was based on nanocomposites synthesized from graphene oxide, conducting polymers (PEDOT: PSS and polypyrrole) and gold nanoparticles (AuNPs) deposited over gold electrodes by drop casting. Impedance spectroscopy measurements revealed the electronic tongue was able to discriminate OPs at concentrations as low as 0.1 nmol L − 1 even in real samples. Authors attributed success of their platform to the composite itself since it made possible to achieve sensing units with dissimilar architectures but maintaining high specific surface area and conductivity [24].

11.4.2   Polymer/Hydrogel Matrices

Conducting polymers (CP) are synthetic polymers with the ability to conduct electrons, while providing flexibility and processability [25]. Due their electric and mechanical properties, they have been applied mostly in conductive and electrochemical sensors, working as signal transduction systems or specialized analyte recognizing components. One of the most important features of CPs is their strongly sensitive response to a wide range of analytes which is directly related with their intrinsic transport properties and polymer structure. In addition, the sensitivity of CPs can be tunable in a simple fashion by adjusting synthesis parameters such as polymerization temperature or counter ions incorporation. Bearing this in mind, CPs have emerged as essential materials for developing high performance and miniaturized supported-sensors for detection of toxic gases and biological species [26]. Recently, some well-studied polymers like polyanilne (PANI) [27], polypyrrole (PPy) [28], and poly(3,4-ethylenedioxythiophene) (PEDOT) [29] have also been explored for construction of sensors based on CPs nanomaterials. The combination between conventional insulating polymers (IPs) and CPs provide a new class of hybrid network, electroconductive hydrogel (ECH) in which IPs provides the 3D aqueous gel and CPs imparts electrical conductivity to the scaffold.

These hydrogel matrices have proven to be very useful for enhancing sensing properties by increasing the effective interface area due to the 3D organic matrix as well as ensuring high density enough to provoke electron set [30]. R.E. Rivero et al., reported the synthesis of a composite formed by CPs (hydrophobic) inside a nanoporous PNIPAm hydrogels matrix [31]. In this case, synergistic effect between both materials provided a nanocomposite which was not only electronically conductive but also elastic enough to deform under pressure allowing them to construct an electronic pressure/force sensor. Going further, development of thermosensitive electrical switch was also possible by inclusion of CPs into hydrogel, since conductive polymer was able to absorb microwaves for driving thermal phase transition and leading a volume change as well as release of loaded solution.

For a better compression, in this chapter all the works reviewed were classified by type of sensor (chemiresistive, optical, gravimetric, etc.) and grouped by material used for developing support or matrix. Before each review, detailed specific information is provided to give a better context about the application.

11.5 Chemiresistive Sensors

11.5.1 Single-Wall Carbon Nanotubes

Chemiresistive sensors works by changing its electrical resistance in response to chemical stimuli from nearby environment. Chemical interaction generated between sensing material and analyte could be by formation of covalent or hydrogen bonding as well as by molecular recognition [32]. Single-walled carbon nanotubes (SWC-NTs) are attractive chemiresistive materials which operate at ambient temperatures [33]. However, synthesis method of SWCNTs produce a mixture of metallic and semiconducting tubes affecting considerably the desirable optoelectronic properties of metallic SWCNTs.

Ishihara et al. [34], reported a study in which two different batches of tubes, metallic or semiconductor, were tested as a mixture and separated to study their effect into SWCNTs-electronic devices. They reported the conjugation of SWCNTs and metallosupramolecular polymer (MSP) to form a composite for detection of diethyl chlorophosphate (DECP), a highly toxic nerve agent simulant. In their study, they developed a chemiresistive gas sensor prepared by wrapping SWCNTs with a square-planar Cu 2+ −based MSP (Fig. 11.3a). Addition of anthracene units to MSP, aligned by Cu 2+ configuration, ensured effective interaction between SWCNTs and MSP through π − π interactions (Fig. 11.3b), leading wrapping and debundling of carbon nanotubes. After exposure to electrophiles, conductivity increased due unwrapping of MSP (Fig. 11.3c) with a detection limit of 0.1 ppm DECP. As mentioned before, to develop and perform the sensor platform, they conveniently separated batches of as-synthesized SWCNTs, demonstrating that S-SWCNTs exhibited the highest sensitivity to target analytes when wrapped with

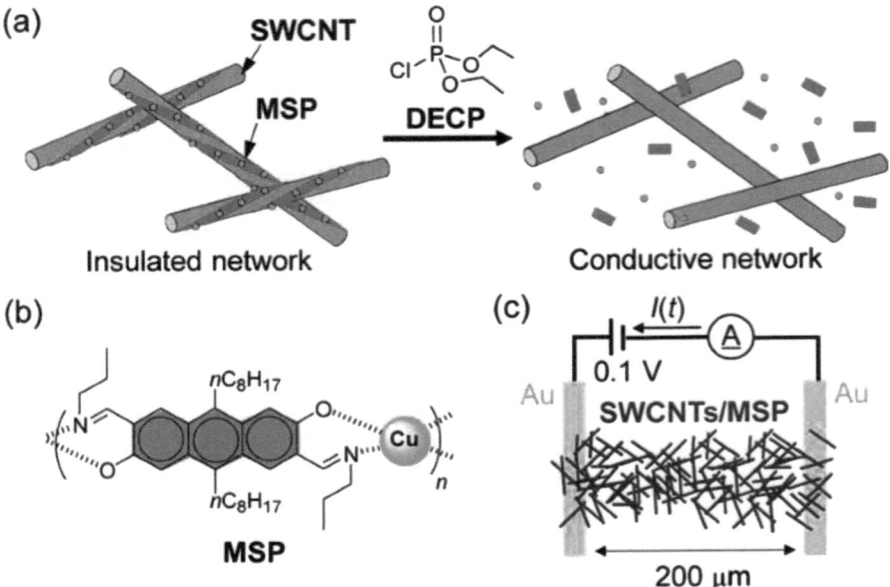

Fig. 11.3 (**a**) Schematic illustration of chemiresistive sensors composed of SWCNTs and metallosupramolecular polymers (MSPs). (**b**) Chemical structure of MSP. (**c**) Sensing device constituted by SWCNTs/MSPs bridging two gold electrodes. Monitoring of electric current (I(t)) under 0.1 V. (Reprinted with permission from [34]. Copyright 2017 American Chemical Society)

MSPs. On the other hand, the responses of M-SWCNT sensors were smaller but more precise in terms of discriminate another stimulus, like saturated water vapor. They concluded also that combination of both, semiconductor and metallic SWCNTs, could be a better option for sensing due its better response to saturated water vapor relative to pure S- and M- SWCNT/MSP sensors.

Yoo and coworkers [35] also explored the conjugation of SWCNTs and polyaniline (PANI) through nanocomposite sensors for detection of DMMP, a typical sarin simulant. Composites were synthesized by drop-cast over SWCNTs polyaniline coated over oxidized Si substrates patterned with palladium electrodes (Fig. 11.4). For measuring sensor response, the composite was exposed to DMMP gas at room temperature and resistance changes were collected. They proposed that DMMP molecules were adsorbed by composite, generating interactions and stimulating electron transfer to the composite. This phenomenon increased the resistance of composite since majority carriers in carbon nanotubes and polyaniline was decreased by transferred electrons. SWCNTs-PANI composite SEM micrographs are shown in the inset of Fig. 11.4a, b, exhibiting a good dispersion among Pd electrodes. Single strand of composite is shown in Fig. 11.4c to demonstrate uniformity of polyaniline coating around carbon nanotube. From this study it was concluded that SWCNTs-PANI composite showed much higher sensitivity with

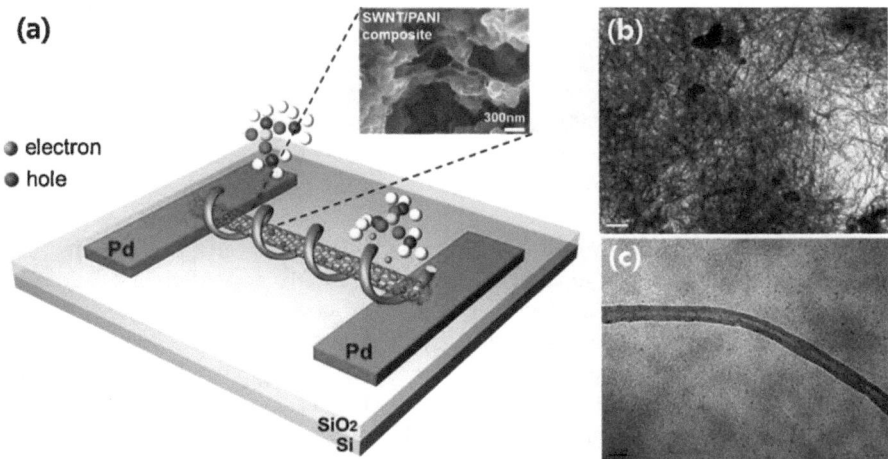

Fig. 11.4 (**a**) A schematic of an SWCNT–polyaniline composite sensor on an oxidized Si substrate with Pd electrodes. The Inset shows an SEM Image of the SWCNT–polyaniline composite. (**b** and **c**) TEM Images of SWCNTs–polyaniline composite. (Reprinted with permission from [35]. Copyright 2015 Elsevier)

better response time when compared with pure SWCNTs or PANI sensor proving that the combination of both had superior carrier collection and transport.

Another important question about the supported sensors is related to degradability of platforms, since these devices are intended to be used in harsh conditions and even maybe stored for prolonged times. In a previous report from Swager et al., it was demonstrated that composites of SWCNTs and polythiophenes (PTs) modified with HFIP can make an effective chemiresistive material for DMMP sensing [36]. Nevertheless, although the PT/SWCNT devices proved to be selective and highly sensitive, they also found significant performance degradation within days of fabrication.

Attending this problem and trying to perform a robust platform, Fennell et al. [17] reported, in a continuation of their previous work, the development and application of three derivatized poly(3,4-ethylenedioxythiophene) (PEDOT) analogs (instead of PT) to wrap SWCNTs to create the same sensitive chemiresistive sensor for detection of DMMP. The sensory response was investigated by measuring the current change between two electrodes at a constant bias voltage of 0.10 V. They proposed that transduction mechanism responsible for the decrease in conductance after exposure to DMMP was the formation of a hydrogen bonding interaction between the DMMP and HFIP (hexauoroisopropyl) moiety incorporated in the derivatized PEDOT sidechain. Strong responses obtained at laboratory environment were as low as 5 ppm DMMP with a calculated detection limit of 2.7 ppm in N2. In a simulated "real world" environment, sensor experienced a strong response at 11 ppm and a detection limit of 6.5 ppm, with no significant response changes after 2 weeks of exposure to 11 ppm DMMP in N2.

11.5.2 Polymer Composites

As established before, one-dimensional conductive polymers are capable to interact in a better way with lower-concentration target analytes to provide a measurable change in conductivity. This phenomenon is possible in CP nanotubes or nanofibers in which charge transport occur along the long-axis direction increasing effective surface area [37].

Trying to use this behavior to construct a sub-ppb concentrations sensor using CPs as transducers, Kwon et al. [38] reported a simple and efficient chemiresistive sensor using carboxylated polyaniline, polypyrrole (PPy) nanotube (cPNT) transducers capable of recognizing the sarin simulant, DMMP. To demonstrate its functionality, an array of interdigitate microelectrodes (IMEs) was fabricated as a sensor substrate through a lithographic process (Fig. 11.5a). An IME consisted of a pair of two gold electrode bands on a SiO2 substrate, with 100 Au/Cr fingers (55-nm thickness); their measurements were 2-μm width and 3.5-mm length (Fig. 11.5b). After deposition, IME substrate was functionalized with aminosilane (APTS) to immobilize cPNTs on surface for further use as resistive transducer. Figure 11.5c

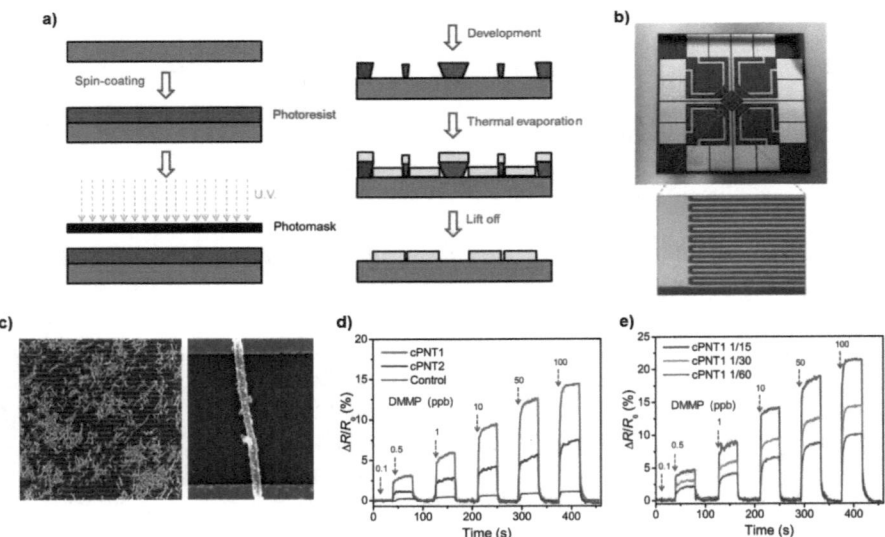

Fig. 11.5 cPNT IME sensor substrate. (**a**) Schematic of the lithographic fabrication process of microelectrode sensor substrate. (**b**) Photo of the chemiresistive DMMP gas sensing substrate, and magnified optical image with finger parts of IMEs. (**c**) Representative SEM images of cPNTs deposited on microelectrode substrate: (left) low and (right) high magnification. Chemiresistive sensing performance. (**d**) Real-time responses of cPNTs with different diameters upon periodic exposure to DMMP (0.1–100 ppb) and (**e**) Real-time responses of cPNT1s with different functionalization degrees of carboxyl group. (Reprinted with permission from [38]. Copyright 2016 PubMed Central)

displays representative scanning electron microscopy (SEM) images of the cPNTs-anchored microelectrode substrate. As can be seen, sensor substrate has numerous cPNTs which clearly bridged microelectrodes (2 μm gap). Excess of nanotubes may enhance interaction between them and target analytes, providing measurable signals even at very low concentrations. Electrically detection of DMMP was first examined on the microelectrode substrate by recorded the real-time change in resistance of cPNTs upon cyclic exposure to DMMP and N2 stream (Fig. 11.5d). When cPNTs were exposed to DMMP, an abrupt rise directly proportional to concentration of DMMP, is shown (lines magenta and blue). Additionally, they found that cPNTs diameter also played an important role increasing resistance change showing small diameter exhibited the bigger increase. In terms of carboxyl group concentration, sensitivity increased as the amount of carboxyl group increased.

11.6 Fluorescent Sensors

Besides the material employed, along last years a range of methods have been designed and employed for CWAs detection, and include not only properly constructed (bio)sensors but also lab techniques such as mass spectrometry, capillary electrophoresis, etc. Nevertheless, use of these instrumental is an expensive and complex task which usually requires specialized training. Additionally, these methods are not designed to be use on field, reducing the possibilities to have a good performance. Among all alternative approaches that can be found in literature, optical sensors and fluorescence probes have emerged as convenient platforms due to high sensitivity, relative low cost and ease of handling [39].

11.6.1 Polymeric Platforms

Smart hydrogels are networks of cross-linked polymer chains that are insoluble, hydrophilic and capable of swelling and soaking up when exposed to aqueous environment. In addition, they can respond to external stimuli and changes in their surroundings [40]. Among the universe of polymers that can be useful for these purposes, polyacrylamide hydrogels have demonstrated to be the best option in terms of the optical transparency they preserve for a vast range of monomer and crosslinker concentrations. By combining these smart hydrogel features with coumarins, a chemical agent marker for OP nerve agents, it was possible to develop a "turn-off" sensor for quick and selective on-site detection of DCP in the aqueous and vapor phases as reported previously.

In this work, Whitaker et al. [41] created a novel fluorescent turn-off sensor supported on a polymeric hydrogel poly (acrylamide) matrix using 6,7-dihydroxycoumarin as a fluorescent probe. Coumarins exhibit strong fluorescence in the UV ($\lambda = 365$ nm) and visible region and also possess very high fluorescent quantum yields, making them ideal candidates to be used in a fluorescence-based sensor. Due to optical transparency of polymer matrix, the strong fluorescence of coumarin and their quenching in presence of DCP, can easily be observed naked-eye by using hand-held UV lamp. In this regard, the research group performed experiments in bulk and thin film hydrogels, finding that in both cases, coumarin integrated efficiently into the matrix without interfering with polyacrylamide structure.

Fluorescence and quenching behavior ware also the same, with a clear expression at $\lambda = 365$ nm. However, for bulk material the rate of fluorescence quenching was very slow (1 up to 12 h), depending on the concentration of the DCP solution (10–1 M to 10–3 M). By the other hand, for thin films the fluorescence was quenched 1000 times faster than bulk hydrogels when exposed to aqueous solutions of DCP (10–1 M). After successful development of their thin hydrogel films, they also investigated their use as chemosensor after dehydration process finding same performance as newly synthesized smart hydrogels. These findings opened the door to solve problems like an effective use even in extreme environments and an easy packaging and adequate storage of hydrated hydrogels.

Although diethyl sulphide (DES), a vapor phase thioether, is as flavoring agent evolved during microbial processes in brewing industry and commonly found in wine and beer, it has also been used as an analog for a constituent of mustard gas, the bis(2-chloroethyl) sulphide [42]. Due lack of sensors for this potentially dangerous compound, Varju et al., [43] developed a "turn-on" vapoluminescent sensor by means of a Cu(I)/Au(I) coordination polymer mix. For measuring vapoluminescent response of Cu2/3Au1/3CN to DES stimuli, the excitation/emission spectra for [Cu2/3Au1/3CN]2(DES) were recorded and compared with previously reported spectrum of ligand-free, Cu2/3Au1/3CN. The emission spectrum of [Cu2/3Au1/3CN]2(DES) exhibited a maximum at 420 nm when sample was excited at 330 nm, while Cu2/3Au1/3CN showed maximum emissions at 380 nm and 560 nm when excited at 315 and 405 nm, respectively. These results demonstrated that changes in emission spectra can be monitored as a function of time and offering proof of correct detection after DES exposure.

Another alternative for improving fluorescent sensors include the incubation of conjugated polymers (CPs) over fibrous nanowebs. Fluorescent intensity of fluorescent sensors based on CPs can be easily altered by binding to an analyte due electron and energy transfer between polymer and quenching species. Specifically, CP dots have proven to have better performance when assembled on the surface of nanofibers than when embedded inside, thanks to quantum yield [44]. Using this previous knowledge, Jo and coworkers [45] explored the synthesis of dots-on-fibers (DoF) hybrid: an organic-inorganic system in which CP dots were dispersed randomly at nanofibers surface (Fig. 11.6a).

Furthermore, using DoF hybrid nanomaterial, they developed a fluorescent "turn-off" sensor based on polyquinoxaline CP dots for the selective detection of DCP (0–1.8 Mm). The fluorescence response of DoF is shown in Fig. 11.6b, where

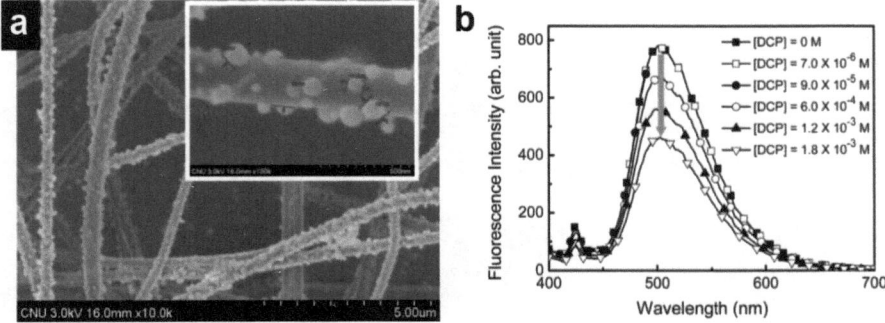

Fig. 11.6 (**a**) SEM micrograph showing distribution of CPdots on the surface of hybrid nanofibers. Inset: High-magnification SEM images of nanofibers. (**b**) Changes in fluorescence spectra of DoF upon exposure to DCP. (Reprinted with permission from [45]. Copyright 2016 American Chemical Society)

quenching was instantaneously observed at 501 nm after exposure to DCP. This successful performance opens the opportunity to prove DoF nanomaterial for detecting other chemicals with same accuracy and simplicity.

11.6.2   Miscellaneous

Climent et al. [46], on the other hand, employed mesoporous silica as a support for immobilization of BODIPY moieties to inhibit their aggregation in water and to facilitate sensitive detection of principal G-agents (sarin, soman, tabun) at picomolar level. BODIPY dyes are commonly used like molecular probes due to their high brightness and photostability [47]. However, self-quenching of BODIPY dyes represents a serious problem when are intend to be used as a fluorescent probe. One strategy employed to solve this drawback consists of embedding BODIPY into polymer matrices (e.g., hydrogels). Nevertheless, this type of support can slow down fluorescent response due to slow diffusion of the CWAs into the matrix. Bearing this in mind, authors of this work used covalent attachment of the probes to the inner walls of mesoporous silica to prevent aggregation and enhancing reaction with the target. When BODIPYs over surface of fluorescent sensor were exposed to nerve agents, a formation of bicyclic ring π-conjugated to the BODIPY occurred, switching off its fluorescence in only a few seconds (Fig. 11.7). The fluorescence spectra were measured under a UV lamp at a $\lambda = 365$ nm and the results clearly showed that these hybrid particles have an enormous potential for CWA detection in field and in real time.

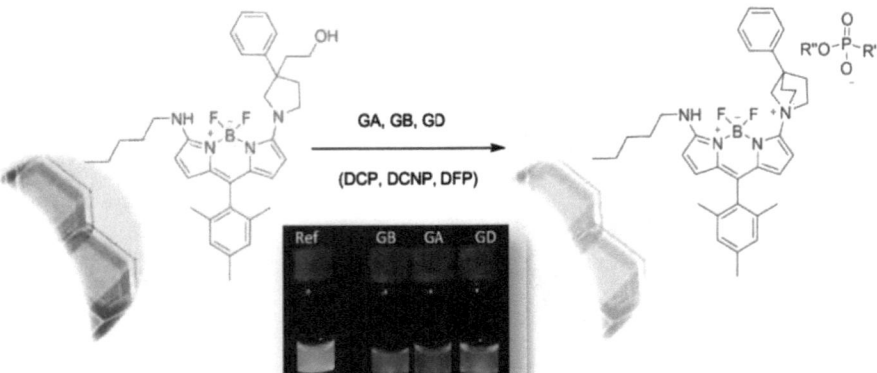

Fig. 11.7 Mechanism and photographs of detection of CWAs Sarin (GB), Soman (GD) and Tabun (GA) via fluorescence quenching response under UV lamp ($\lambda_{exc} = 365$ nm) after 4 min of exposure. (Reprinted with permission from [46]. Copyright 2017 Elsevier)

11.7 Colorimetric Sensors

In the path of trying to develop the optimal sensor whatever its type, not only features of nanomaterial matrix or support must be considered, but also the kind of chemical or agent that would be detected by the sensor. In this sense, dangerous substances like CWAs require an easy, quick and effective method of detection sensitive for very low amounts (ppb level) considering their harmful effects for living organisms when exposed even traces [48]. Colorimetric sensors are designed to be used for instantaneous detection of analyte by means of change color, which can be seen by naked-eye with several advantages associated: detection occur in situ with no further treatment of sample, sensor platforms usually are pretty small (nm or μm size) and for some particular cases (like porous silicon), support can be used again without loss of effectiveness [49].

Nanotechnology plays a key role in current colorimetric sensor development due its compliance with some of the most desirable features for this type of sensor: simplicity, profitability and quick response with an associated relative low cost. Going further, would be a mistake to focus all the efforts on enhancing detection when CWAs can be absorbed through the skin within few seconds. In this sense, nanomaterials can be successfully used for developing mechanical triggered barriers by incorporation into fabrics for protective suits without sacrificing breathability of the fabric [50].

As instance, the work of Belger et al., [51] illustrates the colorimetric response of copolymers prepared by ring-opening metathesis polymerization (ROMP) after addition of chemical warfare agent DCP, at room temperature and environment conditions. By incorporating highly arylated tertiary alcohols as responsive units in the polymers, materials became more robust with an enhancing colorimetric response due excellent reactivity of polymers with DCP. Colorimetric response of

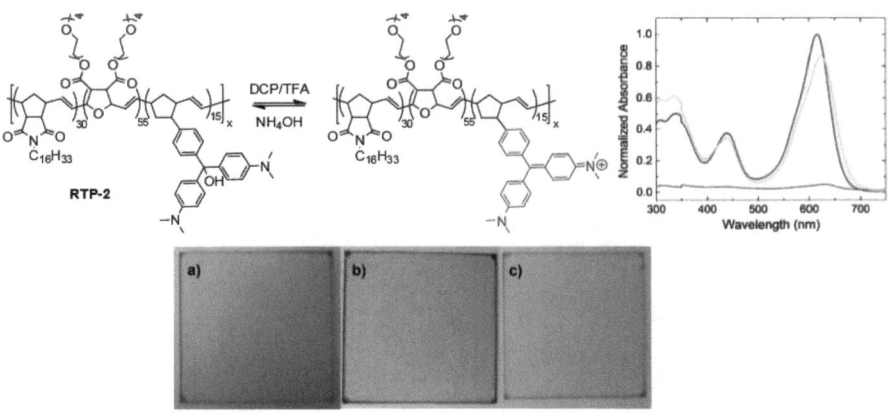

Fig. 11.8 Molecular structure of RTP-2. Spin coated film from chloroform solution before exposure (**a**, grey curve in absorption spectra) and after exposure to DCP (**b**, light green curve in absorption spectra) at room temperature. Regeneration is complete upon exposure to ammonium hydroxide vapor (**c**). (Reprinted adapted with permission from [51]. Copyright 2015 American Chemical Society)

sensor was studied at room temperature and ambient conditions over polymer spin-coated thin films. Different stages of polymer films and its associated absorption spectra are shown in Fig. 11.8. Complex molecular structure of random triple polymer (RTP-2) used for synthesizing the films are also shown. In Fig. 11.8a, is presented the first stage of polymer, before the exposure. As can be seen, its absorption spectrum is plain (gray curve). After DCP exposure (Fig. 11.8b), color visibly changed, and spectrum (light green curve) rise to two absorption bands at 450 nm and 620 nm. As shown in Fig. 11.8c, regeneration of sensor can be easily achieved upon exposure to ammonium hydroxide vapor preserving its activity even after cycle was repeated several times. From this work, authors concluded that incorporation of sensitive probes into random double and triple copolymers generate a highly functional reusable sensor for DCP detection.

Going one step further with designing and development of novel materials, D. A. Giannakoudakis et al. [52], nanoengineered a hybrid smart textile using cotton impregnated with a composite formed by copper based metal organic framework (MOF) and oxidized graphitic carbon nitride (g-C3N4) nanospheres. Conjugation of this nanostructured materials with textile allowed them to create an advanced wearable platform capable to sense, to adsorb and to degrade a surrogate of an organophosphate nerve gas agent, dimethyl chlorophosphate. Firstly, they developed an economically feasible impregnation route to ensure mobile phase was homogeneously deposited over whole textile and they performed experiments to ensure stability of composite over textile fiber. Second step includes the evaluation of multifunctionality of impregnated textiles when exposed to DMCP vapors finding that they became completely yellowish gradually and after 90 min of exposure (Fig. 11.9a). They also tested the textile by applying a droplet of the chemical

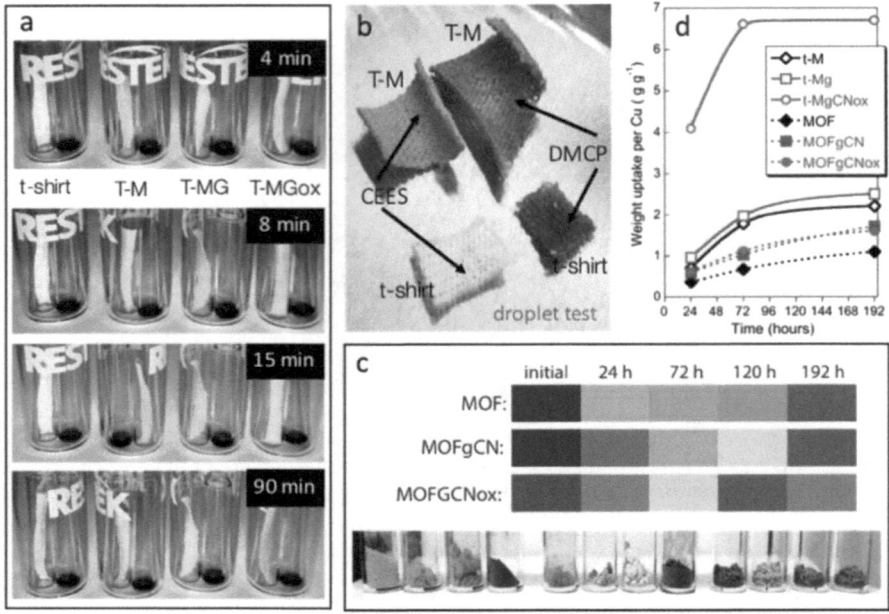

Fig. 11.9 (**a**) Color changes of all studied textiles after various exposure times. (**b**) A pristine T-shirt and T-M after the addition of 4 mL droplets of DMCP and CEES. (**c**) Gradual color changes because of the reactive adsorption process. (**d**) Normalized weight uptakes per gram of Cu after the subtraction of the weight uptake of the unmodified cotton textile (WUpCu). (Reprinted with permission from [52]. Copyright 2017 Royal Society of Chemistry)

(for cases in which CWAs are used as an aerosol) and found that color changed immediately after exposure. This effect only occurred after exposition to DMC and remaining intact when exposed to CEES, a surrogate of mustard gas, assuring calorimetrically selectivity to simulant of nerve agents (Fig. 11.9b). Dried powdered active phases are shown in Fig. 11.9c, in which both MOF and MOFgCN exhibits characteristic dark blue-mauve color, while MOFgCNox has a lighter hue of blue. The color of all phases gradually changed, being dark green in the two first cases after 192 h of exposure. In case of MOFgCNox, dark green color appeared after only 120 h of exposure. At same time of exposition, final color of MOFgCNox was light brown, suggesting important differences in surface reactivity. In Fig. 11.9d, they also shown the weight uptakes due decomposition products generated after up to 192 h DMCP adsorption under ambient light conditions. The main conclusion was that nanocomposite, deposited on the T-shirt cotton fabrics, achieved a DMCP adsorption of ∼7 grams of vapor per gram of copper, while the introduced photoreactivity enhanced the degradation of the vapors.

11.8   Other Type of Sensors

11.8.1   Conductive Sensor: Au/Graphene Oxide

Arsine is not only a well-known dangerous blood agent used as CWA, but also a widely chemical reagent used in semiconductor processes. Their toxicity threshold values were reduced from 0.05 to 0.005 ppmv making mandatory their detection for preserving the safety of industrial semiconductor facilities. Attending this problem, Furue et al. [53], developed a sensor prepared from reduced graphene oxide (rGO) films deposited over a homemade interdigitated electrode (IDE) with lines made of 200-nm-thick Au on 50-nm-thick Cr, both of which were fabricated on a glass substrate by successive RF sputtering and lift-off method. The sensor was placed on a temperature-controlled aluminum block equipped with a heater, thermocouple, and digital temperature controller inside a chamber where AsH3 gas was introduced at a constant flow rate. The conductance of the rGo, Au or Au/rGO sensor was monitored in the presence of flowing AsH3 showing gas sensors fabricated with only rGO or Au did not exhibit AsH3 sensitivity. However, rGO sensors combined with Au exhibited reversible conductivity enhancement with AsH3. The increase in conductivity probably occurred because the AsH3 depleted adsorbed oxygen on the Au islands, resulting in the enhancement of hole conduction in the rGO film. The amounts of gold and rGO, the GO reduction, and the operating temperature were optimized to obtain a detection limit of 0.01 ppmv. Optimization process leading a thickness reduction in rGO and Au films (to set both in 3 nm each) and an increasing operating temperature to 130 °C. Additionally, interference from other gases and vapors was examined founding sensor responded strongly to NO2, but due this is not expected to be present in air-quality-controlled clean rooms, sensor can be used in a semiconductor industry without serious interferences.

11.8.2   Optical Interferometric Sensor: Photonic Crystal Hydrogel

Optical sensors are based on changes of photoluminescence or reflectivity when are exposed to some analytes, which substitutes the air into sensor pores. The effect depends on the chemical and physical properties of each analyte so that sensor can be used to recognize the pure substances. However, due to the sensing mechanism, these kinds of devices are not able to identify components in complex mixtures. That is the reason why in recent researches, some physicochemical surface modifications were proposed [54]. The most common approach is to create covalent bonding between sensor surface and the analytes. In general, sensing principle involves the Fabry-Pérot interference of thins layers [55].

As recent instance, Yan and coworkers [56] provided a research about an interferometric sensor based on a photonic crystal (PhC) hydrogel immobilized with butyrylcholinesterase (BuChE) for sarin sensing. The hydrogel was developed by embedding on into polystyrene colloidal array before BuChE immobilization inside hydrogel matrix via condensation with DEPBT. Morphology of functionalized PhC is shown in Fig. 11.10a. For proving that BuChE was correctly immobilized on the PhC hydrogel, diffraction spectrum was recorded and exhibited in Fig. 11.10b. The unfunctionalized PhC diffracted the light at 632 nm, whilst functionalized PhC rise a peak at 685 nm. This occur because after immobilization of BuChE, hydrogel became more hydrophilic and the free energy of mixing increase leading a red shifting [57]. In addition, other shifts in diffraction were observed when

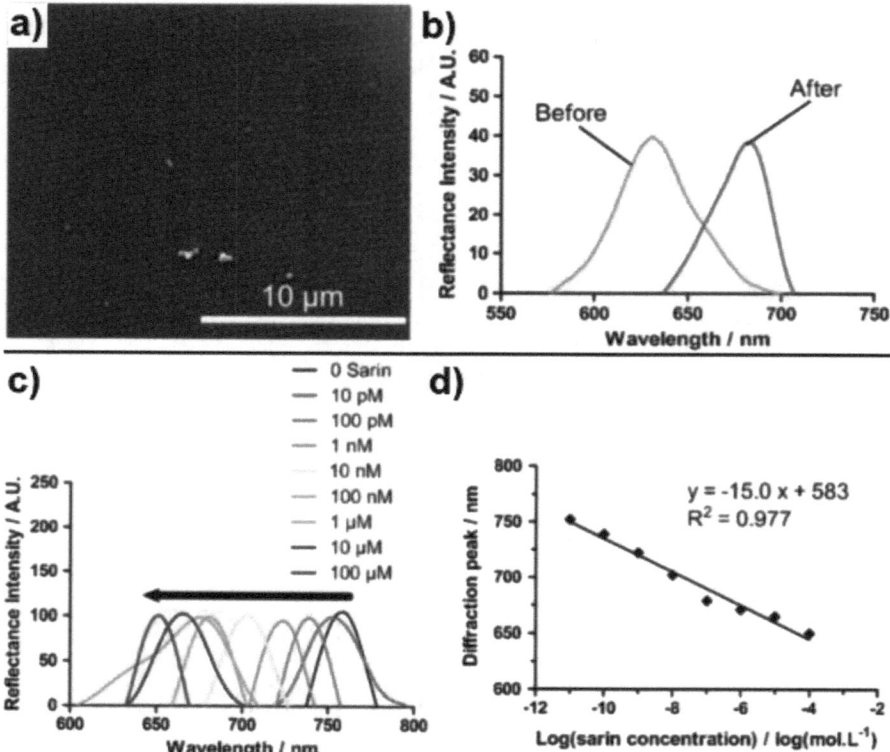

Fig. 11.10 (**a**) SEM of the functionalized PhC, (**b**) diffraction spectrum of the PhC before and after functionalization with BuChE, (**c**) Diffraction spectra of the functionalized PhC at various concentration of Sarin solutions, (**d**) the diffraction peak of functionalized PhC in response to Sarin in aqueous solutions. (Reprinted with permission from [56]. Copyright 2016 Elsevier)

changing the colloidal concentration, the acrylamide fraction and the crosslinking fraction. Firstly, BuChE-free PhC was exposed to sarin solutions and no significant diffraction shift was observed, indicating that resulting diffraction change was not triggered by the amide or carboxyl groups in the polymer backbone but functionalization. Functionalized PhC was exposed to sarin solutions of 10–11 mol L-1-10-4 mol L-1 in sequence showing a diffraction peak blue shifted according with sarin concentration increasing (Fig. 11.10c). A linearity (r2 = 0.977) was observed between amount of diffraction and logarithm of sarin concentration with a limit of detection about 10–15 mol L-1 (Fig. 11.10d). From this work, authors conclude that functionalized PhC could be useful for detection of real NA in a simple accurately way.

11.8.3   Surface Plasmon Resonance (SPR)

Surface plasmon resonance (SPR) has emerged as a suitable detection method for the development of label-free sensors due its unique ability to measure biomolecular interactions in real-time. However, sensitivity of SPR-based sensors is usually insufficient to detect trace amounts of small molecular weight molecules (e.g. CWAs) which are needed for creation of accurately sensors. Fortunately, rapid development of nanotechnology leaded the designed and synthesis of nanomaterials-enhanced SPR sensors for detecting molecules within the concentration range between pmol and amol [58].

Use of SPR-based sensor for identification of extremely dangerous Botulinum Neurotoxin A (BoNT/A) was studied by Tomar and coworkers [59] by means of immobilization of antibody against rBoNT/A-HCC and synaptic vesicles (SV) on a gold chip modified with carboxymethyl dextran. It is known that botulinum neurotoxins (BoNTs) are the deadliest toxins known and among the different protein chip detection methodologies, SPR allowed the detection in real time and without labels of molecular interactions between the immobilized ligand in a sensor chip and analytes injected on surfaces. In this work, the sensor immobilized with BoNT/A antibody was used to detect the BoNT/A antigen with different concentrations of the antigen and they were represented in an SPR sensorgram. The limit of detection (LOD) of the present method was calculated experimentally and found to be 0.045 fM and in addition was the minimum concentration of BoNT/A antigen that showed the response during the interaction of the BoNT/A antigen with its immobilized antibody. Furthermore, the effective interaction between the antigen and the antibody was confirmed based on impedance spectroscopy data. In conclusion, this study provided data for the development of SPR-based sensors that use the interaction antigen and antibody for biological agents.

11.8.4   Impedimetric Sensor: Au-Graphene-Chitosan Composite

Impedimetric immunosensors are a special type of electrochemical biosensors, in which impedance is measured. Impedance is a complex parameter which includes two independent phenomena: resistance and reactance. Resistance measures the extent to which a material opposes the movement of electrons among its atoms while reactance is an expression of the extent to which an electronic component stores and releases energy as the current and voltage fluctuate [59]. Afkhami et al. [61] developed an impedimetric immunosensor by using glassy carbon electrode (GCE) as a working electrode conveniently modified by depositing Au-graphene-chitosan composite. Covalent immobilization of polyclonal antibody will take place; a complete scheme of synthesis process can be observed at Fig. 11.11a. Characterization of composite included UV–vis spectrophotometry to confirm formation of AuNPs (average diameters: 21.63 nm) which was also found through X-ray diffraction (Fig. 11.11b). Topological structure of Gr nanosheets appeared flat and transparent, while large number of AuNPs could be observed spread over the composite indicating successfully formation of composite (Fig. 11.11c). Sensor stability was also studied by monitoring electrochemical response along 1-week period, finding initial resistance remained at least 90% till day fourth. Standard deviation (%)

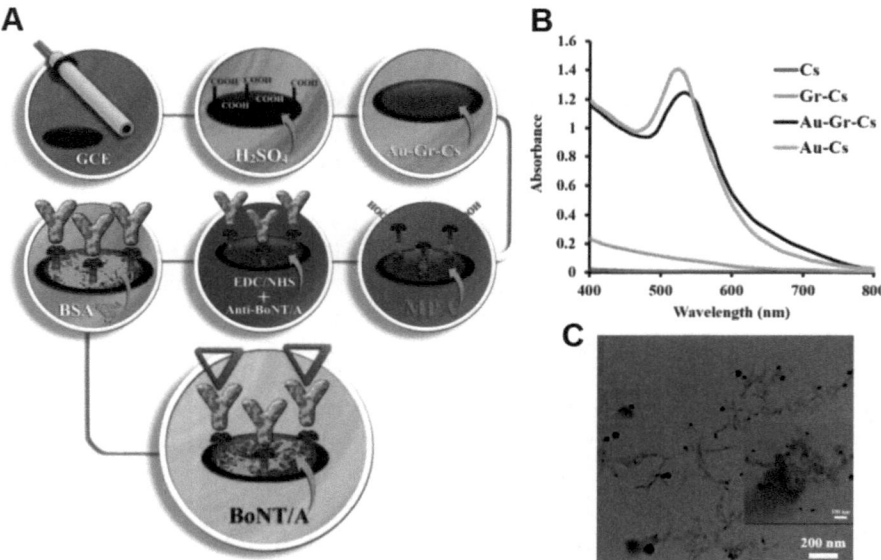

Fig. 11.11 (**a**) Immunosensor construction process. (**b**) UV–vis adsorption spectra and, (**c**) TEM image of Au-Gr-Cs, inset image has higher magnification. (Reprinted adapted with permission from [61]. Copyright 2017 Elsevier)

for three different immunosensors synthesized at the same conditions was 4.73%, which demonstrate excellent reproducibility of proposed immunosensor. Analysis of real samples was conducted in serum, buffer and milk (used as a complex protein abundant matrix to study protein components interference). Results obtained from fabricated immunosensor and ELISA method were in good agreement showing that proposed sensor has a significant response to BoNT/A concentrations in the range of 0.27–268 pg mL-1 when compared to control tests.

11.8.5 Gravimetric Sensor

Among the different transduction technologies known, quartz crystal microbalance (QCM) has recently attracted increasing attention for developing gas sensing platforms due in part to its inherent measurement mechanism which can reach high sensitivity. These transducers work by measuring change of frequency in a quartz crystal resonator when mass variations per unit area occur, (e.g., mass changing before and after analyte capture).

A novel gravimetric sensor example can be found in work reported by Hwang and coworkers [62]. Based on reduced non-stacked graphene oxide (NSrGO)-hexfluorohydroxypropanyl benzene (HFHPB), the sensor allowed the detection of dimethyl methyl phosphonate (DMMP), used as simulant of sarin gas. They reported the synthesis of 3D nanostructured porous NSrGO-HFHPB in which HFHPB worked as receptor for effectively detection of DMMP (Fig. 11.12). The HFHPB receptor is capable to form hydrogen bonding with the phosphonate group present in most nerve agents. The average pore size of composite was expected to be large enough for an easy diffusion of DMMP molecules through matrix, providing a fast response and high sensitivity due to the large exposed surface area. For evaluating performance of sensor, authors explored sensing behavior of GO, NSrGO and the composite hybridized with HFHPB receptors finding that NSrGO-HFHPB film sensor had 13 times greater sensitivity then the other composites. They also demonstrated a synergetic effect among mesoporous composite and receptor by increasing HFHPB amount, which leaded a 3 times faster response and 2 times shorter recovery. Another important feature obtained from hydrophobic HFHPB receptors was maintaining of sensing capability even at 70% humidity. Novel composite developed in this work may guide new strategies of chemical engineering methods for embedding molecules onto carbon-based platforms.

Another recent example of its use as transducer mechanism for highly detection of DMMP was reported by Zhu and coworkers [63]. In their work, they synthesized a mesoporous TiO2-SiO2 platform functionalized with HFIP designed for selective sensing of DMMP due hydrogen bonding formation between analyte and composite. For sensing experiments, compounds were coated over QCM transducers and frequency observed shifted gradually in range from 10 to 60 ppm when exposed to DMMP. Results revealed that sensor decorated with HFIP had better performance at

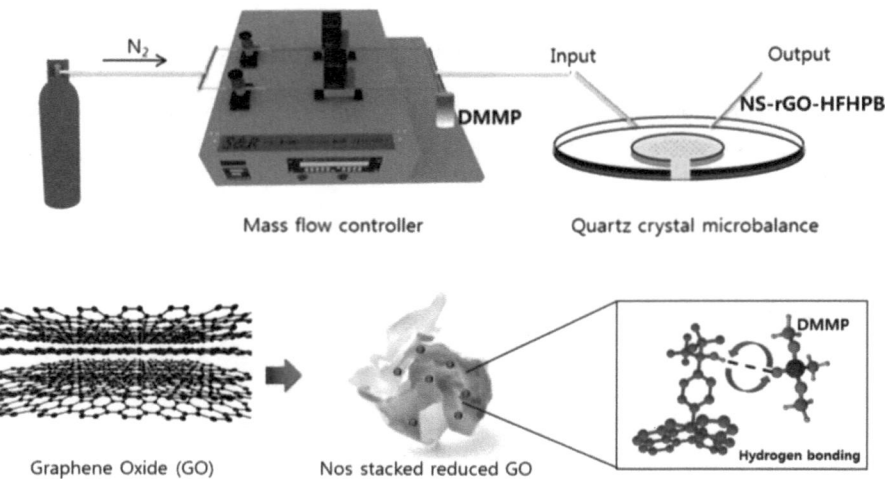

Fig. 11.12 Illustration of the molecular structure of non-stacked, grafted reduce graphene oxide with HFHPB derivative uptake of DMMP. (Reprinted with permission from [62]. Copyright 2016 Nature Publishing Group)

ppb level. Furthermore, 20 wt% HFIP doping led the best sensing response. These findings imply that TiO2-SiO2-HFIP composite could be a good candidate for gas sensing platforms.

11.9 Update

Last year (2018), several investigations were developed with the purpose of implementing new systems of detection for chemical warfare agents; however, only a few works reported the use of nanometric materials-based sensors. These studies explored the sensitivity, selectivity, and stability of the sensors to optimize the efficiency and lifetime detection of simulant nerve agents (NA) and chemical warfare agents (CWAs). As has been explained earlier in this chapter, these compounds are highly toxic even on a trace levels concentration.

For instance, Ali et al. [64] developed a hydrogel film intended for the detection of diisopropyl fluorophosphate (DFP) under conditions potentially low $(6.90 \times 10^{-9}$ mg/cm^3 for 10 min). Remarkably, this sarin simulant deposited on a polyacrylamide (PAAm) hydrogel contained an ionic gradient and the enzyme, diisopropyl fluorophosphatase (DFPase). This last was used to degrade the DFP and release fluoride anions (F-). The internal ionic gradient within PAAm transported the F- to an F-ion-specific sensor engrafted in the polymeric matrix. Hence, the fluoride ions concentrated to an electrochemical sensor minimized the provided amplification of F- signal (an indicator of the DFP exposure). By this method is possible to achieve the detection of CWAs, even ionic species, by using hydrolysis.

In another study, degradation and detection of paraoxon-ethyl (PTL) was carried out by Karthik et al. [65]. PTL is an organophosphate pesticide widely used in the cultivation of beans, wheat, cotton, rice, and fruits. However, by inhalation, the PTL can be entered into the human body leading dangerous health risks, including death in the worst case. Bearing this problem in mind, the synthesis of a novel YbM/f-CNFs composite by wet-chemical and sonication process was performed. In addition to degradation and detection of PTL, the YbM/f-CNFs composite exhibited a great photodegradation and mineralization on the detoxification process preserving at the same time, excellent stability.

A more straightforward detection was studied by Dennison et al. [66], where by means of paper assays of dual-luminescent Ir(III)/Eu(III) generated a qualitative colorimetric analysis under UV irradiation of 254 nm. The results showed that visual comparison of G-series and V-series are different. However, similar results between G-series and their simulants (DMMP, DMF, and GB), and the V-series and their simulants (VX and VO), exhibited the same color emission during the kinetics studied.

Qi et al. [67] developed a two-dimensional photonic crystal (2D PhC) sensor based in a poly(acrylamide-co-acrylic acid) hydrogel functionalized with acetyl-cholinesterase (AChE) for a method efficient and rapid detection of sarin and the simulants, dimethyl methylphosphonate (DMMP,) diisopropyl methylphosphonate (DIMP), isodipropyl methylphosphonate (IDMP). One of the primary results was the optimization of the AChE-functionalized 2D PhC sensor adjusting the amount of monomers in the polymeric 3D network, increasing not only the sensitivity of the sensor but also decreasing the detection limit. A linear relationship between the concentration logarithm of sarin and particle spacing was also observed from 7.1×10^{-17} to 7.1×10^{-4} mol/L.

Barreca et al. [68] reported a synthesis of nano β-MnO_2 by plasma enhanced-chemical vapor deposition (PE-CVD) for the sensing of acetonitrile, a simulant chemical warfare agent (CWA). The results demonstrated that sensors could efficiently detect acetonitrile molecules, with the best response at temperatures over 200 °C. All the features showed by the system, like excellent selectivity and low response time, allow to consider the system as a suitable candidate for practical applications.

11.10 Concluding Remarks

Summarizing, in this chapter we briefly reviewed some of the most recently CWAs sensors based on nanomaterials and described their sensing mechanism with particular emphasis on research performed along past 3 years. Another popular novel materials, like porous silicon, were used in the same direction; however, only a few studies were reported in the early 2000s, so they were not mentioned in this chapter. From the discussed literature, it is clear that carbon-based and polymer materials are the most suitable candidates to design and to develop supported sensors for accurate detection of CWAs due to their inherent properties. In the case of graphene and carbon nanotubes, their high adsorption capacity and sensitive electronic properties, which easily change upon adsorption of other chemical agents, provide the possibility of designing improved electronic and conductive sensors. Besides their primary role as transducers or electrodes, another important application of carbon-based materials is their use as filler into polymer composite matrices for providing connection among other sensing elements, enhancing conductive behavior or increasing mechanical strength. On the other hand, polymeric composites (especially conductive polymers) have been successfully used for detection purposes since their electronic and optical properties are similar to metal or inorganic materials. Advantages of this approach are directly related to their easy methods of synthesis, chemical features, light weight, and low cost production. Nevertheless, further development in the field is required to obtain low cost-efficient portable on-site sensors at industrial scale.

Acknowledgments The authors would like to acknowledge the financial support of CONACyT through PhD grants, No. 385013 and No. 388119.

References

1. Sidell F (1997) Medical aspects of chemical and biological warfare. Textb Mil Med 995: 129–179
2. Ganesan K, Raza SK, R. V (2010) Chemical warfare agents. J Pharm Bioallied Sci 2:166–178. https://doi.org/10.4103/0975-7406.68498
3. Szinicz L (2005) History of chemical and biological warfare agents. Toxicology 214:167–181. https://doi.org/10.1016/j.tox.2005.06.011
4. Riley B (2003) The toxicology and treatment of injuries from chemicalwarfare agents. Curr Anaesth Crit Care 14:173–177. https://doi.org/10.1016/S0953-71
5. Bhaganagar K, Bhimireddy SR (2017) Assessment of the plume dispersion due to chemical attack on April 4, 2017, in Syria. Nat Hazards 88:1893–1901. https://doi.org/10.1007/s11069-017-2936-x
6. López-Muñoz F, Alamo C, Guerra JA, García-García P (2008) The development of neurotoxic agents as chemical weapons during the national socialist period in Germany. Rev Neurol 47:99–106
7. Black R (2016) Development, historical use and properties of chemical warfare agents. In: Chemical warfare toxicology, vol 1: fundamental aspects, pp 1–28
8. Colovic MB, Krstic DZ, Lazarevic-Pasti TD et al (2013) Acetylcholinesterase inhibitors: pharmacology and toxicology. Curr Neuropharmacol 11:315–335. https://doi.org/10.2174/1570159X11311030006
9. Chauhan S, Chauhan S, D'Cruz R et al (2008) Chemical warfare agents. Environ Toxicol Pharmacol 26:113–122. https://doi.org/10.1016/j.etap.2008.03.003
10. Singh VV, Wang J (2015) Nano/micromotors for security/defense applications. A review. Nanoscale 7:19377–19389. https://doi.org/10.1039/C5NR06254C
11. Lévêque C, Ferracci G, Maulet Y et al (2014) Direct biosensor detection of botulinum neurotoxin endopeptidase activity in sera from patients with type A botulism. Biosens Bioelectron 57:207–212. https://doi.org/10.1016/j.bios.2014.02.015

12. Heymann WR (2004) Threats of biological and chemical warfare on civilian populations. Dialog Dermatol:452–453

13. Jang YJ, Kim K, Tsay OG et al (2015) Update 1 of: destruction and detection of chemical warfare agents. Chem Rev 115:PR1–PR76. https://doi.org/10.1021/acs.chemrev.5b00402

14. Yang YC (1999) Chemical detoxification of nerve agent VX. Acc Chem Res 32:109–115. https://doi.org/10.1021/ar970154s

15. Bartelt-Hunt SL, Knappe DRU, Barlaz MA (2008) A review of chemical warfare agent simulants for the study of environmental behavior. Crit Rev Environ Sci Technol 38:112–136. https://doi.org/10.1080/10643380701643650

16. Halliwell J, Savage AC, Buckley N, Gwenin C (2014) Electrochemical impedance spectroscopy biosensor for detection of active botulinum neurotoxin. Sens Bio Sens Res 2:12–15. https://doi.org/10.1016/j.sbsr.2014.08.002

17. Fennell J, Hamaguchi H, Yoon B, Swager T (2017) Chemiresistor devices for chemical warfare agent detection based on polymer wrapped single-walled carbon nanotubes. Sensors 17:982. https://doi.org/10.3390/s17050982

18. Blum AP, Kammeyer JK, Rush AM et al (2015) Stimuli-responsive nanomaterials for biomedical applications. J Am Chem Soc 137:2140–2154. https://doi.org/10.1021/ja510147n

19. Sharma TK, Ramanathan R, Rakwal R et al (2015) Moving forward in plant food safety and security through NanoBioSensors: adopt or adapt biomedical technologies. Proteomics 15:1680–1692. https://doi.org/10.1002/pmic.201400503

20. Li J (2009) Carbon-based sensors. In: Carbon materials for catalisis. Wiley, Hobolen, pp 507–533

21. Basu S, Bhattacharyya P (2012) Recent developments on graphene and graphene oxide based solid state gas sensors. Sens Actuators B Chem 173:1–21. https://doi.org/10.1016/j.snb.2012.07.092

22. Salavagione HJ, Díez-Pascual AM, Lázaro E et al (2014) Chemical sensors based on polymer composites with carbon nanotubes and graphene: the role of the polymer. J Mater Chem A 2:14289–14328. https://doi.org/10.1039/C4TA02159B

23. Mittal G, Dhand V, Rhee KY et al (2015) A review on carbon nanotubes and graphene as fillers in reinforced polymer nanocomposites. J Ind Eng Chem 21:11–25. https://doi.org/10.1016/j.jiec.2014.03.022

24. Facure MH, Mercante LA, Mattoso LH, Correa DS (2017) Detection of trace levels of organophosphate pesticides using an electronic tongue based on graphene hybrid nanocomposites. Talanta 167:59–66. https://doi.org/10.1016/j.talanta.2017.02.005

25. Li L, Shi Y, Pan L et al (2015) Rational design and applications of conducting polymer hydrogels as electrochemical biosensors. J Mater Chem B 3:2920–2930. https://doi.org/10.1039/C5TB00090D

26. Yoon H, Jang J (2009) Conducting-polymer nanomaterials for high-performance sensor applications: issues and challenges. Adv Funct Mater 19:1567–1576. https://doi.org/10.1002/adfm.200801141

27. Baker CO, Huang X, Nelson W, Kaner RB (2017) Polyaniline nanofibers: broadening applications for conducting polymers. Chem Soc Rev 46:1510–1525. https://doi.org/10.1039/C6CS00555A

28. Li M, Li H, Zhong W et al (2014) Stretchable conductive polypyrrole/polyurethane (PPy/PU) strain sensor with netlike microcracks for human breath detection. ACS Appl Mater Interfaces 6:1313–1319. https://doi.org/10.1021/am4053305

29. Sheng G, Xu G, Xu S et al (2015) Cost-effective preparation and sensing application of conducting polymer PEDOT/ionic liquid nanocomposite with excellent electrochemical properties. RSC Adv 5:20741–20746. https://doi.org/10.1039/C4RA15755A

30. Zhao Y, Liu B, Pan L, Yu G (2013) 3D nanostructured conductive polymer hydrogels for high-performance electrochemical devices. Energy Environ Sci 6:2856. https://doi.org/10.1039/c3ee40997j

31. Rivero RE, Molina MA, Rivarola CR, Barbero CA (2014) Pressure and microwave sensors/actuators based on smart hydrogel/conductive polymer nanocomposite. Sensors Actuators B Chem 190:270–278. https://doi.org/10.1016/j.snb.2013.08.054

32. Bai H, Shi G (2007) Gas sensors based on conducting polymers. Sensors 7:267–307. https://doi.org/10.3390/s7030267
33. Fennell JF, Liu SF, Azzarelli JM et al (2016) Nanowire chemical/biological sensors: status and a roadmap for the future. Angew Chemie - Int Ed 55:1266–1281. https://doi.org/10.1002/anie.201505308
34. Ishihara S, O'Kelly CJ, Tanaka T et al (2017) Metallic vs. semiconducting SWCNT chemiresistors: a case for separated SWCNTs wrapped by metallo-supramolecular polymer. ACS Appl Mater Interfaces:7b12992. https://doi.org/10.1021/acsami.7b12992
35. Yoo R, Kim J, Song MJ et al (2015) Nano-composite sensors composed of single-walled carbon nanotubes and polyaniline for the detection of a nerve agent simulant gas. Sens Actuators B Chem 209:444–448. https://doi.org/10.1016/j.snb.2014.11.137
36. Wang F, Gu H, Swager TM (2008) Carbon nanotube/polythiophene chemiresistive sensors for chemical warfare agents carbon nanotube/polythiophene chemiresistive sensors for chemical warfare agents, vol 130, pp 8–10. https://doi.org/10.1021/ja710795k
37. Yoon W, Lee SH, Kwon OS et al (2009) Polypyrrole nanotubes conjugated with human olfactory receptors: high-performance transducers for FET-type bioelectronic noses. Angew Chemie Int Ed 48:2755–2758. https://doi.org/10.1002/anie.200805171
38. Kwon OS, Park CS, Park SJ et al (2016) Carboxylic acid-functionalized conducting-polymer nanotubes as highly sensitive nerve-agent chemiresistors. Sci Rep 6:33724. https://doi.org/10.1038/srep33724
39. Shar M, Khan J, Wang Y-W et al (2017) Sensitive fluorescence on-off probes for the fast detection of a chemical warfare agent mimic sensitive fluorescence on-off probes for the fast detection of a chemical warfare agent mimic sensitive fluorescence on-off probes for the fast detection. J Hazard Mater 342:10–19. https://doi.org/10.1016/j.jhazmat.2017.08.009
40. Kopeček J (2007) Hydrogel biomaterials: a smart future? Biomaterials 28:5185–5192. https://doi.org/10.1016/j.biomaterials.2007.07.044
41. Whitaker CM, Derouin EE, O'connor MB et al (2017) Smart hydrogel sensor for detection of organophosphorus chemical warfare nerve agents. J Macromol Sci Part A Pure Appl Chem 54:40–46. https://doi.org/10.1080/10601325.2017.1250313
42. Raghavender Goud D, Purohit AK, Tak V et al (2014) A highly selective and sensitive "turn-on" fluorescence chemodosimeter for the detection of mustard gas. Chem Commun 50:12363–12366. https://doi.org/10.1039/C4CC04801F
43. Varju BR, Ovens JS, Leznoff DB (2017) Mixed Cu(I)/Au(I) coordination polymers as reversible turn-on vapoluminescent sensors for volatile thioethers. Chem Commun 53:6500–6503. https://doi.org/10.1039/C7CC03428H
44. Son HY, Ryu JH, Lee H, Nam YS (2013) Bioinspired templating synthesis of metal-polymer hybrid nanostructures within 3D electrospun nanofibers. ACS Appl Mater Interfaces 5:6381–6390. https://doi.org/10.1021/am401550p
45. Jo S, Kim J, Noh J et al (2014) Conjugated polymer dots-on-electrospun fibers as a fluorescent nanofibrous sensor for nerve gas stimulant. ACS Appl Mater Interfaces 6:22884–22893. https://doi.org/10.1021/am507206x
46. Climent E, Biyikal M, Gawlitza K et al (2017) Determination of the chemical warfare agents Sarin, Soman and Tabun in natural waters employing fluorescent hybrid silica materials. Sens Actuators B Chem 246:1056–1065. https://doi.org/10.1016/j.snb.2017.02.115
47. Loudet A, Burgess K (2007) BODIPY dyes and their derivatives: syntheses and spectroscopic properties. Chem Rev 107:4891–4932
48. Piriya VSA, Joseph P, Daniel SCGK et al (2017) Colorimetric sensors for rapid detection of various analytes. Mater Sci Eng C 78:1231–1245. https://doi.org/10.1016/j.msec.2017.05.018
49. Betty CA, Lal R, Yakhmi JV, Kulshreshtha SK (2007) Time response and stability of porous silicon capacitive immunosensors. Biosens Bioelectron 22:1027–1033. https://doi.org/10.1016/j.bios.2006.04.022
50. Boopathi M, Singh B, Vijayaraghavan R (2008) A review on NBC body protective clothing. Open Text J 1:1–8. https://doi.org/10.2174/1876520300801010001

51. Belger C, Weis J, Ahmed E, Swager T (2015) Colorimetric stimuli-responsive hydrogel polymers for the detection of nerve agent surrogates. Macromolecules 48:7990–7994. https://doi.org/10.1021/acs.macromol.5b01406
52. Giannakoudakis DA, Hu Y, Florent M, Bandosz TJ (2017) Smart textiles of MOF/g-C3N4 nanospheres for the rapid detection/detoxification of chemical warfare agents. Nanoscale Horiz 2:356–364. https://doi.org/10.1039/C7NH00081B
53. Furue R, Koveke EP, Sugimoto S et al (2017) Arsine gas sensor based on gold-modified reduced graphene oxide. Sensors Actuators B Chem 240:657–663. https://doi.org/10.1016/j.snb.2016.08.131
54. De Stefano L, Rotiroti L, Rendina I et al (2006) Porous silicon-based optical microsensor for the detection of L-glutamine. Biosens Bioelectron 21:1664–1667. https://doi.org/10.1016/j.bios.2005.08.012
55. Lin VS (1997) A porous silicon-based optical interferometric biosensor. Science (80-) 278:840–843. https://doi.org/10.1126/science.278.5339.840
56. Yan C, Qi F, Li S et al (2016) Functionalized photonic crystal for the sensing of Sarin agents. Talanta 159:412–417. https://doi.org/10.1016/j.talanta.2016.06.045
57. Ramu VG, Bardaji E, Heras M (2014) DEPBT as coupling reagent to avoid racemization in a solution-phase synthesis of a kyotorphin derivative. Synthesis 46:1481–1486. https://doi.org/10.1055/s-0033-1341068
58. Nguyen HH, Park J, Kang S, Kim M (2015) Surface plasmon resonance: a versatile technique for biosensor applications. Sensors (Switzerland) 15:10481–10510. https://doi.org/10.3390/s150510481
59. Tomar A, Gupta G (2016) Surface plasmon resonance sensing of biological warfare agent botulinum neurotoxin A. J Bioterror Biodef 7. https://doi.org/10.4172/2157-2526.1000142
60. Marszalek Z, Sroka R, Zeglen T (2017) Multi-frequency conditioning system of the inductive loop sensor – simulation investigations. Methods Model Autom Robot:889–893
61. Afkhami A, Hashemi P, Bagheri H et al (2017) Impedimetric immunosensor for the label-free and direct detection of botulinum neurotoxin serotype A using Au nanoparticles/graphene-chitosan composite. Biosens Bioelectron 93:124–131. https://doi.org/10.1016/j.bios.2016.09.059
62. Hwang HM, Hwang E, Kim D, Lee H (2016) Mesoporous non-stacked graphene-receptor sensor for detecting nerve agents. Nat Publ Gr 6:1–8. https://doi.org/10.1038/srep33299
63. Zhu Y, Cheng Z, Xiang Q et al (2016) Synthesis of functionalized mesoporous TiO2-SiO2 with organic fluoroalcohol as high performance DMMP gas sensor. Sens Actuators B Chem 248:785–792. https://doi.org/10.1016/j.snb.2016.10.080
64. Ali MA, Tsai TH, Braun PV (2018) Amplified detection of chemical warfare agents using two-dimensional chemical potential gradients. ACS Omega 3:14665–14670. https://doi.org/10.1021/acsomega.8b01519
65. Karthik R, Kumar JV, Chen SM, Kokulnathan T, Yang HY, Muthuraj V (2018) Design of Novel Ytterbium Molybdate Nanoflakes Anchored Carbon Nanofibers: challenging sustainable catalyst for the detection and degradation of assassination weapon (Paraoxon-ethyl). ACS Sustain Chem Eng 6:8615–8630. https://doi.org/10.1021/acssuschemeng.8b00936
66. Dennison GH, Curty C, Metherell AJ, Micich E, Zaugg A, Ward MD (2019) Qualitative colorimetric analysis of a Ir (iii)/Eu (iii) dyad in the presence of chemical warfare agents and simulants on a paper matrix. RSC Adv 9:7615–7619. https://doi.org/10.1039/C9RA00824A
67. Qi F, Yan C, Meng Z, Li S, Xu J, Hu X, Xue M (2019) Acetylcholinesterase-functionalized two-dimensional photonic crystal for the sensing of G-series nerve agents. Anal Bioanal Chem:1–9. https://doi.org/10.1007/s00216-019-01700-w
68. Barreca D, Gasparotto A, Gri F, Comini E, Maccato C (2018) Plasma-assisted growth of β-MnO2 Nanosystems as gas sensors for safety and food industry applications. Adv Mater Interfaces 5:1800792. https://doi.org/10.1002/admi.201800792